SCALE EFFECTS IN ANIMAL LOCOMOTION

SCALE EFFECTS IN ANIMAL LOCOMOTION

Based on the Proceedings of
an International Symposium held at
Cambridge University, September 1975

Edited by

T. J. PEDLEY

Department of Applied Mathematics and Theoretical Physics,
University of Cambridge, England

1977

ACADEMIC PRESS

LONDON NEW YORK SAN FRANCISCO
A Subsidiary of Harcourt Brace Jovanovich, Publishers

ACADEMIC PRESS (LONDON) LTD.
24/28 Oval Road
London NW1

United States Edition published by
ACADEMIC PRESS INC.
111 Fifth Avenue
New York, New York 10003

Library of Congress Catalog Number: 76-22865
ISBN: 0-12-549650-8

PRINTED IN GREAT BRITAIN BY
PAGE BROS (NORWICH) LTD, NORWICH

LIST OF PARTICIPANTS

R. McNeill Alexander *Department of Pure and Applied Zoology, The University of Leeds, Leeds LS2 9JT, England.*

R. Bainbridge *Department of Zoology, Cambridge University, England.*

H. C. Bennet-Clark *Department of Zoology, University of Edinburgh, Edinburgh EH9 3JT, Scotland (present address: Department of Zoology, South Parks Road, University of Oxford, England.)*

J. R. Blake *CSIRO, Division of Mathematical Statistics, P.O. Box 1965, Canberra City, A.C.T. 2601, Australia.*

J. Brackenbury *Sub-department of Veterinary Anatomy, Cambridge University, England.*

C. Brennen *Engineering Science Department, California Institute of Technology, Pasadena, California 91109, U.S.A.*

R. H. Brown *Department of Zoology, Cambridge University, England.*

C. A. Cavagna *Istituto di Fisiologia Umana, l'Cattedra, Università di Milano, 20133 Milano, Via Mangiagalli, 32, Italy.*

S. Corrsin *Department of Mechanics and Materials Science, The Johns Hopkins University, Baltimore, Maryland 21218, U.S.A.*

J. D. Currey *Department of Biology, University of York, York YO1 5DD, England.*

P. di Prampero *C.N.R. Centro Studi di Fisiologia del Lavoro Musculare, 20133 Milano, Via Mangiagalli, 32, Italy.*

A. B. Dubois *John B. Pierce Foundation Laboratory, Yale University, 290 Congress Avenue, New Haven, Connecticut 06519, U.S.A.*

H.-R. Duncker *Zentrum für Anatomie und Cytobiologie, am Klinikum der Justus Liebig-Universität, 63 Giessen, den Friedrichstrasse 34, W. Germany.*

C. Ellington *Department of Zoology, Cambridge University, England.*

J. P. Eylers *Department of Zoology, University of Leeds, Leeds LS2 9JT, England.*

M. Fedak *School of Marine and Atmospheric Science, University of Miami, 4600 Rickenbacker Causeway, Miami, Florida 33149, U.S.A.*

G. GOLDSPINK *Department of Zoology, University of Hull, Hull HU6 7RX, England.*

N. C. HEGLUND *Museum of Comparative Zoology, Harvard University, Cambridge, Massachusetts 02138, U.S.A.*

J. HIGDON *Department of Mechanics and Materials Science, The Johns Hopkins University, Baltimore, Maryland 21218, U.S.A.*

D. V. HOLBERTON *Department of Zoology, University of Hull, Hull HU6 7RX, England.*

M. E. J. HOLWILL *Physics Department, Queen Elizabeth College, Campden Hill Road, London W8 7AH, England.*

G. M. HUGHES *Research Unit for Comparative Animal Respiration, University of Bristol, Woodland Road, Bristol, BS8 1UG, England.*

M. JENSEN *Nørrekaer 10, 16, 2610 Rødovre, Denmark.*

T. KAMBE *Department of Applied Mathematics and Theoretical Physics, Cambridge University, England.*

*N. V. KOKSHAYSKY *Academy of Sciences of the U.S.S.R., Institute of Evolutionary Animal Morphology and Ecology, Leninsky Prospekt 33, Moscow W-71, U.S.S.R.*

*JAMES LIGHTHILL (co-chairman) *Department of Applied Mathematics and Theoretical Physics, Cambridge University, England.*

H. LISSMAN *Department of Zoology, Cambridge University, England.*

J. H. LOCHHEAD *Department of Zoology, University of Vermont, Burlington, Vermont 05401, U.S.A. (present address: 49 Woodlawn Road, London SW6, England).*

C. W. MCCUTCHEN *Laboratory of Experimental Pathology, National Institute of Arthritis, Metabolism and Digestive Diseases, National Institutes of Health, Bethesda, Maryland 20014, U.S.A.*

T. A. MCMAHON *Division of Engineering and Applied Physics, Harvard University, Cambridge, Massachussetts 02138, U.S.A.*

A. MONAVON *Mécanique Experimentale des Fluides, Université de Paris— VI, Campus Universitaire, Bâtiment 502, 91405 Orsay, France.*

J. MOORE *Department of Engineering, Cambridge University, England.*

W. NACHTIGALL *Fachbereich 16 der Universität des Saarlandes, Fachrichtung 4—Zoologie, 66 Saarbrücken, West Germany.*

B. G. NEWMAN *Department of Mechanical Engineering, McGill University, P.O. Box 6070, Station 'A', Montreal, Quebec, Canada.*

R. A. NORBERG *Zoologiska institutionen, Göteborgs universitet, Fack, 400 33 Göteborg 33, Sweden.*

U. M. NORBERG *Zoologiska institutionen, Göteborgs universitet, Fack, 400 33 Göteborg 33, Sweden.*

H. OEHME *Akademie der Wissenschaften der DDR, Forschungsstelle für Wirbeltierforschung (im Tierpark Berlin), 1136 Berlin, den Am Tierpark 125, East Germany.*

J. W. M. OSSE *Vakgroep Experimentele Diermorfologie en Celbiologie, Landbouwhogeschool, Hollandseweg 13, Wageningen, Netherlands.*

*T. J. PEDLEY (Secretary) *Department of Applied Mathematics and Theoretical Physics, Cambridge University, England.*

C. PENNYCUICK *Department of Zoology, University of Bristol, Woodland Road, Bristol BS8 1UG, England.*

J. PIIPER *Max-Planck-Institut für experimentelle Medizin, Abteilung Physiologie, D-34 Göttingen, Hermann-Rein-Strasse 3, West Germany.*

H. PRANGE *Department of Zoology, University of Florida, Gainesville, Florida 32611, U.S.A. (present address: Physiology Section, Indiana University School of Medicine, Bloomington, Indiana 47401, U.S.A.)*

J. W. S. PRINGLE *Department of Zoology, Agricultural Research Council Unit, Oxford University, South Parks Road, Oxford OX1 3PS.*

J. M. V. RAYNER *Department of Applied Mathematics and Theoretical Physics, Cambridge University, England.*

G. RÜPPELL *Institut für Zoologie, der Friedrich-Alexander-Universität, Erlangen-Nürnberg, 852 Erlangen, Bismarckstrasse 10, West Germany.*

G. W. SCHAEFER *Ecological Physics Research Group, Cranfield Institute of Technology, Cranfield, Bedford, MK43 0AL, England.*

*K. SCHMIDT-NIELSEN *Department of Zoology, Duke University, Durham, North Carolina 27706, U.S.A.*

M. A. SLEIGH *Department of Biology, University of Southampton, Southampton SO9 3TU, England.*

C. R. TAYLOR *Biological Laboratories, Harvard University, 16 Divinity Avenue, Cambridge, Massachussetts 02138, U.S.A.*

Y. TOSHEV *Bulgarian Academy of Sciences, Centre for Biomechanical Research, Sofia 1, 7 Noemvri Street, Bulgaria.*

V. A. TUCKER *Department of Zoology, Duke University, Durham, North Carolina 27706, U.S.A.*

J. J. VIDELER *Zoölogisch Laboratorium, der Rijksuniversiteit te Groningen, Kerklaan 30, Haren (Gr.), Netherlands.*

S. VOGEL *Department of Zoology, Duke University, Durham, North Carolina 27706, U.S.A.*

C. S. WARDLE *Department of Agriculture and Fisheries for Scotland, Marine Laboratory, P.O. Box 101, Victoria Road, Aberdeen AB9 8DB, Scotland.*

P. W. WEBB *School of Natural Resources, University of Michigan, 430 E. University St., Ann Arbor, Michigan 48104, U.S.A.*

D. WEIHS *Department of Aeronautical Engineering, Technion, Haifa, Israel.*

*TORKEL WEIS-FOGH† (co-chairman) *Department of Zoology, Cambridge University, England.*

D. R. WILKIE *Department of Physiology, University College, Gower Street, London W.C.1, England.*

*T. Y. WU *Engineering Science Department, California Institute of Technology, Pasadena, California 91109, U.S.A.*

---

* Member of the Scientific Committee
† Now deceased

DEDICATED TO THE MEMORY OF
TORKEL WEIS-FOGH

(Deceased 13 November 1975)

All who took part under Torkel Weis-Fogh's inspiring chairmanship in the 1975 Symposium on Scale Effects in Animal Locomotion wish to dedicate these Proceedings to the memory of a supreme explorer of those fields where zoology and mechanics meet.

Professor Torkel Weis-Fogh

TORKEL WEIS-FOGH

(1922–1975)

Torkel Weis-Fogh was born in Aarhus, Denmark on 22 March 1922. In 1947, after wartime studies at Copenhagen University, he joined the laboratory of August Krogh, specialising in research on the desert locust *Schistocerca gregaria*. Weis-Fogh soon began to complement Krogh's fine work on locust metabolism and respiration with investigations of striking originality into locust musculature, cuticle and flight mechanics. After Krogh's death in 1949 he became head of the laboratory for four years, during which he began a long and effective collaboration with the engineer Martin Jensen. Their great series of papers under the heading "Biology and physics of locust flight" began to be published in the Philosophical Transactions of the Royal Society in 1956 (when it already occupied 170 pages devoted to analysis based, above all, on their remarkable wind-tunnel studies). In the meantime, Weis-Fogh was appointed successively to a lectureship in the Copenhagen Institute of Neurophysiology, a Rockefeller Fellowship, and a Balfour Studentship which involved him in work at Cambridge University under Professor Sir James Gray, the famous leader in animal-locomotion research.

In 1958, Weis-Fogh became Professor of Zoophysiology in Copenhagen University. This was the period of his work leading to the discovery of resilin, the extraordinary elastomer, with properties similar to those of vulcanized rubber, which enables many insects to put their wings into tuned oscillation with very low internal damping. He visited Harvard University as Prother Lecturer in 1961. Then in 1966 he succeeded Sir James Gray as Professor of Zoology in the University of Cambridge. He died suddenly on 13 November 1975.

During his time in Cambridge, Torkel Weis-Fogh made massive extensions to both major themes of his previous research. Using many powerful techniques he probed deeply into the structural features underlying the remarkable mechanical properties of certain biological materials. These included not only resilin, but also elastin (the very different elastomer found in the vertebrates), and the contractile "spasmoneme" found in certain protozoans. In addition, he embarked on a fundamental and comprehensive investigation of how animals hover in still air. In papers published from 1972 onwards he gave profound analyses both of what he named "normal hovering" and of various exceptional modes of hovering. One of the most striking "scale effects in animal locomotion" was brought to light when he showed that insects of sizes below about 2 mm are prevented by the viscosity of the air from using normal hovering. By cinephotography at 7000 frames a second, he proved that they use a mechanism of lift generation previously unknown to aerodynamicists: the "clap and fling". He was actively engaged in research on exceptional modes of hovering up to the time of his death.

Weis-Fogh always retained his Danish citizenship. He was a Fellow of the Royal Danish Academy, and of the Academy of Technical Sciences. He was a member of the Danish State Research Foundation, the Chairman of its Natural Sciences Committee. A terrible tragedy for him was the car accident in Denmark in 1971, in which his wife was killed (he had married Hanne Heckscher in 1946). He was himself seriously injured in this accident, but stoically bore the continued disabilities that resulted.

During the last three years of Torkel Weis-Fogh's life I enjoyed a close collaboration with him in researches on animal flight and in the planning of the Cambridge Symposium whose proceedings form the contents of this volume. His great width of understanding, inventiveness, enthusiasm, integrity and determination were mingled with personal charm in a way that made working with him an unforgettable experience. Just before his death, we had completed plans for a seven-year programme of collaborative research in biological fluid dynamics.

There is a terrible sense of loss in Cambridge as a result of the sudden death of Torkel Weis-Fogh. At least one aspect of it, the knowledge that much of what he would have been able to do in his future research work is irrecoverably lost because he was unique, is felt all over the world.

James Lighthill

EDITOR'S PREFACE

This book concentrates on just one aspect of animal location: the effect of scale, or the relationship of locomotory function to animal size. However, all types of locomotion are included—swimming and flying; walking, running and jumping—and all factors involved in that locomotion are examined: metabolic rate, respiration, muscle power, skeletal strength; the dynamical forces exerted by leg muscles when a foot is on the ground; the fluid dynamical forces exerted between an animal and its aquatic or aerial surroundings; and the solid dynamical response of that body or limb to the same propulsive forces. Many animals and many mechanisms are considered, but the concentration on scale effects confers an overriding unity to the work and makes a coherent whole out of subjects as disparate as the aerodynamics of hovering insects and the swimming of crustacea, size limitations on flying birds and on swimming spermatozoa, elastic energy storage in a fish's tail and in a kangaroo's Achilles tendon.

Animal locomotion is a truly interdisciplinary field, in which the work of the mechanical scientist impinges directly onto that of the biologist, because the mechanical forces of propulsion interact directly with the animal itself; and this is a truly interdisciplinary book. The authors are drawn from many fields—biophysics, physiology and zoology on the one hand, engineering and applied mathematics on the other—but most have a wide experience of working with practitioners of the other disciplines, and all have written their chapters with the aim of making them comprehensible to both physical and life scientists. The biologists have tried to avoid all but the most pertinent anatomical or taxonomic jargon, and the theoreticians have concentrated on the physical understanding of their work, not on the mathematical details. The result is a book which is accessible to almost any graduate scientist (and to many undergraduates too), and it is hoped that many will be stimulated to involve themselves in the field, and discover, as the authors of this book have already discovered, how profitable and enjoyable it is.

As explained in the Chairman's Foreword, the contents of this book are the Proceedings of an International Symposium held at Cambridge in

September 1975. It was a great honour to me to be invited by Sir James Lighthill and Torkel Weis-Fogh to be the scientific secretary of the Symposium and to edit these Proceedings. Since my own work is not in the field of animal locomotion, but rather in the application of fluid mechanics to circulatory and respiratory systems, I approached the task with some trepidation but great enthusiasm. As the great fluid dynamicist, Theodore von Kármàn, said when confronted with an unfamiliar problem, my imagination was "unfettered by any knowledge of the facts", and I found the whole venture enormously stimulating. I am sure that the person who has learnt most from it is myself, and I greatly appreciate the opportunity offered by it. The editorial burden was eased by the ready cooperation of all the authors, who showed considerable tolerance of my ignorance, and by the constant encouragement of Torkel Weis-Fogh (until his death) and of Sir James Lighthill. I am extremely grateful to them all. I am grateful, too, to Celia Whitchurch, Maggie Brown and Anne Rickett, who gave me invaluable assistance in the organisation of the Symposium and in all the correspondence which both preceded and followed it. I would also like to thank my wife, Avril Pedley, without whom the Index to this book could not have been compiled. I gratefully acknowledge all publishers and authors who gave permission for their original figures to be used in this volume (as recorded in the appropriate captions), and finally I would like to express my thanks to Academic Press who have made order out of the apparent chaos of thirty edited manuscripts.

Cambridge   T.J.P.
February 1977

FOREWORD

Professor Sir James Lighthill, F.R.S.

The International Symposium "Scale Effects in Animal Locomotion" was held at Cambridge University in September 1975, under the co-chairmanship of Torkel Weis-Fogh and myself. It was, however, two years earlier, in September 1973 at Duke University, North Carolina, that the desirability of holding such an International Symposium had first been demonstrated and agreed. Professor Knut Schmidt-Nielsen had invited a few people to take part in a working conference, to be held at Duke University with support from the Cocos Foundation, on scaling problems in biology. This was attended by nine persons, all of whom also attended the later Cambridge Symposium and are authors of papers in this volume. They comprised Professor Schmidt-Nielsen and two other members of the Duke University Department of Zoology (V. A. Tucker and S. Vogel), together with four more life scientists (R. M. Alexander, C. R. Taylor, T. Weis-Fogh and D. R. Wilkie) and two mechanical scientists (T. A. McMahon and myself).

A clear conclusion from the (unpublished) proceedings of that conference was the decision to hold in Cambridge during 1975 a major international meeting devoted to scale effects (that is, dependence on size) in animal locomotion. The subject matter of such a meeting would include terrestrial, aquatic and aerial locomotion; metabolism, musculature and skeletal strength; solid dynamics and fluid dynamics; all treated with the aim of elucidating in different kinds of animal locomotion how various key parameters may be related to size (as measured by the animal's mass or linear dimensions), or how there may exist critical limitations on size.

It was already envisaged that, for such a Symposium spanning mechanics, biophysics and zoology, sponsorship should be sought from three International Unions: those for Theoretical and Applied Mechanics, for Pure and Applied Biophysics, and for Biological Sciences. We were delighted that all of them gave us their strong support. In the meantime, it was arranged that participants in the symposium would stay in Christ's College Cambridge (of which Torkel Weis-Fogh was a Fellow) and that the lectures would be held in the Cambridge University Zoology Department.

In July 1974, Weis-Fogh and I both attended the important Symposium on Swimming and Flying in Nature, held at the California Institute of Technology, under the chairmanship of Professor T. Y. Wu. (Published as "Swimming and Flying in Nature," (Wu, T. Y., Brokaw, C. J. and Brennen, C. (eds.) Plenum Press, 1975)). Our attendance at this Symposium greatly helped us to identify persons concerned with aquatic or aerial locomotion who would be able to make a valuable contribution to a meeting on scale effects.

Later, we were further assisted in identifying those who should be invited to give papers or otherwise participate in the International Symposium "Scale Effects in Animal Locomotion" by our fellow-members of the Symposium's Scientific Committee: R. M. Alexander, S. S. Grigoryan, R. D. Keynes, N. V. Kokshaysky, K. Schmidt-Nielsen, T. Y. Wu, and T. J. Pedley (Scientific Secretary and Editor of these Proceedings). After the resulting invitations had been sent out, we were overjoyed that the vast majority of the recipients were able to accept and, furthermore, agreed wholeheartedly with the aims of the meeting as described above.

The meeting was attended by 58 invited participants (1 from Australia, 1 from Bulgaria, 1 from Denmark, 1 from France, 4 from the Federal Republic of Germany, 1 from the German Democratic Republic, 1 from Israel, 1 from Italy, 2 from the Netherlands, 2 from Sweden, 24 from the UK, 16 from the USA, and 1 from the USSR). Of these, 40 had their main educational background in the life sciences, and 18 in the mechanical sciences. All those invited, however, had had previous experience in interdisciplinary collaboration. As a result, a good standard of mutual intelligibility between the disciplines was achieved both in the papers and in the discussions.

The papers given at the meeting are collected together in this book. I believe that readers interested in quantitative aspects of animal locomotion, and coming from backgrounds either in the mechanical sciences or the life sciences, will find this book a clearly intelligible and comprehensive account of what is known about scale effects.

The structure of the book, besides being a natural one, reflects the structure of the meeting. Parts II, III and IV are concerned with terrestrial, aquatic and aerial locomotion respectively, just as were the second, third and fourth days of the meeting. (The only papers given at the meeting, but not included here because the authors were unable to prepare a manuscript by the time the book had to go to press, were those by Professor S. Corrsin, in the session on terrestrial locomotion, and by Professor G. W. Schaefer, in the session of aerial locomotion.) Papers on certain topics important to locomotion in general are grouped together in Part I.

There was lively and constructive discussion throughout the Symposium. Much of this led to the clarification of points touched on in the papers;

often, in such cases, authors took the opportunity after the Symposium to amend their papers accordingly for publication in this book. A few discussion contributions were selected to appear separately in the Proceedings.

Various pleasant evening occasions added to the enjoyment of participants in this Symposium but, in the nature of things, cannot be communicated to readers of the book. Besides a Cocktail Party and a Symposium Dinner, they included a showing of films on animal flight by R. A. Norberg, U. M. Norberg, G. Rüppell and T. Weis-Fogh. They included also a helpful informal discussion on dimensional analysis. Although we have not tried to summarize this in the book, we believe that it influenced the final published form of many authors' contributions.

Conscious of the great influence on research in animal locomotion of the work of Sir James Gray (born 14 October 1891, died 14 December 1975; Professor of Zoology in the University of Cambridge 1937–59), members of the Symposium collectively signed and sent to his nursing-home bedside "A warm greeting to the Doyen of Animal Locomotion, Sir James Gray". He expressed his great appreciation at receiving this.

The untimely death of Torkel Weis-Fogh prevented him from carrying out the thorough revision of the text of his paper which he is known to have planned. The Editor and I have made a few revisions in matters of detail, mainly at points where he had indicated to us his plans for emendation. Nevertheless, we emphasize that the paper as published does not take the precise final form that the author had wished, although we believe that readers will still find it of absorbing interest.

Just before his death, however, Weis-Fogh had been working actively with Professor R. M. Alexander to produce a definitive paper based on the very lively Concluding Discussion at the end of the Symposium. This Concluding Discussion was mainly concerned with problems of estimating (and scaling) the maximum sustained power output from muscle in general and flight muscle in particular. We publish this paper by Alexander and Weis-Fogh, summarizing their agreed view of that Concluding Discussion, as an epilogue to these Proceedings, rounding off this tribute to the memory of Torkel Weis-Fogh with his last piece of scientific work.

CONTENTS

Part I

GENERAL CONSIDERATIONS

Part II

TERRESTRIAL LOCOMOTION

Part III

AQUATIC LOCOMOTION

Part IV

AERIAL LOCOMOTION

Part V

PART I

GENERAL CONSIDERATIONS

1. Problems of Scaling: Locomotion and Physiological Correlates*

KNUT SCHMIDT-NIELSEN

Department of Zoology, Duke University, Durham, North Carolina, U.S.A.

ABSTRACT

This is an elementary introduction to the theme of structural and functional consequences of changing body size. I discuss lower and upper limits to body size, the use of graphs and simple equations in examining scaling problems, and the existence of constraints and discontinuities in both structure and function. Other functions that must be scaled to body size include metabolic maintenance rates, muscle power, and the supply of oxygen and fuel to the tissues. These functions in turn are the foundation for the main theme of this volume, the energetics of animal locomotion as related to body size.

INTRODUCTION

When I was invited to address this conference, and indeed to have the honor of giving the opening lecture, I accepted in spite of considerable hesitation. Obviously, the many distinguished participants have made and will make substantial contributions to our knowledge in the field covered by the conference title, and I have no hope of telling much that is new. Rather than attempt to predict what will come, I decided to speak about what biologists have done in the past, what problems have been raised, what has been achieved, and to some extent what problems seem to remain unsolved.

First of all, when we consider how animals are designed, move about, and function, we find similarities as well as differences. One difference, so obvious

* Support for this work was provided by the Cocos Foundation, and by the U.S. National Institutes of Health Grants No. HL-02228 and 1-K6-GM-21, 522.

that we often give little attention to it, is the enormous differences in size
among living organisms. A few examples of the size of organisms are listed in
Table I. Each single step in this list represents a thousand-fold difference in
size, for a total difference between the smallest and largest of 10^{21}. I have
stopped the list at the largest living animal, well knowing that the giant
Sequoia trees of California can exceed the largest whales by 10 or 15-fold.

TABLE I. Size range of living organisms, listed in steps approximately
a thousand-fold apart.

Organism		Mass
Blue whale	> 100 tonne	$> 10^8$ gram
Human	100 kg	10^5
Hamster	100 g	10^2
Bee	100 mg	10^{-1}
Large amoeba	0·1 mg	10^{-4}
Tetrahymena (ciliate)		10^{-7}
Average bacteria		10^{-10}
Mycoplasma (PPLO)		$< 10^{-13}$

BODY SIZE-LIMITS

These enormous size differences bring up two questions:
1. What are the size limits of living organisms, and
2. for an organism of a given structure, what are the consequences of a
   change in size?

Let us first examine the lower limits to size. The *Mycoplasma*, also known
as PPLO (pleuropneumonia-like organism), is the smallest organism known
to live and reproduce by itself in an artificial medium, and hence is not a virus.
Its non-aqueous cell mass is less than 10^{-14} g, and Morowitz[30] calculated
that if the inside of the cell were at neutral pH, it would contain, on the average,
no more than two hydrogen ions. Within this small volume the cell must contain the necessary metabolic equipment for its life processes, including proteins and enzymes, as well as the complete genetic information necessary for
reproduction of itself. Since the size of these essential molecules probably
cannot be scaled down, the *Mycoplasma* may very well represent an ultimate
lower size limit for a living organism.

At the other end of the size range, more than 10^{21} times larger than
Mycoplasma, we find the giant blue whale. It is commonly said that this animal
can reach its giant size only because it is aquatic and its enormous bulk is
supported by water.

The fact that there are structural constraints on the size of animals has been known since the time of Galileo.[9] Galileo understood the simple geometrical consequence of increased body size, that, as a linear dimension is increased, the mass of a similarly shaped animal increases by the cube of the linear dimension, and that the strength of the supporting structures must be correspondingly increased. It is a minor matter that Galileo, judging from his drawing of a scaled-up bone, made an arithmetical mistake and increased its diameter by the square of its length, instead of the 1·5 power.

FIG. 1. An artist's interpretation of the common (although untenable) view that the extinct giant dinosaurs were of such immense size that they could not move freely on land and therefore led a semi-aquatic life. From Gregory.[14]

The largest living land animal, the elephant, attains only about five tonnes, and it is frequently stated that the much larger extinct giant dinosaurs lived a semi-aquatic life because their great bulk prevented them from moving freely on land. The largest dinosaur, *Brachiosaurus*, weighed about 80 tonnes, and the smaller long-necked *Brontosaurus* about 30 tonnes. Figure 1 shows how these animals are supposed to have grazed at the bottom of water and used their long necks as snorkels which, according to paleontologists, simplified the problems of locomotion as well as those of breathing.[33] Anyone who has tried to walk on a muddy bottom with the water at the level of his neck knows that locomotion is not precisely simplified. Furthermore, at the indicated depth the chest would have to support an excess pressure of some 5000 kg m^{-2}, a pressure that would make inhalation at atmospheric pressure impossible.

The largest land mammal that has ever lived was the *Baluchitherium*, an extinct relative of the modern rhinoceros. It stood over 5 m high at the shoulder and weighed about 30 tonnes, as much as the *Brontosaurus*.[10] No paleontologist doubts that *Baluchitherium* was a fully terrestrial plant-eating mammal. Since well preserved skeletal material is available, we can use the size of the bones to estimate their strength and thus evaluate whether they could support this bulky animal on land. From the size of the metacarpal bone it can be calculated that its compressive strength must have been about 280 tonnes, which for a 30 tonne animal gives a safety factor of about 10. It seems to be no coincidence that this is roughly the same as the safety factor for static loads on the bones of man. We must conclude that the strength of the skeleton to support static loads is not an ultimate limit to the size of terrestrial animals. However, the ability of the skeleton to support static loads is actually irrelevant, for the stresses on the bones are much greater in locomotion when forces of acceleration and deceleration predominate. The fact that such forces bring the organism close to its design limits is witnessed by the many pulled muscles, torn ligaments, and sprained bones suffered by human athletes in competitive sports.

SCALING—USE OF ALLOMETRIC EQUATIONS

As we consider a group of reasonably similar animals, such as land mammals ranging from shrews to elephants, they cover more than a million-fold difference in size, from a few grams to several tonnes. What are the scaling effects over this size range?

In my use of the word scaling, *I define scaling as the structural and functional consequences of a change in size or in scale among similarly shaped animals.* A question may be raised about what scale we should use. Undoubtedly, for many purposes, the quantity of mass is both the most convenient and the most meaningful. For certain purposes other quantities may be more useful, for example, for similarly shaped fish, the quantity of length may be convenient and indeed more informative (cf. Ref. 2). However, for animals as different as a rhinoceros and a giraffe, what measurements of length would be meaningful? Mass is easily measured with accuracy, and the use of mass has another advantage: the density of virtually all animals is close to 1·0, and mass is therefore also a good measure of total volume, a fact which greatly simplifies many functional considerations.

A moment ago, I discussed the need for a change in the relative dimensions of bones as the size of an animal increases. The available data on the dimensions of bones are rather inadequate, but we can nevertheless find some useful information. In Fig. 2 the mass of the skeleton of some mammals is plotted on logarithmic coordinates against body mass. If all mammals were isometric,

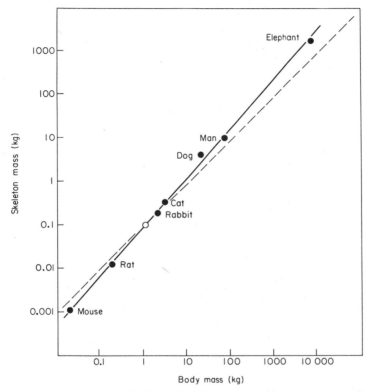

FIG. 2. The mass of mammalian skeletons (plotted on logarithmic coordinates) against body mass. The solid regression line has a slope of 1·13, and the dashed line 1·0. Data from Kayser and Heusner.[22]

that is, scaled to the same linear proportions, the mass of their skeletons would be proportional to body mass and the points would fall along the dashed line with a slope of 1·0. In reality, the skeleton of the elephant is 2·5 times as heavy as this, in accord with what was discussed above. The fully drawn line in Fig. 2 shows the empirically observed mass of the mammalian skeleton (m_{sk}) in relation to body mass (m_b), and this regression line can be represented by the equation

$$m_{sk} = 0 \cdot 1 \, (m_b)^{1 \cdot 13},$$

the exponent 1·13 indicating the slope of the regression line. Should the mass of the skeleton be scaled in proportion to the static load of the body mass, however, it should be scaled with an exponent of 1·33.* Thus, if static loads were the only consideration, the skeleton of the elephant must be considered underdimensioned. However, as I said, static loads are not the ultimate consideration, and if an elephant were to jump like a mouse, he might break the bones on take-off and totally collapse on impact.

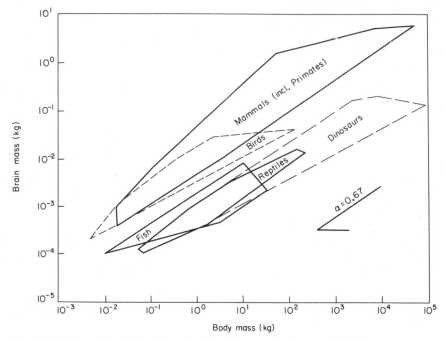

FIG. 3. It has been claimed that the extinct giant dinosaurs had very small brains relative to their body size, but their actual brain sizes fall within an extension of the range of modern reptiles. Redrawn from Jerison.[21]

The equation shown above is commonly known among biologists as an *allometric equation* (alloios = different), because it expresses in a simple form the deviation from geometric similarity, or simple isometric scaling. An amazing number of biological measurements fit such allometric, or logarithmic, equations. The value of this approach to the problem of making comparisons over a wide range of body size was emphasized by Julian Huxley,[20]

* Let the compressive strength of the supporting column increase in proportion to body mass; thus its cross-sectional area (a_{sk}) should increase in proportion to the body mass (m_b). Also, let the length of the supporting column (l_{sk}) increase in proportion to a characteristic linear dimension, i.e. in proportion to $m_b^{1/3}$. The mass of the supporting column (m_{sk}) (the product of its cross-sectional area, a_{sk}, and its length, l_{sk}) will then be: $m_{sk} \propto m_b^{1·33}$.

originally for the purpose of describing differential growth. Since then this approach has become an important tool in biology, especially when making interspecific comparisons over a wide size range. Let me mention an example.

It has been stated that the reason that dinosaurs became extinct is that their brains were exceptionally small for their large bodies, and that they therefore lost out in competition with the smaller but brainier mammals. This question has been examined by Jerison.[21] His compilation (Fig. 3) shows that the brain sizes of reptiles and fish fall within ranges well below those of mammals and birds, and without overlap with them. By making casts of the brain cavity of dinosaurs Jerison obtained data to show that the brain size of extinct dinosaurs was perfectly within the range that can be expected for reptiles of their size. Since modern reptiles with their relatively small brains have survived for hundreds of millions of years, it raises serious doubts as to the argument about small brains and dinosaur extinction.

TABLE II. The average brain size of mammals in general and of major primate groups, expressed as a function of body mass (m_b in kg). Data from Ref. (39).

Animal	Brain mass (kg)
Mammals	$0\cdot01\ m_b^{0\cdot70}$
Monkeys	$0\cdot02$ to $0\cdot03\ m_b^{0\cdot66}$
Great apes	$0\cdot03$ to $0\cdot04\ m_b^{0\cdot66}$
Humans	$0\cdot08$ to $0\cdot09\ m_b^{0\cdot66}$

Allometric scaling does not need to use graphs, we can merely use the corresponding allometric equations as a basis for comparison. As an example, Table II shows the relative brain sizes of mammals in general and of the major groups of primates. We note that in all the groups the brain size varies with nearly the same power of body mass, about $0\cdot67$ (incidentally, the same slope as that which predominates in Fig. 3). When the exponent in these equations is the same, the numerical factor preceding the exponential term directly expresses the relative magnitude of the variable in question. Looking at these factors we see that monkeys have brains that are 2–3 times larger than mammals in general, the great apes twice as large as monkeys, and humans twice as large again. These equations simply tell us the essential difference in brain size between monkey and man, and this comparison we can make at a glance although there is no overlap in body size between the two groups.

This approach to the study of size, or the effects of scale, has been a powerful tool in biology, but it is not without limitations. In a recent paper allometric scaling was used to estimate the wing span of a pterosaur, a fossil flying reptile.[27] In this case the length of the bone in the upper arm, the humerus, was used to estimate a total wing span of 15·5 m, "undoubtedly the largest flying creature presently known". The extrapolation was based on the proportion between the short humerus and the total wing length in other kinds of pterosaurs of much smaller body size, from which all the wing bones are preserved. In a letter to the editor of Science, Greenewalt[13] criticized this extrapolation and suggested that a more realistic calculation could be based on the equations for observed wing lengths of birds. The wing span would then be only one-third of the estimated 15 m, or 5·25 m. Both calculations are based on extrapolations, and it is probably unwise to use allometric scaling for any extrapolation, in particular when limits and constraints are unknown and discontinuities may exist.

LIMITATIONS AND CONSTRAINTS

The question of limitations or constraints is very important. When size is changed, a limit may be reached which can be overcome only by changes in design, in materials, or in both. Biology is full of examples of such discontinuities. It is equally familiar to engineers. For example, the limit to the length of a bridge built as a stone arch is limited by the compressive strength of stone. By changing the design and supporting the bridge on tensile elements, and by changing the material to steel which is strong in tension, the engineer can readily increase the span by perhaps 100-fold.

Analogous examples of changes or discontinuities can be found in biology, in regard both to design and to materials. When the size of a small organism increases to a point where oxygen cannot be supplied at a sufficient rate by diffusion alone, convection, a new principle, is added. When convection takes place in the external medium, whether water or air, we call it respiration or ventilation. Inside the body the addition of a convective system to facilitate oxygen transport is, of course, known as circulation. Another discontinuity occurs when an oxygen carrying compound, such as hemoglobin or hemocyanin, is added to the circulating blood, thus augmenting its oxygen carrying capacity.

Insects use an entirely different design to solve the problem of rapid internal distribution of oxygen; air-filled tubes lead from the body surface directly to the organs and cells. The diffusion coefficient for gases in air is 10 000 times higher than in water, and in insects the blood plays no role in the distribution of oxygen. However, even so, highly active insects have convection added to the already rapid diffusion, for they actively ventilate their tracheal system, thus providing the flight muscles with oxygen at rates that permit these to be the most active animal tissues known.[47] It is, in fact, most likely that the upper limit on the power output of muscle is inherent in the muscle itself, and not limited by the rate of oxygen supply. This latter is an important consideration, for it has been suggested that the possible body size of insects is limited by the need for an adequate supply of oxygen. Since active ventilation is common in insects, it is most unlikely that oxygen supply is a constraint on size. However, the design of the skeleton as an external hard cuticle, as opposed to the internal skeleton of the vertebrates, has been discussed by Currey[5] as a possible limitation on their size, a subject to be discussed elsewhere in this volume.

We know that animals must live within the laws of physics and chemistry. In the examples I have mentioned, the solution to a problem was found by changes in scale, in design, or in material. Animals are unable to manipulate physical constants such as diffusion coefficients in air and water; in other words, these are scale-independent constants. Similarly, the physical strength of structural materials in the bridges, stone and steel, and the bones and cuticles of animals, are probably also scale independent. Animals can use different principles of design, and different materials such as bone or chitin, but they must still function within the constraints of the physical world they live in. However, by finding novel solutions to their problems they can overcome some of the limitations that are imposed by scale.

METABOLIC RATE

So far I have mainly discussed the scaling of structure, rather than function. However, animals characteristically are alive, move about, feed, mate, pursue prey, and escape from predators. The mechanical energy for locomotion is derived from chemical energy, which comes from the oxidation of suitable fuel. This we frequently refer to as metabolism, and for reasons that I shall not discuss here, it is convenient, as well as reasonably accurate, to measure the rate of metabolism as the rate of oxygen consumption. In fact, the two concepts have, in the minds of many biologists, almost become synonymous.

In the absence of external activity the rate of metabolism, or of oxygen consumption, continues at some rate which could be called the resting or maintenance rate. In cold blooded animals this level is ill-defined, it changes drastically with temperature and also with such factors as feeding or merely the presence of food and its amount, the previous history of the animal, and innumerable other environmental and internal factors. In the so-called warm-blooded animals, mammals and birds, the resting metabolic rate is relatively well-defined at a given level which I shall call the maintenance level. A major contributing factor to this constancy is that the body temperature of these animals remains relatively constant, with variations of only a few degrees. The various groups of warmblooded vertebrates maintain their body temperatures at characteristic although somewhat different levels (Table III), independently of whether the animal lives in hot or cold climates, on land or in the sea. Within each group, the normal body temperature appears to have no relation to the body size of the animal, and this situation prevails although the conditions for heat loss depend on the surface area of the animal and its insulation, factors which differ very much with body size and thus are scale dependent.

TABLE III. The body temperature of higher vertebrates is, within each group, independent of body size, i.e. it is a scale independent invariable. Within each major group the lethal temperature is approximately 6°C above the "normal" temperature range of that group. Data quoted from Ref. (37).

Animal	Approximate normal core temp.	Approximate lethal core temp.
Monotremes (echidna)	30–31	37
Marsupials	35–36	40–41
Insectivore (hedgehog)	34–36	41
Eutherian mammals	36–38	42–44
Birds, non-passerine	39–40	46
Birds, passerine	40–41	47

That body temperature is a scale independent quantity is an empirical finding which we must accept, although it is difficult to find convincing rational explanations for this constancy, as it is for the characteristic group-to-group differences. Equally difficult to explain is the apparent constancy in the range of body temperature that can be tolerated, for it appears that the lethal limit is reached when the body temperature is raised to about 6°C above the "normal" body temperature.

Let us return to the maintenance metabolic rate, which can also be expressed as heat production. The well known "mouse-to-elephant curve", reproduced in Fig. 4, shows that, when plotted on logarithmic coordinates, the heat production of birds and mammals increases linearly with increasing body size. A great mass of additional data has now been accumulated to show that the slope of the regression line in this plot is 0·75 or very close to this value. It was earlier believed that, for animals to maintain a constant body temperature, their heat production should be proportional to their relative surface area.[34] This would lead to a slightly lower slope, 0·67, and a great deal of effort went into attempts at fitting the observed data to this concept. The relationship was given names such as "the surface law", and led to the unfortunate practice of expressing metabolic rates relative to body surface area.

It is to the credit of Max Kleiber[23] that he expressed the "mouse-to-elephant curve" as an allometric equation, thus emphasizing observed factual information rather than preconceived explanations. As additional information has accumulated, and has been carefully reviewed,[4, 16, 24] we can now

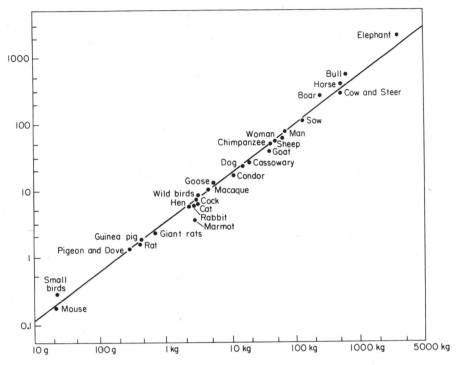

FIG. 4. Metabolic rates (in watts) of mammals and birds plotted on logarithmic coordinates against body mass (in kilograms).

summarize our information about maintenance metabolic rates of birds and mammals in allometric equations as shown in Table IV. Since the body size expression in these equations occurs with nearly the same exponent, we can again compare directly the preceding numerical factors. In a few words, these say that, with allowance for individual variations in both directions, marsupial mammals have lower metabolic rates than ordinary placental mammals, that birds have metabolic rates similar to placental mammals of the same size, and that passerine birds, that is sparrows and finches and crows and starlings, have metabolic rates consistently higher than other birds. These few numbers actually put in a nutshell the essence of nearly a century of accumulated knowledge of the maintenance metabolic rates of birds and mammals.

As we see, the exponents, representing the slope of the regression lines, are close to 0·75. If the heat production per unit body mass were the same in the mouse and the elephant, the regression line would have the slope of 1·0, and the total heat production of the elephant would be 20 times higher than in reality it is. Why is heat production not proportional to body mass? Small and large animals have cells that are roughly of the same size, within an order of magnitude of 10 μm.[43] The large organism is therefore not made up of larger cells, but of a larger number of cells of about the same size. It seems, thus, that although the large animal consists of similar-sized cells, these somehow are metabolically more dilute.

TABLE IV. Metabolic rates (in watts) of warm-blooded vertebrates as a function of body mass (m_b in kg). Data recalculated from Refs. (6, 26, 28).

Marsupial mammals	$2\cdot36\ m_b^{0\cdot737}$
Eutherian placentals	$3\cdot34\ m_b^{0\cdot75}$
Non-passerine birds	$3\cdot79\ m_b^{0\cdot723}$
Passerine birds	$6\cdot25\ m_b^{0\cdot724}$

It has long been understood that mammals and birds could not have metabolic rates in proportion to their mass for the simple reason that heat exchange with the environment is a surface-dependent process. This was succinctly expressed by Max Kleiber in the following way: If a steer were designed with the same specific metabolic rate as a mouse, it could dissipate the metabolic heat only if its surface temperature were well above the boiling point. Conversely, if a mouse had the same low specific heat production as a steer, it could keep warm only if it were insulated with a fur at least 20 cm thick. This convincingly demonstrates that the metabolic rate and problems of heat dissipation must be considered in relation to body surface area. However, if this were the only consideration, we might expect metabolic regression lines with the slope 0·67 and not 0·75. The interesting fact is that several groups of cold blooded animals, which have no problem of heat regulation, and even some plants, display metabolic regression lines relative to body size that have the same slope, 0·75, or nearly so.[16] We should, however, note that this is not a universal biological rule, for there are many exceptions with both higher and lower slopes.

A great deal of effort and time have gone into attempts at "explaining" the deviations from a surface-related exponent, often with arguments that have a certain metaphysical quality. Recently, a more rational analysis of this exponent was presented by McMahon,[29] who started with the well-known fact that vertebrates are not geometrically similar or isometric. He suggested a model based on the analysis of the functional requirements of elastic similarity between animals. This model requires a distortion from geometric similarity, which led McMahon to the conclusion that metabolically related variables should be scaled according to elastic similarity, with body mass raised to the power 0·75 and biological frequencies inversely as the body mass to the power 0·25.

Although a wealth of information is available about the maintenance metabolic rates of mammals and birds, our knowledge of maximum performance or peak power is only rudimentary. We know that in some animals the peak power output during maximum exertion in locomotion, such as running or flying, may exceed the maintenance level by a factor of 10 or 15, or perhaps even 20. Hart and Berger[15] suggested that power output of birds in flight exceeds their maintenance rate by a constant factor of about ten, independently of body size. However, the available information is insufficient to say with any certainty that this factor is really scale independent. For mammals, the best information is available for man and dogs, but these represent roughly the same body size. Several investigators have been unable to induce small rodents to achieve peak power of more than 6 or 8 times their maintenance rate, only one-half of that in dog and man.[32, 38, 48] The fact that peak power output in small birds can be very high could be related to the design of their respiratory system, which involves a much more effective air flow in the lung than in mammals. On the other hand, the mammalian design undoubtedly permits very high peak power, even in those of small body size, for bats in flight reach a power output similar to that of birds of the same body size.[44] We can therefore suggest that peak power may exceed the maintenance power by a factor of 10 to 15 in both birds and mammals, but whether or not this factor actually is body size-independent remains to be determined.

MUSCLE POWER

In locomotion the power is produced by the muscles, and in this field more precise information is available. It appears that the muscles make up about the same fraction of the total body mass of all mammals, around 45%, irrespective of their body size.[31] Much more detailed information is available for birds, and the extensive material compiled by Greenewalt[11, 12] is particularly useful. It seems that the large pectoral muscle, the main flying muscle of birds, makes up about 15% of the body mass. Although there are variations in this figure, related to the flight habits of the bird in question, this percentage remains a fairly constant fraction, irrespective of the body size. It is interesting that the flight muscles of hummingbirds make up a larger fraction of their body mass, some 25 to 30 %. This is in accord with the greater power requirements for hovering flight than for flapping flight. Also, in humming-birds, the muscle responsible for the upstroke of the wing makes up about one-third of the total mass of the flight muscles, while in other birds these muscles are only one-tenth of the total. Without any knowledge of other observations, the availability of this large muscle for the upstroke would by itself suggest that lift is provided during both upstroke and downstroke. This, of course, is confirmed by the aerodynamic analysis of the wing movements of hovering hummingbirds.[41]

As for muscle as a tissue, it appears that the maximum tension per unit area, or stress, that can be delivered by muscle is roughly 300–400 kN m^{-2} (3–4 kg cm^{-2}). The amazing fact is that this maximum stress appears to be body size independent, not only in higher vertebrates, but also in many other organisms, including at least some invertebrates. In vertebrate skeletal muscle the maximum relative shortening, or strain, also seems to be rather constant at around 0·3. The conclusion is that the maximum work (force × distance) performed in one contraction, when calculated per unit volume of muscle, is invariable and independent of size. This generalization was recognized by Hill[17] and is well supported by later evidence.

The conclusion that the maximum work per contraction is scale-independent is fully in accord with our present knowledge of the structure of skeletal muscle. In widely different organisms the number of filaments per cross-sectional area of muscle is the same (i.e. filament thickness is constant), and the sarcomere length, the length of the filaments, and thus the maximum overlap between thick and thin filaments, are all of the same magnitude in small and large animals. This uniformity of filament thickness, length, and overlap explains the constancy of stress and strain (although it does not answer the question of why these parameters remain constant).

If the work per contraction is constant, the power output during contraction will be a direct function of the speed of shortening, or the strain rate. For repeated contractions, as in running or flying, the average power output of a muscle will be directly proportional to the frequency of its contraction. It should be noted, however, that the very high frequency of contraction in flying insects does not imply power inputs at inordinately high levels, for the flight muscles of insects (due to the peculiarities of the flight apparatus) do not shorten by more than perhaps a couple of per cent.

In summary then, we can say that skeletal muscle is uniformly of a similar structure, and is scale-independent in regard to maximum stress and strain. In regard to strain rate, or speed of shortening, there are vast differences, and the frequency of contraction increases regularly with decreasing body size. Thus, power output is scaled to increase with decreasing body size.

FUEL AND OXYGEN

For a steady state power output, muscle must be supplied with fuel and with oxygen at the appropriate rate. To provide oxygen, the respiratory systems must be scaled accordingly, whether they be gills, lungs, or tracheae, as must the circulatory systems, including the heart and its pumping capacity.

I shall only say a few words about the scaling of these systems and their function, and refer to a previous review for further details.[35] In mammals, the lung volume seems to be a constant fraction of body size. The relative lung volume is thus body-size-independent, and a logarithmic plot of lung volume against body mass gives a regression with a slope of nearly 1·0. The total area of the alveoli of the lung, however, is directly proportional to the rate of oxygen consumption, with a slope of 1·0 in the regression line between these two variables (or 0·75 when related to body mass).

TABLE V. Relationships between variables related to oxygen supply and body mass (m_b in kg) of mammals. Data from Refs. (1, 7, 40)

O_2 consumption (litre min^{-1})	$= 0{\cdot}0116\, m_b^{0{\cdot}76}$
O_2 consumption per kg (litre min^{-1} kg^{-1})	$= 0{\cdot}0116\, m_b^{-0{\cdot}24}$
Lung ventilation rate (litre min^{-1})	$= 0{\cdot}334\, m_b^{0{\cdot}76}$
Lung volume (litre)	$= 0{\cdot}063\, m_b^{1{\cdot}02}$
Tidal volume (litre)	$= 0{\cdot}0062\, m_b^{1{\cdot}01}$
Respiration frequency (min^{-1})	$= 53{\cdot}5\, m_b^{-0{\cdot}26}$
Blood volume (litre)	$= 0{\cdot}055\, m_b^{0{\cdot}99}$
Heart weight (kg)	$= 0{\cdot}0058\, m_b^{0{\cdot}99}$
Heart rate (min^{-1})	$= 241\, m_b^{-0{\cdot}25}$

A large amount of data has been accumulated on the scaling of other variables in the oxygen transport system, such as ventilation of the lungs, circulation, heart size, heart frequency, blood volume, dimensions of blood vessels, and so on (see Table V). In the tissues the oxygen must be delivered at the high rate required in the small animal, and variables such as diffusion distances and diffusion gradients are also scaled to body size. The hemoglobin content of mammalian blood is body size independent, but the unloading pressure for oxygen is higher in the small animal, thus facilitating the delivery of oxygen. Further aids to a faster delivery of oxygen are provided by a more pronounced Bohr effect in the small animal and by the presence of the enzyme carbonic anhydrase.

Similar information on a wide range of respiratory variables in birds has been analyzed, from the viewpoint of scaling, by Lasiewski and Calder[25] and Hinds and Calder,[18] and the dimensions of fish gills have been subject to a series of studies by Hughes and Morgan[19] and others.

The completely different system for delivery of oxygen in insects, the subject of careful analyses by Weis-Fogh and others, was mentioned above. Although insect blood plays no direct role in the delivery of oxygen, blood is necessary for the fuel supply to the tissues. In order to supply fuel at the required high rate over the available diffusion distances, it seems that the most active insects, for example honeybees, depend on sugar concentrations in the blood as high as 5 or 10%, or 100 times as high as in man.[49]

With an adequate supply of oxygen and fuel assured, are these factors eliminated as a posible constraint to the power of insect muscle? If so, could the power output of insect muscle be further increased? There may be another design feature that puts an absolute limit to the power output. The most active insect flight muscles have as much as 30% of their volume occupied by mitochondria, and if a further increase in power were to be achieved by increasing the density of mitochondria, the volume of contractile elements would decrease, thus defeating the intended increase in power.

A recent report suggests that the glucose concentration in mammalian blood is also a function of body size, with higher concentrations observed in mammals of small body size.[46] This is analogous to the situation in insects, and would be in accord with the higher specific rate of oxygen consumption in the small mammal.

Although I have now discussed a number of variables that appear to be scaled in accord with the actual needs of animals of various sizes, the system must also provide power for peak requirements (except that, for very brief periods, the muscles may work without being in steady state in regard to oxygen and fuel supply).

As I said before, the information we have about peak power for maximum sustained exertion is relatively meagre, and this makes it difficult, at this time, to make any even reasonably definite statement about how peak power may be scaled. All we can say is that the information at hand does not contradict the tentative assumption that both birds and mammals can maintain, for some time, peak power output of the order of 15 times their maintenance rate.

LOCOMOTION AND BODY SIZE

With the preceding information on the scaling of power output, it is useful to proceed to the problems of scaling in locomotion. This, in fact, is the subject to be dealt with in the remainder of this volume. When animals swim or run or fly or jump, they perform activities that in various ways are size or scale dependent. Obviously, they are bound by and must act within the physical laws of mechanics and fluid dynamics, and they encounter limitations and constraints that put limits on their activities. Most of us have encountered discussions of the largest possible size for an animal that moves in flapping flight, and perhaps also the interesting case in which an apparent lower size limit for a jumping animal can be overcome by a novel design.

Let us examine the latter case as an example of a discontinuity in design. A very small animal, such as a flea, of necessity has a very short take-off distance. It must therefore provide all the energy and acceleration for the jump in a very short time, shorter than even the fastest muscle could contract. The flea has overcome this constraint by a novel design; the necessary energy for the jump is stored in a pad of the elastic material, resilin, and is released on take-off so that the animal is thrown into the air as from a catapult.[3]

During the last several years much new material has been accumulated as to the energy requirements for swimming, for flying, and for running. The understanding of swimming, and especially of flying, has advanced faster than that of running. In swimming and flying, the laws of fluid dynamics permit a detailed analysis of the energy expenditure required to overcome drag and to provide lift. In contrast, when an animal moves horizontally over land, practically no external work is performed, yet the cost of moving a unit of body mass over a unit distance is higher than for both swimming and flying.[36,45] Information about the cost of running for animals of various body sizes has been accumulated by Taylor and his collaborators.[42] A few examples from their findings will show that their observations still seem to lack adequate mechanical and physiological explanations.

As an animal runs on a horizontal treadmill, its oxygen consumption increases with the running speed (Fig. 5). What is surprising, and could not easily be predicted, is that the increase in the rate of oxygen consumption (power) is linear. The increase in power is much steeper for the small than for the large animal. Since the regression lines are straight, the increment cost of running for an increment in speed is speed independent. If we plot the slopes of these lines relative to body mass on logarithmic coordinates, we obtain the straight regression line shown in Fig. 6. The ordinate in this graph gives the slopes from Fig. 5 expressed as the amount of oxygen needed to move one gram of the animal over 1 km. The slope of the regression line in Fig. 6 is -0.4.

It is interesting that all available information for running humans seems to fall outside the regression line for animals that run on four legs. This observation led Fedak et al.[8] to examine other bipedal runners, and since few mammals run on two legs, he chose birds which range in size from small quail (30 g) to the 1000 times larger South American rhea (30 kg). Again, the observations fell on a straight regression line, but in this case with a different slope, -0.2. The data for man fall directly on an extension of the regression line for the birds, but whether this expresses a fundamental difference between quadrupedal and bipedal locomotion remains uncertain.

As more information is accumulated on the power requirements for terrestrial locomotion, it becomes increasingly unsatisfactory that our under-

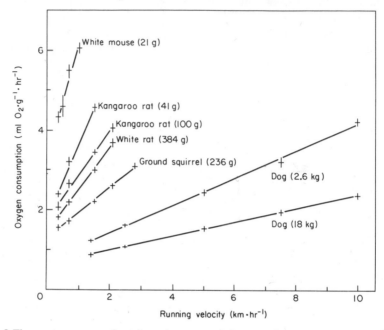

FIG. 5. The oxygen consumption of running mammals increases linearly with running velocity, the increase being higher in small than in large animals. From Taylor *et al.*[42] (1 ml $O_2 \cdot g^{-1}$ hr^{-1} ≈ 5.6 J kg^{-1} s^{-1}; 1 km. hr^{-1} ≈ 0.28 m s^{-1}).

standing is incomplete. Although we have reasonably good information about how power requirements are scaled to body size, we would also like to know why this is so. A major difficulty in this interpretation is that we lack an understanding of how energy is used when no external work is performed. We are still uncertain about the relative uses of energy in the acceleration and deceleration of the limbs, the movement of the body mass in the field of gravity, the importance of elastic storage in the recovery of kinetic energy, and in what physiologists call "negative work", that is, when a tensed muscle is being stretched and work thus is done on the muscle.

Many of these problems will be discussed by later contributors to this volume. I have had the honor of presenting an elementary introduction to the general theme of scaling and animal function, and I am eagerly looking forward to the contributions that are yet to come.

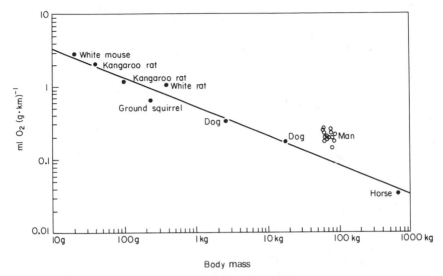

FIG. 6. The slopes of the regression lines in Fig. 5 plotted against body mass. These slopes represent the increment cost of moving one unit of body mass over one unit distance. From Taylor *et al.*[42] (1 ml $O_2 g^{-1} km^{-1} \approx 20 \cdot 1$ J kg^{-1} m^{-1}).

REFERENCES

1. Adolph, E. F. Quantitative relations in the physiological constitutions of mammals. *Science*, **109**, 579–585 (1949).
2. Bainbridge, R., The speed of swimming of fish as related to size and to the frequency and amplitude of the tail beat. *J. Exp. Biol.* **35**, 109–133 (1958).
3. Bennet-Clark, H. C, and Lucey, E. C. A. The jump of the flea: a study of the energetics and a model of the mechanism. *J. Exp. Biol.* **47**, 59–76 (1967).
4. Brody, S. "Bioenergetics and Growth. With Special Reference to the Efficiency Complex in Domestic Animals." Reinhold Publ. Co., New York (1945). Reprinted Hafner Publ. Co., Darien, Conn. (1964) 1023 pp.
5. Currey, J. D. The failure of exoskeletons and endoskeletons. *J. Morphol.* **123**, 1–16 (1967).
6. Dawson, T. J. and Hulbert, A. J. Standard metabolism, body temperature, and surface areas of Australian marsupials. *Am. J. Physiol.* **218**, 1233–1238 (1970).
7. Drorbaugh, J. E. Pulmonary function in different animals. *J. Appl. Physiol.* **15**, 1069–1072 (1960).
8. Fedak, M. A., Pinshow, B. and Schmidt-Nielsen, K. Energy cost of bipedal running. *Am. J. Physiol.* **227**, 1038–1044 (1974).
9. Galilei, G. "Dialogues Concerning Two New Sciences." (1637). Translated by Henry Crew and A. De Salvio. Macmillan, New York (1914).
10. Granger, W. and Gregory, W. K. A revised restoration of the skeleton of *Baluchitherium*, gigantic fossil rhinoceros of central Asia. *Am. Mus. Novitates*, No. 787, 1–3 (1935).

11. Greenewalt, C. H. Dimensional relationships for flying animals. *Smithsonian Misc. Coll.* **144**, 1–46 (1962).
12. Greenewalt, C. H. The flight of birds. *Trans. Am. Phil. Soc., New Ser.* **65**, Pt. 4, 1–65 (1975).
13. Greenewalt, C. H. Could pterosaurs fly? *Science,* **188**, 676 (1975).
14. Gregory, W. K. "Evolution Emerging. A Survey of Changing Patterns from Primeval Life to Man." Vols. 1–2. Macmillan Company, New York (1951) 736 and 1013 pp.
15. Hart, J. S. and Berger, M. Energetics, water economy and temperature regulation during flight. *In* "Proc. 15th Int. Ornith. Congr." pp. 189–199 (1972).
16. Hemmingsen, A. M. Energy metabolism as related to body size and respiratory surfaces, and its evolution. Copenhagen, *Reports of the Steno Memorial Hospital and Nordisk Insulinlaboratorium,* **9**, 1–110 (1960).
17. Hill, A. V. The dimensions of animals and their muscular dynamics. *Proc. Roy. Inst. G.B.* **34**, 450–471 (1950). Also published in *Sci. Prog.* **38**, 209–230 (1950).
18. Hinds, D. S. and Calder, W. A. Tracheal dead space in the respiration of birds. *Evolution,* **25**, 429–440 (1971).
19. Hughes, G. M. and Morgan, M. The structure of fish gills in relation to their respiratory function. *Biol. Rev.* **48**, 419–475 (1973).
20. Huxley, J. S. Constant differential growth-ratios and their significance. *Nature,* **114**, 895–896 (1924).
21. Jerison, H. J. Gross brain indices and the analysis of fossil endocasts. *In* "Advances in Primatology," Vol. 1: "The Primate Brain" (C. R. Noback & W. Montagna, eds), pp. 225–244. Appleton-Century-Crofts, New York (1970).
22. Kayser, Ch. and Heusner, A. Étude comparative du métabolisme énergétique dans la série animale. *J. Physiol. (Paris),* **56**, 489–524 (1964).
23. Kleiber, M. Body size and metabolism. *Hilgardia,* **6**, 315–353 (1932).
24. Kleiber, M. "The Fire of Life. An Introduction to Animal Energetics." John Wiley, New York (1961) 454 pp,
25. Lasiewski, R. C. and Calder, W. A., Jr. A preliminary allometric analysis of respiratory variables in resting birds. *Resp. Physiol.* **11**, 152–166 (1971).
26. Lasiewski, R. C. and Dawson, W. R. A re-examination of the relation between standard metabolic rate and body weight in birds. *Condor,* **69**, 13–23 (1967).
27. Lawson, D. A. Pterosaur from the latest Cretaceous of West Texas: Discovery of the largest flying creature. *Science,* **187**, 947–948 (1975).
28. MacMillen, R. E. and Nelson, J. E. Bioenergetics and body size in dasyurid marsupials. *Am. J. Physiol.* **217**, 1246–1251 (1969).
29. McMahon, T. Size and shape in biology. *Science,* **179**, 1201–1204 (1973).
30. Morowitz, H. J. The minimum size of cells. *In* "Principles of Biomolecular Organization" (G. E. W. Wolstenholme and M. O'Connor, eds) pp. 446–459. J. & A. Churchill Ltd., London (1966).
31. Munro, H. N. Evolution of protein metabolism in mammals. *In* "Mammalian Protein Metabolism," Vol. 3, 133–182. Academic Press, New York (1969).
32. Pasquis, P., Lacaisse, A. and Dejours, P. Maximal oxygen uptake in four species of small mammals. *Resp. Physiol.* **9**, 298–309 (1970).
33. Romer, A. S., "Vertebrate Paleontology." 3rd Edition. University of Chicago Press, Chicago, Ill. (1966) 468 pp.
34. Rubner, M. Ueber den Einfluss der Körpergrösse auf Stoff-und Kraftwechsel. *Z. Biol.* **19**, 535–562 (1883).

35. Schmidt-Nielsen, K. "How Animals Work." Cambridge University Press, England (1972) 114 pp.
36. Schmidt-Nielsen, K. Locomotion: energy cost of swimming, flying and running. *Science*, **177**, 222–228 (1972).
37. Schmidt-Nielsen, K. "Animal Physiology. Adaptation and Environment." Cambridge University Press, England (1975) 699pp.
38. Segram, N. P. and Hart, J. S. Oxygen supply and performance in *Peromyscus*. Metabolic and circulatory responses to exercise. *Can. J. Physiol. Pharmacol.* **45**, 531–541 (1967).
39. Stahl, W. R. Organ weights in primates and other mammals. *Science*, **150**, 1039–1042 (1965).
40. Stahl, W. R. Scaling of respiratory variables in mammals. *J. Appl. Physiol.* **22**, 453–460 (1967).
41. Stolpe, M. und Zimmer, K. Der Schwirrflug des Kolibri im Zeitlupenfilm. *J.f. Orn.* **87**, 136–155 (1939).
42. Taylor, C. R., Schmidt-Nielsen, K. and Raab, J. L. Scaling of energetic cost of running to body size in mammals. *Am. J. Physiol.* **219**, 1104–1107 (1970).
43. Teissier, G. Biométrie de la cellule. *Tabulae Biologicae*, **19**, Pt. 1, 1–65 (1939).
44. Thomas, S. P. and Suthers, R. A. The physiology and energetics of bat flight. *J. Exp. Biol.* **57**, 317–335 (1972),
45. Tucker, V. A., Energetic cost of locomotion in animals. *Comp. Biochem. Physiol.* **34**, 841–846 (1970).
46. Umminger, B. L. Body size and whole blood sugar concentrations in mammals. *Comp. Biochem. Physiol.* **52A**, 455–458 (1975).
47. Weis-Fogh, T. Diffusion in insect wing muscle, the most active tissue known. *J. Exp. Biol.* **41**, 229–256 (1964).
48. Wunder, B. A. Energetics of running activity in Merriam's chipmunk, *Eutamias merriami. Comp. Biochem. Physiol.* **33**, 821–836 (1970).
49. Wyatt, G. R. The biochemistry of sugars and polysaccharides in insects. *In* "Advances in Insect Physiology," Vol. **4** (J. W. L. Beament, J. E. Treherne, and V. B. Wigglesworth, Eds), 287–360. Academic Press, London (1967).

2. Metabolism and Body Size

D. R. WILKIE

Department of Physiology, University College, London, England

ABSTRACT

I. Statement of the problem. The regular variation of metabolism with body size. The inadequacy of the classical "surface law" and of considerations of heat balance to explain this regularity.

II. Optimal design. Animal design has been highly optimised by millions of years of natural selection. What are the key factors that must be optimised to ensure survival? What limits are set by the physical and chemical properties of matter to the possible variations of animal design? Can these explain problem *I*?

III. To what extent is the metabolism of the whole animal determined by the properties of its cells? Apparently not very much, though far more experimental evidence is needed. The properties of the cells *in vivo* are evidently strongly conditioned by control mechanisms that in turn depend on animal size. This leads to interesting problems of considerable practical importance when a foetus becomes a free-living organism.

IV. Dimensional analysis in biology. The theoretical basis of dimensional analysis. Does it enable us to arrive at general conclusions of biological interest?

INTRODUCTION

At this Symposium we are discussing a topic that forms a bridge between biology and engineering. We believe that living creatures operate subject to the laws of physics and chemistry, partly from our historical experience that in the past *some* parts of biology have been thus explained, and partly as an act of faith. Further development in physical science may well be needed in order to explain biological function: forty years ago physics could not explain how a modern pocket calculator works. We may need to know very much more to explain the brain itself.

Living creatures resemble man-made machines in very many ways, and they have been highly optimised by the harsh conflict for survival. How do they differ? There are superficial differences, for example in the materials employed and in the absence of freely-rotating parts (with the exception, apparently, of some bacterial flagella). At a deeper level there is a difference in that a machine is specified by a complete blue-print showing every single nut and bolt whereas a living system must be one that can be specified by genetic instructions of more general character. This leads to a certain plasticity in structure which, moreover, is retained through life. Not only are living things self-organising: they are also to a remarkable degree self-repairing.

Other similarities and differences will no doubt be discussed later. I wish to draw attention to some features of animal design which are well established experimentally, but which nevertheless lack a satisfactory physical explanation.

METABOLISM AND BODY SIZE, THE PROBLEM

Living creatures capable of sustaining an independent existence are found in an astonishingly wide range of sizes. The smallest unicellular organisms weigh less than 10^{-15} kg, the largest about 10^{-6} kg: multicellular organisms span the range from 10^{-8} kg to 10^5 kg. Obviously the large organisms are not merely magnified versions of the small ones. Design varies radically with size and, at least in the middle size range (say 10^{-7} to 10^2 kg), there are usually, at a given body weight, several radically different types of design (for example, in different phyla) that have been efficient enough for the species involved (of similar size) to have remained competitive with one another over many millions of years.

Notwithstanding the astonishing variety of size and design there is an even more astonishing regularity. A very large number of physiological parameters vary in a regular way with body mass: not only this but the mathematical relationship is very frequently of the same mathematical form

$$P = aM^b, \tag{1}$$

where P is the physiological parameter in question, M is total body mass; and a and b are constants,

A wide range of such equations, mainly concerning locomotion, are described by other contributors to this Symposium. In several instances explanations have been given, in the satisfactory terms of engineering, for the basis of the relations observed experimentally.

I propose to discuss the relation between resting metabolic rate and body size, where the experimental evidence is more extensive but the theoretical situation is frankly confused. By drawing attention to this unsatisfactory state I hope at least to indicate the lines along which an explanation might be sought and to urge the compilation in one place of the requisite experimental information, much of which probably already exists in fragmentary and dispersed form.

It has been known for a very long time, certainly since the work of Sarrus and Rameaux[16], that small warm-blooded creatures have a more intense metabolism (power output, measured in $W\,kg^{-1}$) than large ones. They proposed the classical "surface law" according to which the total metabolic rate of an animal should be proportional not to its body mass but to its surface area. The surface area for bodies of similar shape and density is proportional to $M^{2/3}$ and the surface area does indeed vary in this way, at least in vertebrates, over a size range of almost 10^7 (Ref. 6, Fig. 10).

If the proposal by Sarrus and Rameaux were true it would follow that

$$P = aM^{2/3}, \tag{2}$$

where P is the resting metabolic rate expressed in watts and a is a constant expressed in $W\,kg^{-2/3}$. Equations such as (1) and (2) are commonly transformed into the linear form:

$$\log P = \log a + b \log M. \tag{3}$$

The "two-thirds power law" is indeed followed quite closely so long as one does not examine too wide a range of size. Figure 1 shows how accurately it may be obeyed *within* a single species (the guinea pig) over a mass range of 10:1. The guinea pig is particularly suitable for this study since the young are born in a very mature state.

Acceptance of the "surface law" has had at least one unfortunate consequence. Sarrus and Rameaux had argued, quite correctly, that since heat is lost largely from the body surfaces it was necessary for purposes of temperature regulation that the metabolism of different animals should vary in proportion to their surface area. This explanation was held for many years to be sufficient (it is still often encountered at the present time) and this almost certainly delayed a deeper investigation of the matter.

The simple explanation in terms of heat balance is unsatisfactory for at least three reasons:

1. Most of the measurements on which the "surface law" was founded were made on animals under resting or basal conditions; this was the case for the results shown in Figs. 1–5 of this paper. In order to make such measurements it is essential to adjust the environmental temperature in relation to the animal's thermal insulation, so the problem of thermal balance is taken care of by the experimenter. Under natural conditions too it has been established[17] that animals living in cold climates adjust their thermal insulation at rest so that they do not normally need to produce extra heat merely to maintain their body temperature. It is under quite different circumstances, for example

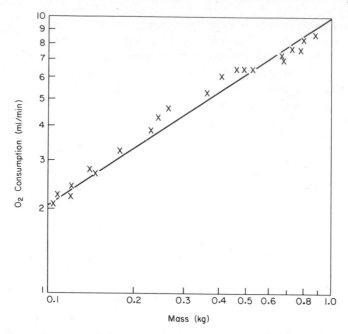

FIG 1. The resting metabolic rate of guinea pigs of varying size. The line has been fitted by the equation: O_2 consumption (ml min^{-1}) = 9·55 $M^{2/3}$ (kg). (Hill, June, R., unpublished results). *Note*: 1 ml O_2 min^{-1} ≈ 0·35 W.

immediately after birth, or during maximal exercise (when a human athlete may be obliged to dispose of more than 1 kW of heat) that the relation between metabolism and surface area becomes critical for survival or performance.

Thus for warm-blooded terrestrial creatures the surface area does impose lower and upper limits on metabolism, but they are very wide ones.

2. More precise and more extended measurements on warm-blooded verte-brates, such as the well-known "mouse to elephant" series by Kleiber[10] (Fig. 2) have shown that the resting metabolic rate (unit W; M is expressed in kg) is given by

$$P = 3·416 \, M^{0·734},\tag{6}$$

where the exponent, 0·734, is significantly different from the 0·667 required by the surface law. The metabolic *intensity*, expressed in W kg^{-1}, is conse-quently given by

$$P/M = 3·416 \, M^{-0·266}.\tag{7}$$

3. Far more telling than either of these objections was the accumulation of evidence from other than warm-blooded species, many of them aquatic and even unicellular. For such creatures the problems of losing heat or of retaining it can only seldom be critical. Nevertheless a similar regular relationship between metabolic rate and body mass is observed, Fig. 3.

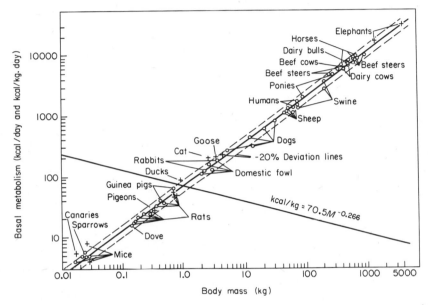

FIG. 2. The basal metabolic rate of warm-blooded creatures as a function of body mass. Logarithmic scales on both axes. The rising line and the experimental points show metabolic rate in kcal day^{-1}; the falling line shows the metabolic intensity in kcal day^{-1} kg^{-1}. From Kleiber.[10] 1 kcal day^{-1} = 48·4 m W.

The evidence is accumulated and critically evaluated in the classic monograph by Hemmingsen (Ref. 6). The conclusion is astonishing: when fair allowance is made for the different temperatures at which the different measurements were made (the standard rule is that P increases by a factor of 2·5 for each 10°C increase in body temperature) then over a mass range of 10^{18} we have:

$$P = aM^{0·75}$$

where the exponent, 0·75, is not significantly different from that of Kleiber.

Thus we find ourselves with a biological law (and biology is not exactly rich in laws!) that encompasses an almost astronomical range of phenomena. Admittedly, when plotted on this widely ranging scale quite large departures

from the line can occur without appearing dramatically discrepant. This
adds to, rather than diminishes from, the importance of the overall regularity.
Yet this "biological law" totally lacks an adequate explanation in terms of
mechanism and its very existence is unknown to many biologists: for example,
it is not even mentioned in most textbooks of physiology.

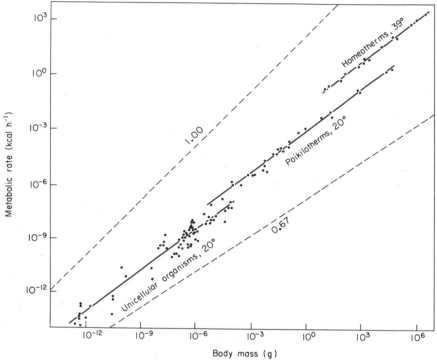

FIG. 3. Metabolic rates of various organisms in relation to body mass, both plotted on logarithmic
scales. The full lines have slope 0·75. (1 kcal h^{-1} = 1·16 W). Data from Hemmingsen[6] redrawn
by Schmidt-Nielsen. The measurements have been corrected either to 20° or to 39°, as shown.

OPTIMAL DESIGN

We can take it for granted that the design of living creatures has been optimised
by the competition of many millions of years of natural selection, and our
problem is to disentangle the key factors in this optimisation process which
have constrained metabolic rate in the observed fashion. Some features,
notably the relations between lengths, areas and volumes, change radically
with size and design; many others do not. The composition of living creatures
does not vary much; they are mostly water so the properties of water and of

dilute aqueous solutions may be dominant. Density, diffusivity (both of solutes and of heat) electrical conductivity, viscosity are thus roughly indepen-dent of size. The constituents other than water are also fairly similar, so the intrinsic strength of skeletal tissues is invariant. At the biochemical level the substrates such as ATP are identical and the enzymes catalysing their reac-tions are consequently very similar in creatures of different size. Most enzymes become denatured at temperatures not much above 40°C, which limits the permissible operating range from about 45°C down to a little below 0°C. Those organisms that have become adapted to life in hot springs form an interesting exception. They seem to have modified existing enzymes to work at higher temperatures, rather than merely managing to contrive metabolic pathways that use only those normal enzymes that happen to be intrinsically heat-resistant. Their capacity to adapt in this way poses the interesting question of why it is that warm-blooded creatures have settled on maximum operating temperatures no higher than 40°C. Perhaps some enzymes that are essential to more complex organisms cannot be adapted to higher tempera-tures.

I have listed above some of the factors that set bounds on the way in which animals can be designed. None of them taken by itself seems to impose the behaviour shown in Fig. 3, as Hemmingsen himself concludes. No doubt we shall at some future time know enough to be able to "design" a simplified ani-mal on paper (or computer) and, by varying systematically those parameters that can be varied, discover which fundamental factors actually govern animal design. Presumably (since animal design is already optimised) we would find that is is not possible to design animals that depart very far from what we actually observe in Nature. In the process we ought to be able to discover what the critical limitations on design are in each size range—and why they happen to lead to regularity of behaviour over all sizes.

TO WHAT EXTENT IS THE METABOLISM OF THE WHOLE ANIMAL DETERMINED BY THE PROPERTIES OF ITS CELLS?

As we have seen, the characteristic difference between large creatures and small ones is that the latter have a much higher metabolic intensity (expressed in $W kg^{-1}$) than the former. The same rule must be expressed at cellular level, yet it remains unclear to what degree the metabolism of an individual cell in a multicellular organism is limited by its own anatomical and biochemical constitution and to what extent it is subject to control mechanisms exercised by the organism as a whole. This topic has been ably investigated and deeply discussed for many years (for bibliography see Ref. 13) yet much remains unclear.

The size and microscopic appearance of cells from corresponding organs probably do not vary regularly with size of the whole animal though I have not been able to trace a comprehensive review of this topic covering a wide range of tissues. Certainly the size of red blood cells has been studied in a wide range of species (Ref. 7, kindly provided by Dr D. S. Parsons) and although fairly large differences are found, the size of the cell is not related in any intelligible fashion to the size of the organism from which it came.

The abundance of some organelles, notably the mitochondria, must have a bearing on the maximal oxidative metabolism of the cell but I have not come across any systematic study of mitochondrial concentration in tissues with an intense oxidative metabolism, for example the cardiac muscle of animals of different size. Such studies may exist, and we should know of them.

In seeking biochemical correlates of metabolic intensity both striated and unstriated muscle should provide a useful object of study, as it has become clear from the work of several laboratories[15] that the maximal speed of shortening and hence the maximal mechanical power output of muscles is directly proportional to the activity of the actomyosin ATP-ase in each case.

These studies were made on *different types* of muscle but until now there seem to have been no comparable systematic investigations on the *same type* of muscle taken from animals of different size. For that matter, even the mechanical properties of muscles have not been examined in this way (Ref. 3 p. 181), although there is good reason to believe that the intrinsic speed of shortening should vary systematically with animal size[8].

There have been many experimental studies of the relationship between the metabolic rate in isolated tissue slices and that of the whole animal from which they are taken. If these slices metabolise *in vitro* in the same way as *in vivo*, and if the proportions of the different tissues are similar in animals of differing size then we should expect the metabolic intensity of tissue slices to vary with body mass as indicated by Eqn (7). Early experimental results were conflicting (see Ref. 13, p. 19): more recent work, and especially the authoritative paper by Krebs,[11] seems to be in agreement that metabolic intensity of different isolated tissues is less if the samples are taken from larger animals but that the variation is not nearly so great as is that of the intact animals. Krebs suggested that these facts could be reconciled if the larger animals contained a higher proportion of accessory structures of low metabolic rate such as blood vessels, glandular ducts and connective tissues. However, it has still not been shown that this explanation will stand up to quantitative test.

Krebs's measurements were made in incubation media that might encourage high rates of metabolism (abundant and varied substrates) and might also diminish control (the medium that gave the highest metabolic intensity contained no calcium). Using less exotic media, Martin and Fuhrman[13] obtained *in vitro* measurements that were generally lower than those of

Krebs.[11] Combining this information with careful measurements of organ sizes in mouse and dog these authors were able to account for roughly 70% of the total metabolic rate of the mice and 100% of that of the dogs. Clearly it would be illuminating, albeit tedious, if a similar investigation could be extended over a wider range of species,

At present it thus appears that the maximal metabolic intensities of the tissues may be set by intrinsic factors such as the abundance of mitochondria and of key enzymes. However within the body other, central, control mechanisms must play an important part in determining that the individual never departs very much from the metabolic intensity appropriate to its size.

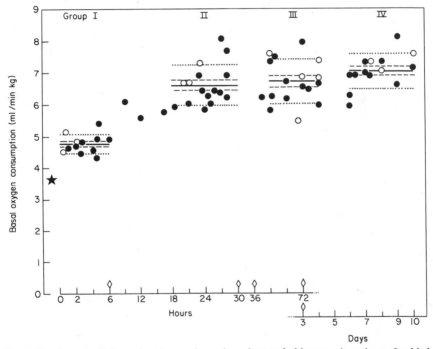

FIG. 4. Basal metabolic intensity of normal new-born human babies at various times after birth expressed in ml O_2 min^{-1} kg^{-1}. (*Note*: 1 ml O_2 min^{-1} kg^{-1} ≈ 0·35 W kg^{-1}). Note the non-linear time scale: the breaks in it are indicated by diamond top-marks. The star indicates the basal metabolic intensity in normal young women.[19] For further details see Hill and Rahimtulla[9].

This is shown especially clearly by the changes observed shortly after birth (see Figs. 4 and 5). The measurements of metabolic intensity during the first hours after birth (Fig. 4) rise rapidly from the point (indicated by a star) which represents the probable metabolic intensity in the mother. Similar conclusions can be drawn from direct measurements of metabolic rate (in animals) made while the foetus remained *in utero*.[1, 18]

Thus the foetus behaves like a typical organ in the mother, with the low metabolic intensity corresponding to her weight; but after 36 hours the baby has achieved the appreciably higher metabolic intensity characteristic of a free-living mammal of its own weight. This is shown by Fig. 5. At birth (line *a*) the baby is appreciably below the "Kleiber line" (shown dotted; slope taken as 0·75) but after 18–30 hours of life (line *b*) it has practically reached the dotted line. Such a rapid change can hardly result from an equally rapid proliferation of enzymes and mitochondria—almost certainly it shows the operation of central control of metabolism.

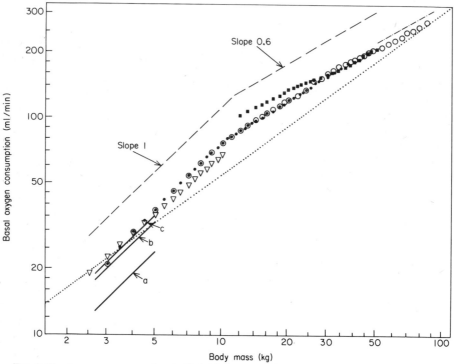

FIG. 5. Basal oxygen consumption (ml/min) in humans related to body mass (kg) for the entire period from birth to adult life (in double logarithmic coordinates). (1 ml O_2 min^{-1} ≈ 0·35 W). The points, drawn from various sources, could designate the increase in metabolic rate with body mass of a growing individual. The lines show how metabolic rate is related to body mass in different babies at various ages: *a*, 0–6 hours; *b*, 18–30 hours; *c*, 6–10 days. The baby's metabolic rate is proportional to its mass at birth, but this rate increases rapidly *without* increase in the mass of the individual baby; i.e. an individual baby rises almost vertically from *a* to *c* during the first ten days of its life. The dotted line shows the "mouse–elephant" line of Kleiber, from Fig. 2. For further details see Hill and Rahimtulla.[9]

Figure 5 illustrates several other features of interest. During the first year of life (roughly up to 10 kg) metabolic intensity stays constant (slope of line $\simeq$ 1), but then the slope changes fairly suddenly to approximately 0·6 for the remainder of the growth period; this is perhaps associated with a larger metabolic requirement for rapid growth in the early part of life. (Fish show a similar discontinuity, with a slope of 3 at the early stages: Hughes, discussion). Thus the experimental values never have the slope of 0·75 found when one makes comparisons over the full range of warm-blooded creatures; they wander from one side to the other of this line without, it seems, ever being able to depart far from it. Why this should be so remains the central unsolved problem which this article has attempted to illustrate, if not to illuminate.

DIMENSIONAL ANALYSIS IN BIOLOGY

The regular variation of metabolic rate with body size is but one example among many. A good deal of this Symposium is devoted to an attempt to understand why so many biological parameters are governed with such regularity. In a few cases—notably those involving locomotion—we can begin to identify the physical restrictions at work. In the majority of instances, varying from metabolic parameters to the duration of life or the frequency of the heart beat, we still have very little idea of the underlying mechanism. The wide variety of such phenomena is described in a recent review by Günther.[4]

Not only is the variation with body size a regular one: in the vast majority of cases it follows with reasonable accuracy the same type of *Allometric Equation* given as Eqn (1) above. The reason for this is itself obscure and deserves study. The explanation put forward by Von Bertalanffy in 1968[2] (see Günther p. 28), that both P and M are exponential functions of time, is quite simply not true in the majority of cases.

Dimensional analysis. Can we learn anything from the physical theory of dimensional analysis? Since the details of this theory may not be familiar to all biologists I will attempt a brief explanation of the usefulness and the limitations of this theory as applied in biology. The first postulate is that every physical quantity can be expressed in terms of physical quantities which are regarded as dimensionally independent. At present, by international agreement, seven such quantities are recognised[12, 14]: mass, M; length, L; time, T; electric current, I; thermodynamic temperature T; amount of substance, n; and luminous intensity I_v. To a fair extent, this choice is arbitrary; for example, electromotive force could have been chosen instead of electric current. Equally obviously, not *any* choice of seven quantities would be useful for dimensional analysis: they must be independent quantities, in that it is possible to change the units in which one is measured without affecting the others. For our present purpose the problem can be much simplified since in a purely mechanical situation we may restrict attention to only M, L and T.

Any mechanical quantity y can be expressed in terms of M, L and T by substitution in the following equation

$$y = kM^{\alpha}L^{\beta}T^{\gamma} \tag{8}$$

where k, α, β and γ are constants. For example, if y is velocity, $\alpha = 0$, $\beta = 1$ and $\gamma = -1$; if y is pressure or stress, $\alpha = 1$, $\beta = -1$, $\gamma = -2$. This equation can be put to good use when comparing a model (say of an aeroplane) with the thing itself. We are doing something exactly analogous when we compare animals of different size with one another. We can simplify Eqn (8) into the form of the allometric Eqn (1)

$$y = aM^{b}$$

if we are prepared to assume, or preferably to demonstrate experimentally, the way in which two suitable physical quantities differ as a function of body mass.

One such similarity condition is easy to establish and is roughly true: if the animals of different size are of similar shape, and if they have the same average density, we can write:

$$M = k_{1}L^{3}. \tag{9}$$

Where L is a characteristic length. Thus

$$L = M^{1/3} k_{1}^{-1/3}. \tag{10}$$

Substituting in Eqn (8) we have

$$y = kM^{\alpha}k_{1}^{-\beta/3}M^{\beta/3}T^{\gamma} = k_{3}M^{(\alpha + \beta/3)} T^{\gamma} \tag{11}$$

The term in T can also be eliminated if we can find a physical quantity (which must contain the dimension T) which varies in a known way with size or else remains invariant. Out of many possible examples one of the most illuminating is to suppose that mechanical stress and pressure are independent of size (a particularly interesting alternative will be proposed by McMahon in his paper). The physical meaning of this is simply that the internal as well as the external proportions of the different animals are similar and that they are constructed of the same materials stressed to an equal fraction of what they can bear. These assumptions are close to reality.[10]
Since stress $= S = k_{4}ML^{-1}T^{-2} =$ constant, $\tag{12}$

we can now extend Eqn (11) to

$$y = k_{5}M^{(\alpha + \beta/3 + \gamma/3)}. \tag{13}$$

Equation (13) can now be used to predict how various other quantities should vary with size. For example power or metabolic rate ($\alpha = 1$, $\beta = 2$, $\gamma = -3$) should be proportional to:

$$M^{(1 + 2/3 - 1)} = M^{2/3}. \tag{14}$$

We have seen from Fig. 1 that this is accurately true over small ranges of size, and from Figs 2 and 3 that it is only approximately true over wide ranges.

Running speed ($\alpha = 0$, $\beta = 1$, $\gamma = -1$) should not vary with size, which is also approximately true.[8] Time ($\alpha = 0$, $\beta = 0$, $\gamma = 1$) should vary as $M^{1/3}$. Clock time is obviously the same for all animals but here we are concerned with the notion of 'Biological time' which passes more swiftly for small animals than for large ones. Quantitatively the time unit may be expressed as the duration of a heart beat or a whole lifetime (or in various other ways), and it found to vary as $M^{0.27}$ to $M^{0.29}$, only slightly less than predicted. Dimensional arguments were used to confirm that man-powered flight was possible,[20] some three years before it was actually achieved; many other examples of successful application will emerge in the course of this Symposium.

Günther and Guerra[5] have shown that the predictive power of Eqn (13) for a wide variety of phenomena is improved if it is slightly modified to:

$$y = k_5 M^{(\alpha + \beta/3 + 0.31\gamma)}. \tag{15}$$

This probably means (see Ref. 4, p 27) that biological systems are to some degree "mixed regimes" whose behaviour is affected by physical quantities other than those assumed to be dominant. The same problem arises in hydrodynamics where the various numbers used in scaling (Reynolds's, Froude's, Weber's) may each be dominant in different situations; no one such number can be used to scale the wide diversity of fluid behaviour.

There seems to be some fundamental factor determining biological time scales which we do not fully understand. This brings me back to the debate about the basis on which dimensional reasoning rests. Some hold that the dimensions of a quantity refer to its essential physical nature; others that they merely arise from the conventions employed in making *physical* measurements. If this latter view is correct it may well be that the application of dimensional ideas in biology should be restricted to mechanical events such as muscular movement or blood flow. However, a statement of this viewpoint by one of its most distinguished exponents, P. W. Bridgeman, seems to leave a good deal of leeway for experiment and to attach a good deal of credit to success: ". ... dimensional analysis is an analysis of an analysis: that is, an analysis of the implications of the fact that methods of analysing experience have been found profitable which employ certain types of measuring process and certain methods of mathematical treatment of the results of the measurements. There is nothing absolute here, but a great deal of flexibility."

REFERENCES

1. Acheson, G. H., Mott, J. C. and Dawes, G. S. Oxygen consumption and the arterial oxygen saturation in foetal and new-born lambs. *J. Physiol (London)* **135**, 623–642 (1957),
2. Bertalanffy, L. Von. "General Systems Theory. Formulation, Development, Application". New York, George Braziller (1968).
3. Close, R. I. Dynamic properties of muscle. *Physiol. Rev.* **52**, 129–197 (1972).
4. Günther, B. On theories of biological similarity. *Fortschritte der experimentellen und theoretischen Biophysik.* **19**, 1–11 (1975). See also: Dimensional Analysis and Theory of Biological Similarity, *Physiol. Rev.* **55**, 659–699 (1975).
5. Günther, B., and Guerra, E., Biological similarities. *Acta Physiol. Latinoamericana.* **5**, 169–186 (1955).
6. Hemmingsen, A. M. Energy metabolism as related to body size and respiratory surfaces, and its evolution. *Rep. Steno Hosp. Copenh.* **9**, Part 2, 7–110 (1960).
7. Hewson, W. *In* "The Works of W. Hewson", (ed. G. Gulliver), Sydenham Society, London (1846), pp. 236–244.
8. Hill, B. V. The dimensions of animals and their muscular dynamics. (a) *Proc. Roy. Inst.* **34**, 450–473; (b) *Sci. Progr. Twent. Cent.* **38**, 209–230 (1950),
9. Hill, June R. and Rahimtulla, Kulsam A. Heat balance and the metabolic rate of new-born babies in relation to environmental temperature and the effect of age and of weight on basal metabolic rate. *J. Physiol.* **180**, 239–265 (1965).
10. Kleiber, M. Body size and metabolic rate, *Physiol. Rev.* **27**, 511–541 (1947).
11. Krebs, H. A. Body size and tissue respiration, *Biochim. et Biophys. Acta* **9**, 249–269 (1950).
12. McGlashan, M. L. "Physiochemical Quantities and Units". Roy. Inst. Chem. London. Monographs for teachers, No. 15, 1–117 (1971).
13. Martin, A. W. and Fuhrman, F. A. The relationship between summated tissue respiration and metabolic rate in the mouse and dog. *Physiol. Zoology* **28**, 18–34, (1955).
14. Royal Society of London. "Quantities, Units and Symbols" (1971).
15. Ruegg, J. C. Smooth muscle tone, *Physiol. Rev.* **51**, 201–248 (1971).
16. Sarrus and Rameaux. Rapport sur un memoire adressé à l'Académie royal de medecine. *Bull. Acad. Med., Paris.* **3**, 1094–1100, (1839).
17. Scholander, P. R., Hock, R., Walters, V., Johnson, F. and Irving, L. Heat regulation in some arctic and tropical mammals and birds. *Biol. Bull., Wood's Hole*, **99**, 237–258 (1950).
18. Smith, C. A. "The Physiology of the Newborn Infant". 3rd Ed. pp. 203–204, Springfield, Illinois, Chas. C. Thomas (1959).
19. Van Döbeln, W. Human standard and maximal metabolic rate in relation to fat-free body mass. *Acta Physiol. Scand.* **37**, suppl. 126, 7–79 (1956).
20. Wilkie, D. R. The work output of animals: flight by birds and by manpower. *Nature*, **183**, 1515–1516 (1959).

3. Mechanics and Energetics of Muscle in Animals of Different Sizes, with Particular Reference to the Muscle Fibre Composition of Vertebrate Muscle

G. GOLDSPINK

Muscle Research Laboratory, Department of Zoology, University of Hull, England.

ABSTRACT

The inherent mechanical and thermodynamic characteristics of muscle will be discussed in relation to bird flight, fish swimming and terrestrial locomotion.

Force versus velocity plots and thermodynamic efficiency versus velocity plots will be discussed in relation to flapping frequency and the design of muscles in large and small birds. The relationship between body size and flapping frequency will be explained in terms of the maximum permissible strain rate for the tendons. Gliding will be used as an example of isometric contraction and the economy of this type of muscular activity will be discussed for muscle fibres with different rates of contraction.

Swimming in fish involves two types of muscle: slow red muscle, which is concentrated down each side of the fish near to the lateral line, and the bulk of musculature which is white and fast contracting. This, therefore, represents a two-geared system which is important in the fish because tail beat frequency, and hence rate of shortening of the muscle, increases with an increase in swimming speed.

Muscle action during terrestrial locomotion is more complex because of the influence of gravity and the storage of energy in the elastic components of the limbs. In running we have isotonic, isometric and negative work contractions during each step. We find, therefore, that the limb muscles in terrestrial vertebrates are made up of several different kinds of muscle fibres which have different mechanical and bioenergetic characteristics. Evidence for the hierarchy of activation and division of labour of the different kinds of muscle fibres during locomotion will be presented and an attempt will be made at explaining why the rate of energy usage during running is dependent on body size and stepping frequency.

STRUCTURE OF MUSCLE

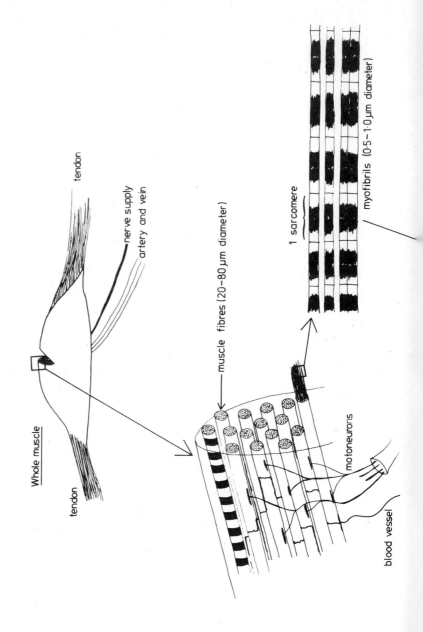

Whole muscle

tendon

tendon

nerve supply

artery and vein

muscle fibres (20–80 μm diameter)

motoneurons

blood vessel

1 sarcomere

myofibrils (0·5–1·0 μm diameter)

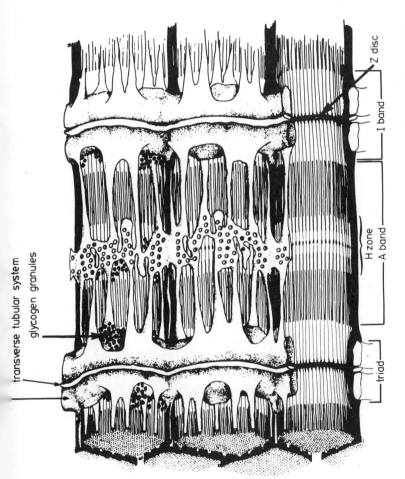

FIG. 1. Diagram summarising the structure of muscle. Muscle fibres contain contractile material or myofibrils. The myofibrils are made up of sets of thick (myosin) and thin (actin) filaments which are arranged in units called sarcomeres. Not shown in this diagram are the myosin cross-bridges which generate force by interacting with the actin filaments. These cause the thin filaments to slide between the thick filaments to produce shortening of each sarcomere and hence of the muscle fibre as a whole. The lower part of the diagram is taken from Peachey.[17]

In this paper I shall attempt, in a general way, to bridge the gap between classical muscle physiology and vertebrate locomotion and to pay attention to the relationship between animal size and locomotion. I shall also mention some of the ways in which muscles from animals of different sizes differ and how this is related to the efficiency of locomotion and to maintaining posture. I shall avoid any detailed discussion of the mechanics of locomotion because this is dealt with by the other contributors.

THE STRUCTURAL BASIS OF MUSCULAR CONTRACTION

Most biologists know something about the structure and function of muscle. This section is therefore for the benefit of the mathematicians, engineers and physicists. As muscle is a mechanical tissue it is, of course, important that we know something about the moving parts; indeed the special mechanical characteristics of the tissue can only be understood if one has some structural knowledge.

The muscles of the skeleton are made up of cellular units called muscle fibres. These are about 50 μm in diameter and are attached to the bones by tendons. The tendons, as we shall learn, play a passive but important part in certain kinds of locomotion because they are able to store energy. Muscle fibres however have the ability to develop force actively and to shorten in length because they contain contractile elements called myofibrils (Fig. 1). The myofibrils are about 1 μm in diameter and are made up of two kinds of protein filament which are arranged in a highly ordered fashion. In longitudinal section it can be seen that the filaments are arranged in structures called sarcomeres. The sarcomere is the basic unit, repeated along the length of the myofibril and hence of the muscle fibre. Each sarcomere consists of one set of thick filaments and two sets of thin filaments which overlap with the thick filaments. The system can generate force because the thick filaments have protruding cross-bridges which, when activated, interact with the thin filaments. The activity of the cross-bridges is a cyclical one of attaching, generating force and detaching. Each cycle of a cross-bridge requires a packet of energy. Before attaching to the actin filament the cross-bridge has to be reprimed each time with a high energy phosphate supplied by ATP. When the cross-bridges generate force in this way, the thin filaments are pulled "over" the thick filaments, unless of course the opposing force is equal to or greater than the force they are generating. When a muscle fibre shortens, the absolute rate at which it shortens is determined by the rate at which the individual sarcomeres shorten (this is called the intrinsic rate of shortening and it is determined by the rate at which the cross bridges work) and the number of sarcomeres in series; in other words, the effects of all the sarcomeres shortening in phase are added together. The force per unit cross sectional area that a muscle can generate depends on the number of filaments in parallel and as the filament packing and myofibril packing is about the same in most species, this is scale independent. When muscle fibres exert a constant force while shortening this is referred to as an *isotonic* contraction. When muscle fibres develop force (tension) without an appreciable change in their length, this is referred to as an *isometric* contraction. Under certain circumstances muscle fibres may be extended whilst they are still generating force and trying to shorten. This is called an eccentric or *negative work* contraction, because, from a mechanical point of view, the muscle is then doing a negative amount of work on its environment. Nevertheless, the metabolic cost to the musculature under these circumstances is positive; mechanical scientists must bear this contrast in mind when studying biological systems. Most of our knowledge of muscle physiology has been obtained from muscles maintained *in vitro* and made to contract either isotonically or isometrically. Needless to say, when the muscle is working *in situ* during locomotion the type of contraction is often not so easy to define.

During locomotion it is unusual for all the muscle fibres to contract simultaneously. In vertebrate muscle the fibres are arranged (not necessarily spatially) in groups called motor units. A motor unit comprises all those fibres that are activated by the same nerve fibre (neuron). The intensity of the contraction of a muscle is regulated by recruiting more or fewer motor units. However, these motor units are not necessarily of the same type and therefore contractile parameters other than force may also be changed by recruiting more or fewer motor units. This will be explained in more detail in the discussion on terrestrial locomotion. Some details of the different kinds of muscle fibres found in skeletal muscle are given in Fig. 2.

BIRD FLIGHT

Let us look at the way muscles work during bird flight because this kind of locomotion is, in some ways, the most easily understood. There are basically two kinds of flight: *flapping flight*, which involves isotonic contractions of more or less constant velocity, and *gliding flight* which involves mainly isometric contractions. Bird flight muscles are made up of fibres which all have about the same intrinsic rate of shortening therefore they have essentially a one-geared system. It seems that the power requirements for flight are such that it would not be feasible to have more than one type of fibre otherwise the pectoral muscles would be far too bulky.

During flapping flight there are several ways that the bird could alter the speed at which it is flying. However, as a general rule it does not change to any great extent the rate at which the pectoral muscle fibres contract, so that wing beat frequency remains more or less the same. It chooses instead to alter the

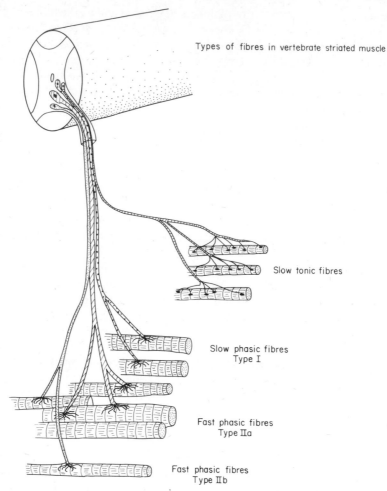

Types of fibres in vertebrate striated muscle

Slow tonic fibres

Slow phasic fibres
Type I

Fast phasic fibres
Type IIa

Fast phasic fibres
Type IIb

Fig. 2. Diagramatic representation of the different types of fibres found in vertebrate skeletal muscle. The characteristics of the different types and their role in locomotion and posture are given below.

Main types of muscle fibre found in vertebrate skeletal muscle

True slow or tonic fibres

Very slow contracting fibres. "En grappe" innervation. Poor propagation of action potential. Myofibrils not very discrete—Z discs zig-zag across the fibre. Very good isometric economy can hold isometric tension for long periods without fatigue. Postural role only, i.e. not used for locomotion.

Slow twitch or slow phasic fibres

(Type I, S.O., S, β, red fibres)

Slow contracting fibres with high isometric economy and high isotonic efficiency. 'En plaque' innervation. High levels of oxidative enzymes. Resistant to fatigue. Thicker Z discs, more mitochondria. Low myofibrillar ATPase but relatively acid stable. Recruited frequently for both isometric and isotonic contractions.

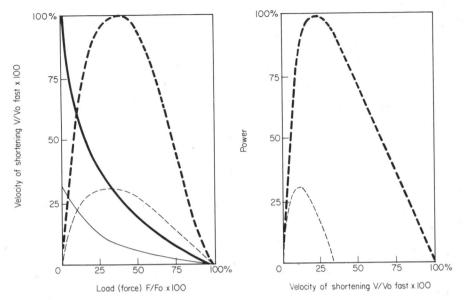

FIG. 3. Relationships between the velocity of shortening, force and power developed by fast and slow muscles. The values are expressed as a percentage of the maximum value for the fast muscle. ——— Velocity of shortening of fast muscle, ——— Velocity of shortening of slow muscle, ----- Power developed by fast muscle, ----- Power developed by slow muscle.

shape, and the force of the stroke (i.e. the number of muscle fibres recruited) and it may also alternate between gliding and flapping, or in the case of smaller birds may undergo intermittent flight.

Let us consider why it is inadvisable for the bird to regulate its speed by changing the wing beat frequency. Figure 3 shows the relationship between the force developed by the muscle and the velocity of shortening. Using the same data we can also derive the power developed by the muscle at the different shortening velocities; this is given by the dotted lines. I have included plots for

---

Fast twitch glycolytic fibres
(Type IIb, FG, FF, α_1, white fibres)
Fast contracting fibres with high power input but with low isometric economy and fairly low isotonic efficiency. Thin Z discs, few mitochondria. Metabolism mainly glycolytic. High myofibrillar ATPase which is stable at alkaline pH. Used mainly for rapid isotonic contractions but also involved in large isometric contractions. Recruited infrequently.
Fast twitch oxidative fibres
(Type IIa, FOG, FR, α_2, pink fibres)
Similar to type IIb but contain more mitochondria and oxidative enzymes. High myofibrillar ATPase, but less acid stable than either Type I or Type IIb. Fairly fatigue resistant. Recruited fairly frequently.

fast contracting fibres and slow contracting fibres in Fig. 3 because these are relevant to the discussion of fish swimming and terrestrial locomotion. From these shortening velocity plots it will be seen that there is an optimum rate of shortening at which maximum power is produced. This means that for each kind of bird there is a very definite optimum frequency of wing beat.

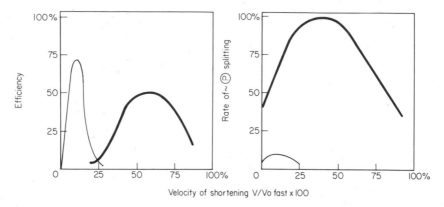

Velocity of shortening V/Vo fast x 100

FIG. 4(a). *Left* The relationship between the velocity of shortening and the mechanochemical (thermodynamic) efficiency of fast and slow muscles. Velocity values are given as a percentage of the maximum velocity for the fast muscle (between 20 and 30 muscle lengths per second). The efficiency values are the percentage of free energy from high energy phosphate that is converted into work. In other words the efficiency values do not include the efficiency of the reactions which replenish the high energy phosphate levels.

4(b). *Right* The relationship between the velocity of shortening and the rate of energy utilization (high energy phosphate splitting) in fast and slow muscles. Values are given as a percentage of the maximum values for the fast muscle. ━━━ Efficiency and rate of energy utilization for fast twitch muscle, ─── Efficiency and rate of energy utilization for slow twitch muscle.

Of perhaps even greater importance is the relationship between the velocity of shortening and the thermodynamic (or mechanochemical) efficiency of the muscle. This relationship is shown in Fig. 4(a). From this Figure it will be seen that the mechanochemical efficiency peak tends to be sharper than the power peak, and this makes it more important that the wing beat frequency is maintained at the optimum rate. Figure 4(a) shows efficiency curves for slow muscle fibres as well as for fast muscle fibres and it will be seen why in certain circumstances it is an advantage to have two types of fibre in a muscle as this widens the range of velocities over which it can contract efficiently. However, as mentioned above, it would be too expensive for the bird to carry two muscle systems. Incidentally, the efficiency values given in Fig. 4 are for the contractile process including activation. In other words, the values cover the ATP used for moving the Na^+, K^+ and Ca^{++} ions used in activating the contractile process, and the transduction of ATP into work, but not the replenishment reactions. The reactions involved in replenishing the ATP, i.e. glycolysis and oxidative phosphorylation, are about 45 % efficient, so if one includes these in muscle efficiency then the maximum values obtained are of the order of 30 %.

Wing Beat Frequency and Size of Bird

Although individual birds do not vary their flapping frequency to any great extent, flapping frequency varies considerably between different species. This is related to size. Large birds, such as the Stork, flap at about two beats per second, whereas a small bird such as a Wren has a wing beat frequency of 10 beats per second.[13]

Why is wing beat frequency related to body size? It is true of course that large birds do not neet to flap their wings as rapidly as small birds in order to achieve the necessary power for flight. Power is force times displacement over time, and as the greater wing span of the larger birds means an increased displacement, the time factor may also be increased, in other words they can have a longer wing beat cycle. However, why cannot large birds have a high wing beat frequency and thus generate much more power and hence travel at much greater speeds?

As mentioned above, the rate at which the muscle shortens and develops force is dependent on the number of sarcomeres in series and the intrinsic velocity of each sarcomere. The muscle fibres in large birds are very long with a lot of sarcomeres and if they had the same intrinsic speed of shortening they would shorten (i.e. develop strain) at a very high rate. The material of which muscle fibre and tendons are made cannot apparently withstand such high strain rates.[15] Figure 5 shows a plot for the absolute rate of extension and breaking force of a whole muscle preparation, including muscle, tendon and tendon insertions. From this plot it will be seen that the force necessary to break the preparation initially rises as the extension rate is increased. This is a property of many polymers including biological polymers. However, when the rate of extension of this preparation exceeds 80 mm s^{-1} the force required to break the preparation decreases. This is presumably because the forces imposed on the muscle and tendon cannot be dissipated rapidly enough. Although the larger birds have longer and thicker tendons it may be the absolute rate of extension that is the important factor. Therefore the larger birds may have no alternative but to have a lower intrinsic speed of shortening in order to stay on the rising phase of the breaking force: extension rate curve. The same arguments can incidentally be used for explaining why in terrestrial locomotion the larger animals are limited to lower stepping frequencies than smaller animals, and why larger species of fish have lower tail beat frequencies.

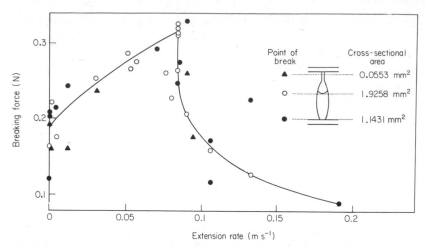

FIG. 5. Shows the force necessary to break a muscle preparation using different rates of extension. The muscle preparation used was the biceps brachii of the mouse; length of preparation (not accurately measured) was 1–2 cm. The different rates of extension were applied using a Levin–Wyman ergometer fitted with a 7 kg weight. The Levin–Wyman apparatus is a moving beam and dashspot which is used in muscle physiology for producing isovelocity movements. The breaking force was measured using a transducer fitted to the end of the beam. The slack in the system was taken up by putting 0·25 N of initial tension on the muscle. At the higher velocities the dashspot was used without fluid. The point of break is signified by the following symbols; ▲ Origin tendon, ○ Origin tendon/muscle fibre junction, ● Insertion tendon.

There are probably other mechanical reasons why it is undesirable to have high flapping frequencies or stepping rate for the larger birds and animals, such as the problem of the inertia associated with long limbs, and the optimum resonance frequency of the limbs and trunk as discussed below by McMahon. However, the mechanical properties of the tissues certainly seem to be among the limiting factors, although of course they may not be the main ones. More studies on the dynamic properties of muscle and tendon need to be carried out so that we know what sort of constraints and safety limits are involved.

Economy during Gliding

The muscles involved in gliding are primarily the triceps and deltoid groups which hold the wings forward and outstretched, and the pectorals which maintain the horizontal position of the wings. The energy needed for gliding is known to be considerably less than that needed for flapping flight. Baudinette and Schmidt-Nielsen[4] have shown that the metabolic rate during gliding flight is only about twice the resting metabolic rate, whereas in flapping flight it is about seven times the resting metabolic rate. From measurements of the electrical activity of the muscles it is apparent that fewer groups of muscle fibres (motor units) are being recruited than is the case in flapping flight. Thus one would expect less energy to be used. Also the type of muscle activity involved in gliding is primarily isometric contraction, as the muscles are maintaining posture rather than doing work, and it is known that the energy turnover in muscle is lower in isometric contractions than in isotonic contractions.[14] This is probably because the cross-bridge cycle time is longer during an isometric contraction.[8] This can also be seen if one looks at the plots for rate of high energy phosphate splitting against the velocity of shortening (Fig. 4(b)) and compares those for zero velocity (isometric contraction) with those for other velocities. The reason why the turnover of energy is greater when the muscle is shortening seems to be due to the fact that the thin filaments are moving over the thick filaments and there are more active sites on the thin filaments presented per unit time. However, if the active sites move by too quickly then the cross-bridges are not able to make contact and the splitting rate goes down again,

There are several other factors that would be expected to influence the economy of gliding in birds. As will be seen from Table I the economy of an isometric contraction is greater, the longer the duration of the contraction. This is because the initial phase, during which tension is being developed, is relatively expensive.[2, 3] Therefore the longer the muscle fibres can maintain tension without being reactivated, the more economical the flight will be. Birds such as the albatross, which are very well adapted for gliding, may glide for very long periods with very little expenditure of energy. It will also be seen from Table I that the slow muscles are considerably more economical in maintaining isometric tension than fast muscles.[2, 11] The reason why slow muscle fibres are able to maintain tension more economically is believed to be due to their longer cross-bridge cycle time. Cross-bridges are believed to need energy only when they disengage and re-engage and not during the time they are attached and generating force. Therefore the longer the cycle time the

TABLE I. Expenditure of high energy phosphate and economy for isometric contractions of different durations in hamster fast (biceps brachii) and slow (soleus) muscles. From Awan and Goldspink.[2]

	Duration of contraction	1·2 s	30 s	60 s
Expenditure (µmol high-energy phosphate/g)	Fast	1·5 + 0·06	2·2 + 0·05	2·5 + 0·05
	Slow	1·1 + 0·03	1·5 + 0·02	1·8 + 0·01
Economy (g.s) per µmol high-energy phosphate)	Fast	99 + 7	2483 + 110	3635 + 180
	Slow	143 + 8	6377 + 250	9012 + 230

more economically the muscle can maintain tension. Birds that glide for long periods of time are all relatively large, and as already stated they need to have fairly slowly contracting fibres which are a very definite advantage when it comes to gliding because of their superior economy. There may, of course, be aerodynamic reasons why a small bird with the same relative dimensions cannot glide as well as a large bird. However, one can certainly argue, albeit in a teleological way, that large birds have retained the ability to glide for long periods of time because they have relatively slow muscles and thus they are the only ones which could exploit gliding as an economical means of locomotion.

FISH SWIMMING

It is well known that both the amplitude and the frequency of tail beat increase with increase in swimming speed. Hence fish swimming differs from bird flight both in that the rate at which the muscles are required to shorten changes, and in that there is no isometric component except that involved in maintaining the position of the fins. However, we find that fish have two main types of muscle fibre, one fast and one slow (Fig. 6). Thus in effect they have a two-geared system. The slow aerobic red fibres make up 5 to 20% of the musculature and in most fish are found as a discrete red band which runs down each side of the body near to the lateral line. The bulk of the musculature consists of

TABLE II. Statistical analyses of the effects of different types of forced swimming on the concentrations of free creatine, glycogen and lactate in the muscles compared to non-exercised fish.

| Type of swimming | Level of significant difference compared with non-exercised fish | | | | | |
| | Free creatine | | Lactate | | Glycogen | |
	Red muscle	White muscle	Red muscle	White muscle	Red muscle	White muscle
Low speed sustained (200 min)	NS	NS	$P < 0.01$	NS	$P < 0.05$	NS
High speed sustained (200 min)	NS	$P < 0.05$	$P < 0.001$	$P < 0.001$	$P < 0.01$	NS
Vigorous swimming (15 min)	NS	$P < 0.001$	NS	$P < 0.001$	NS	$P < 0.01$

NS = Not significant at $P < 0.05$ level.

FISH RED AND WHITE MUSCLE

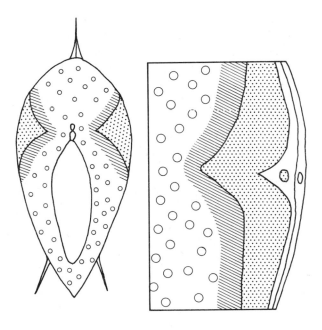

▢ Superficial red muscle

▢ White muscle   ▨ Pink muscle

FIG. 6. A schematic diagram of the distribution of the different kinds of muscle fibres found in fish. The red muscle fibres are slow fibres whilst both the pink and the white are fast fibres. The pink fibres, however, contain more mitochondrial enzymes and are less anaerobic than the white fibres. The red fibres are believed to be recruited at all swimming speeds, the pink fibres are believed to be recruited at intermediate and faster swimming speeds, whilst the white fibres are recruited only at the faster swimming speeds.

fast anaerobic white fibres. There is another kind of fibre which is pink in colour. This pink fibre is another kind of fast fibre which contains high levels of mitochondrial enzymes and therefore it is reasonably aerobic and fairly resistant to fatigue. By studying the change in levels of metabolites during different swimming speeds, Johnston and Goldspink[16] were able to show that the fast fibres are only recruited at the higher speeds (see Table II). Some recent work by Davison and Goldspink (unpublished findings) which involved measuring the depletion of glycogen and lipid in individual muscle fibres of the three types has indicated that, at reasonably slow speeds in the trout (1 body length s^{-1}), the red fibres are very active, using both glycogen and lipid. The pink fibres are also quite active metabolically, but use glycogen only. It seems therefore that the pink fibres are the ones that are recruited after the red fibres; they have a reasonable supply of mitochondria so that they can be recruited fairly frequently. From Figs 3 and 4(a) we can see why, from the mechanical and thermodynamical points of view, the fish should have fast and slow muscle fibres. When the fish is cruising slowly it is using only its slow muscle fibres and these are contracting at about the optimum velocity. When the fish wishes to put on a burst of speed it turns on its fast fibres. The fast fibres have a different optimum velocity, and therefore the fish can change into higher gear and propel itself more rapidly without a great loss in thermo-dyanamic efficiency. However, at these higher speeds the fish fatigues very rapidly because oxygen cannot possibly be made available at the rate at which the fast white fibres hydrolyse ATP.

We may look on fish as having a two-geared system with the lower gear being used most of the time. However, this analogy is somewhat misleading because it is not a question of switching over from slow muscle to fast. The evidence suggests that the slow muscle is not switched off; it continues to be activated but mechanically it is not effective at the higher speeds. We can see from the data in Fig. 4(b) that there is no great need for the central nervous system to switch off the slow fibres as the speed is increased because the amount of energy they are consuming decreases markedly. Concerning the question of swimming and body size, it would seem that the advantage that a large fish has over a small fish is higher absolute speed. Relative swimming speeds do not seem to change very much so that we may assume that the increase in absolute speed is due to the lengthening of the muscle fibres, giving them more sarcomeres in series.

TERRESTRIAL LOCOMOTION

Muscle action during terrestrial locomotion is more complex than during flying or swimming. In terrestrial locomotion we have to give more consideration to the effects of gravity. In swimming and flying these are reduced by buoyancy and lift respectively. Associated with the greater effect of gravity, energy is stored in the elastic elements of the limbs (see the papers in Part II of this volume). It seems that one of the main muscle actions during terrestrial locomotion is in fact to get the limbs in the correct position so that the elastic elements—the tendons—are tensioned so that there is maximum transfer of stored energy. During rapid locomotion the muscles may be required to help "rewind" the tendons as there is insufficient time for them to recoil of their own accord.

During running and walking there are at least three different kinds of muscle activity. We have *isotonic* contractions in swinging the limbs forward, and also in tensioning the tendons when the animals is running at speed. We have *isometric* or nearly isometric contractions involved in tensing the tendons so that they store the maximum amount of energy before impact. Just after impact we also have some *negative work* contractions as the muscles are stretched whilst they are still generating force. This is particularly true if the animal is running down a slope or climbing down a tree. Because of this diversity of muscle function we should not be surprised to find that the muscles of terrestrial vertebrates are not homogeneous and are in fact made up of several kinds of muscle fibres. These can be distinguished histochemically because they possess different amounts and different kinds of enzymes.[7, 10] They can also be examined electrophysiologically if the individual nerve cell or nerve fibre which innervates a group of like muscle fibres (motor unit) is stimulated, and the contractile characteristics measured.[6]

Although the evidence regarding the kinds of muscle fibres (motor units) that are found in the limb musculature is somewhat confusing, the generalized scheme shown in Fig. 2 seems to apply. There are two main types that are readily distinguished; the tonic or true slow muscle and the twitch or phasic muscle. The former may not concern us here because it is not involved in locomotion but only in maintaining posture (the 'postural cost' of locomotion appears to be independent of speed; see the paper by Taylor below). The twitch of phasic muscle is also made up of different kinds of fibre. There is the slow contracting fibre which has high levels of oxidative enzymes and which is resistant to fatigue, and a fast contracting fibre which has low levels of oxidative enzymes and fatigues rapidly. There is yet a third type which is also a fast contracting fibre but which has high levels of oxidative enzymes and is fairly resistant to fatigue.

There is evidence that the slow muscle fibres are always recruited first in mammals as in fish,[12] and we may presume that the fast oxidative (fatigue resistant) fibres are recruited next. During slow walking we may expect that only the slow fibres and possibly some fast, fatigue resistant fibres will be recruited. When the speed of locomotion is increased, more fibres are recruited and these are the fast fibres as they are required to produce the faster movements. Fast fibres are used not only to swing the limbs forward and backwards more rapidly but they are needed to help rewind the tendons because there probably is insufficient time for the tendons to recoil completely

This of course would be more important in smaller animals with their higher stepping frequency. At optimum galloping velocities it seems that the tendons may recoil at the same rate at which the muscle fibres develop force and therefore the muscle will be contracting isometrically. As already stated isometric contraction is less costly than isotonic contraction and thus there should be a considerable saving in the energy required for locomotion.

Considering the different functions of the muscle fibres during walking and running, it is perhaps not surprising to find that the ratios of the different types of fibres differ from muscle to muscle according to the predominant kind of contractile activity of the muscle. For example, the soleus muscle in animals that run on their toes has a very high percentage of slow muscle fibres. This muscle is attached to the long Achilles tendon which is one of the main elastic storage elements for locomotion, therefore one of its main functions is to hold the tendon in the tensed state. Muscles that swing the limbs forward, such as the vastus group in the thigh, usually have all three types of fibre, no doubt so that they can move them either slowly or quickly with reasonable efficiency.

One of the problems of terrestrial locomotion that has so far defied proper explanation is the finding that the net cost of transport is lower for larger animals than for small animals (as remarked by Schmidt-Nielsen in Chap. 1, see also the paper by Taylor). Larger animals take longer strides; however, it is not immediately obvious why fewer long strides are less costly than a greater number of short strides. Taylor and coworkers[18] have also carried out work which suggests that the energy that goes into swinging the limbs is relatively small, so presumably most of the energy goes into rewinding the tendons and holding them in the rewound state. In the smaller animals with their high stepping frequency, the tendons will be rewound more times per second, and because of the short time interval it is likely that fast fibres will be required to assist in the rewinding. This may entail some shortening of the fibres. In larger animals there is more time for the tendons to recoil on their own accord and the muscles can contract isometrically. Another aspect is that because of the longer duration of the stride the muscles are maintaining isometric tension rather than developing it. Thus the larger animals are able to save energy on several scores: (1) They need fewer fast fibres and it is known that fast fibres are more expensive than slow fibres in both isotonic and isometric contraction. (2) The elastic storage of energy enables the muscles of larger terrestrial animals to contract isometrically rather than isotonically, and this also represents a large saving in energy. (3) The longer stride duration means that the muscles maintain isometric tension for a greater proportion of time, and maintaining isometric tension is inexpensive compared with developing tension. In terrestrial locomotion muscle efficiency is therefore not a very meaningful parameter; what we really should look at is the rate at which energy is being used (Fig. 4(b)) and the effectiveness of the type of locomotion.

COMPOSITION OF MUSCLES FROM ANIMALS OF DIFFERENT SIZES

The difference in size of muscles in large and small animals is mainly accounted for by the difference in the number of fibres and the length of the fibres. Larger animals tend to have larger muscle fibres, but there are certain physical limits on the size which muscle fibres can attain and remain viable. For example, it is important that the fibres have a high surface area to volume ratio to allow for rapid diffusion of oxygen. As the muscles of large animals are generally less metabolically active than those from small animals, it is permissible that they should have a slightly lower surface to volume ratio.

Black-Schaffer et al.[5] carried out an interesting study on cardiac muscle for which the same situation must presumably hold. They concluded that a rat weighing 0·0014 kg contains about 9×10^7 fibres whilst the heart of a blue whale which weighs approximatly 290 kg contains about 2×10^{13} fibres. Thus the increase in fibre number is almost proportional to the weight difference.

The ratios of the numbers of the different types of fibre also vary with animal size.[9] In larger animals there are more slow fibres, which is expected in view of their longer stride duration. Also, as the linear dimension increases the weight of the animal increases as length cubed, and therefore larger animals have a greater postural problem. On this basis therefore, we would expect them to have more slow fibres as these are the ones that are economical in maintaining isometric tension, i.e., maintaining posture. We should perhaps mention the tortoise here because this animal also has a postural problem as it carries a heavy protective carapace and plastron. It therefore is also obliged to have slow muscle fibres, and as it has not many effective predators, speed is not a priority and it does not need any fast fibres. The near relative of the tortoise—the turtle—has less of a postural problem because it spends most of its time in water and we find that this animal has more fast muscle fibres than the tortoise. A similar situation holds true for some animals which rely on camouflage such as the chameleon and sloth. These animals have gone in for thermodynamic efficiency rather than for a high power output, because a rapid means of escape is no longer required. Incidentally, the part of the chameleon which can move very rapidly is its tongue. This rapid movement is achieved because the tongue, which is a fluid filled tube, has a series of circular muscle bands and these contract in a rapid series thus displacing the fluid and extending the tongue forward. The tongue can also be rapidly withdrawn back into the mouth by long muscles running down its length. This movement is quite rapid because the longitudinal muscles have many sarcomeres in series. To return to the postural problem of larger terrestrial animals; it seems that although the large animals have the advantage of a lower net cost of transport, they have to devote a much larger proportion of their musculature to postural activities. This means that the very large terrestrial mammals are really obliged to be herbivores.

As well as possessing more fast fibres, the muscles of small animals are metabolically more active. Gauthier and Padykula[10] found that the fibres of the diaphragm in small animals had higher levels of mitochondrial enzymes than those in larger animals. The diaphragm in small animals, like the skeletal muscles, has to contract more often per unit time than in larger animals, and thus the fibres have to be more metabolically active. If we make the hypothesis that the metabolic rate can only be increased by a certain amount as activity builds up from rest, it seems plausible that muscle, which makes the greatest energy demand, sets the pace for the rest of the body. Thus, as the muscles of smaller animals work at higher frequencies during locomotion, their basal metabolic rate must be higher so that the increment due to activity is sufficient to satisfy the big energy demands of the active muscles. According to this hypothesis, the respiratory and cardiovascular system and indeed the metabolism of virtually every cell has to be faster in the small mammal even at rest. This topic is discussed in the paper by Wilkie.

REFERENCES

1. Alexander, R. McN. The mechanics of jumping by a dog. *J. Zool. Lond.* **173**, 549–573 (1974).
2. Awan, M. Z. and Goldspink, G. Energetics of the development and maintenance of isometric tension by mammalian fast and slow muscles. *J. Mechanochem. Cell Motility*, **1**, 97–108 (1972).
3. Aubert, X. "Le Couplage Energetique de la Contraction Musculaire". Editions Arscia Brussels (1956).
4. Baudinette, R. V. and Schmidt-Neilsen, K. Energy cost of gliding flight in herring gull. *Nature*, **248**, 83–84 (1974).
5. Black-Schaffer, B., Grinstead, C. E. and Braunstein, J. N. Endocardial fibroelastosis of large mammals. *Circ. Res.* **16**, 383–390 (1963).
6. Burke, R. Motor unit types of cat triceps surae muscle. *J. Physiol.* **193**, 141–160 (1967).
7. Close, R. Dynamic properties of mammalian skeletal muscle. *Physiol. Rev.* **52**, 129–183 (1972).
8. Curtin, N. A., Gilbert, C., Kretzschmar, K. M. and Wilkie, D. R. The effect of performance of work on total energy during muscular contractions. *J. Physiol.* **238**, 455–472 (1974).
9. Davies, A. S. and Gunn, H. M. A comparative histochemical study of the mammalian diaphragm and m. semitendinosus. *J. Anat.* **110**, 169 (1971).
10. Gauthier, G. F. and Padykula, H. A. Cytological studies of fibre types in skeletal muscle. A comparative study of the mammalian diaphragm. *J. Cell. Biol.* **28**, 333–354 (1966).
11. Goldspink, G., Lawson, R. E. and Davies, R. E. The immediate energy supply and the cost of maintenance of isometric tension for different muscles in the hamster. *Z. Vergleich. Physiol.* **66**, 389–397 (1970).

12. Gollnick, P. D., Piehl Karin and Saltin, B. Selective glycogen depletion patterns in human muscle fibres after exercise of varying intensity and at varying pedalling rates. *J, Physiol*, **241**, 45–58 (1974).

13. Greenewalt, C. H. Dimensional relationships for flying animals. *Smithsonian misc. Collns.* **144** (2), 1–46 (1962).

14. Hill, A. V. The heat of shortening and the dynamic constants of muscle. *Proc. Roy Soc. B*, **126**, 136–195 (1938).

15. Hill, A. V. The dimensions of animals and their muscular dynamics. *Science Progress*, **38** (150), 209–231 (1950).

16. Johnston, I. A. and Goldspink, G. A study of the swimming performance of the Crucian carp Carassius Carassius (L) in relation to the effects of exercise and recovery on biochemical changes in the myotomal muscles. *J. Fish Biol.* **5**, 249–260 (1973).

17. Peachey, L. D. The sarcoplasmic reticulum and transverse tubules of the frog's sartorius. *J. Cell Biol.* **25** (no. 3, pt. 2) 209–232 (1965).

18. Taylor, R. C., Shkolnik, A., Omiel, R., Baharav, D. and Borut, A. Running in cheetahs, gazelles and goats: energy cost and limb configuration. *Am. J. Physiol.* **227**, 848–850 (1974).

4. Dimensions and the Respiration of Lower Vertebrates

G. M. HUGHES

Research Unit for Comparative Animal Respiration, University of Bristol, England

ABSTRACT

Scaling of important parameters of the respiratory systems of fish, amphibians and reptiles in relation to body mass is discussed. Particular attention is given to the dimensions of the gas exchange surfaces, and relationships between these parameters and the amount of oxygen transferred. It is shown that, although the exponent relating standard oxygen consumption to body mass is usually about 0·75, this may vary from 0·5 to more than 1·0 for individual species. There are also variations in the corresponding exponents for respiratory areas.

In some organisms the ratio between oxygen consumption and respiratory area increases with body mass and consequently the respiratory surface may be limiting. In others there is an increase in surface area relative to the oxygen demand and consequently a possible increase in scope for activity with body size.

The general conclusion is that, while overall studies involving a range of species of different body sizes have provided valuable guidelines, intraspecific studies may not necessarily show the same relationships between body mass and the respiratory parameters.

INTRODUCTION

In his essay "On being the right size" J. B. S. Haldane[21] rightly remarked that the size of animals is often neglected in many zoological text-books. Unfortunately there is still some truth in such a view today, but thanks to studies on allometry (Huxley;[36] D'Arcy Thompson[55]) and the relationship between size and metabolism, at least some aspects of animal size are now frequently mentioned. The overall surveys of Kleiber,[37] Benedict,[5] Brody,[8] Hemmingsen,[23] Zeuthen,[61] Schmidt-Nielsen,[47] and others have

57

illustrated the general value of such studies by the use of examples ranging from the need to relate drug and anaesthetic dosage to metabolism rather than body mass, to the Lilliputians' problem when feeding Gulliver! Changes in size between humans and animals have fascinated many authors of English literature, and their illustrators have usually shown isometric changes in body parts. However, Alice during her surprising growth, and when she utters the famous "curiouser and curiouser", is illustrated by John Tenniel with a changing proportion between her neck and body! Changes in proportion are a very important factor in animal evolution, the adaptive radiation of a basic plan of organisation frequently being achieved by an increase in size associated with a given allometric relationship (Carter[10]). Palaeontologists have long used bi-logarithmic plots of the dimensions of one part of an organism relative to another (Romer,[45] Joysey[36a]) and have recognised that changes in the slope of these lines may occur during the evolution of a particular group. Perhaps this more general viewpoint can help our understanding of the relationship between interspecific and intraspecific studies concerned with metabolic rate (MR) and body mass (m). On the one hand we have a fairly definite interspecific relationship $MR \propto m^{0.75}$ whereas intraspecifically, $MR \propto m^{0.5-1.0}$. Perhaps the latter represent stages in the evolution of the allometry of morphological and physiological characteristics, whereas the former represents an overall trend. This paper emphasises the variation between species and groups of animals and at the same time recognises the general validity of overall interspecific relationships. By drawing attention to these differences it is hoped to reduce the dangers which sometimes arise because those with less experience may expect to find that relationships in intraspecific studies are the same as the better-known overall relationships. Furthermore it will emphasise the value of intraspecific studies, using a range of body sizes, to functional interpretations of the respiratory and circulatory systems.

In the context of the present Symposium, it is clear that even the finest adaptations at the muscular and skeletal levels cannot be effective if insufficient oxygen is transferred across the respiratory surfaces, and/or the circulatory system is unable to transport sufficient oxygen to the cells. Certainly muscular systems can function under anaerobic conditions, but oxygen is ultimately required for the oxidative mechanisms and aerobic metabolism is normally preferred for constant cruising activity.

It is impossible to cover all aspects concerning the dimensions of these systems in fish, amphibians and reptiles; in many cases because insufficient information is available. Figure 1 and Table I indicate some of the main relationships that could be considered in a survey of gas exchange across a respiratory surface from water or air to blood. Flows of the external fluid and blood may be in the same, opposite, or cross-current directions. The first

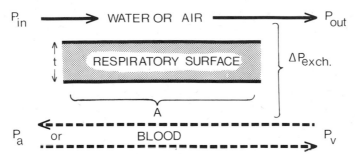

FIG. 1. Diagram to show the basic parameters in relation to oxygen exchange at a respiratory surface. Table I summarises some of the chief oxygen transfer equations and the dimensions that can be measured in these organs. Several ratios between parameters in common use by respiratory physiologists are also included. P represents the partial pressure of oxygen.

two equations were originally given in relation to fish gills (Hughes and Shelton[37]) and indicate that oxygen uptake may be considered in relation either to its removal from the medium or to the capacity of the blood to transport it away. Diffusing capacity is defined by physiologists as the ratio between the oxygen transferred and the mean difference in oxygen

TABLE I

Equations	Quantities to be measured	Ratios
1. *Ventilation*		
	Ventilatory Pump $\big\langle$ Stroke / Frequency	
$\dot{V}_{O_2} = \dot{V}_E \alpha_w (P_{in} - P_{out})$	Length and cross section of water or air flow resistances	$\dot{V}_E/\dot{V}_{O_2}$ (ventilation requirement)
2. *Perfusion*		
$\dot{V}_{O_2} = \dot{Q} \alpha_b (P_a - P_v)$	Cardiac Pump $\big\langle$ Stroke / Frequency	$\dot{V}_E/\dot{Q}$ (ventilation-perfusion ratio
	Length and cross section of blood flow resistances; heart mass; blood O_2 capacity; viscosity etc.	$\dot{V}_E \alpha_w/\dot{Q}\alpha_b$ (capacity rate ratio)
3. *Diffusing capacity* D_{O_2} (*Physiol*) = $\dot{V}_{O_2}/\Delta P_{exch}$ D_{O_2} (Morphometric) = $(A/t) K$	Area and thickness of exchange surfaces; capillarisation	$\dot{V}_{O_2}/A$ or $A/\dot{V}_{O_2}$
4. *Exchange capacity* $E_c = A \cdot p_{max} \Delta P_{exch}$	Permeability characteristics	

tension (ΔP) between the external medium and blood. From a morphometric point of view this can also be estimated by measurements of surface area (A) and thickness (t) of the respiratory surface; K is Krogh's permeation coefficient; α is the oxygen solubility coefficient. Exchange capacity is a more recent term in which the overall permeability of the animal (p) is equivalent to K/t. Some of the quantities which should be measured in relation to these equations are also indicated. For relatively few of them is sufficient data available to give good interspecific comparisons, but for certain species the data is quite good. The detailed nature of these quantities varies according to the particular organism and whether its respiratory surface is gill, skin or lung. The final column in Table I indicates certain ratios that are often used in respiratory physiology, some of which will be referred to in the present paper.

There is very little data, compared with that available for birds and mammals, for the three classes of animal that I shall consider. These animals are largely cold-blooded, although in recent years it has become clear that some fish, e.g. Tuna (Carey *et al*[9]) and reptiles such as sea snakes (Graham[19]) maintain a constant difference between their body temperature and that of the environment. Within the three groups there are large variations in the basic design and adaptations of animals' respiratory systems. An example is the use of different types of surface with different relative importance with respect to oxygen or carbon dioxide exchange. The size range among present day forms is not great, and many studies have been over a relatively small range, so that the slopes of the log–log regression lines cannot be obtained accurately. Measurements of some of the parameters is also fairly difficult and the conditions under which they are carried out vary between species; once more this makes interspecific comparisons more difficult. Nowadays, however, there is greater standardisation of the conditions under which standard oxygen consumption, and active metabolism at different levels of locomotory activity, are determined. There still remains a great need for further measurements over a range of body sizes.

RESPIRATORY SYSTEMS OF FISH

Although studies of metabolism in relation to body mass have been made for many years under varied conditions (see Winberg[59]) it is only relatively recently that any interest has developed in the dimensions of respiratory surfaces. Furthermore, in more recent studies (e.g. Barlow[1], Beamish[3], de Silva[14], Rao[44], Zeisberger[60]) the variability in the exponent relating metabolism to body mass has been shown to be even greater than that indicated by Winberg. The exponent may range from less than 0·5 to slightly greater than 1·0. Although studies of the dimensions of fish respiratory

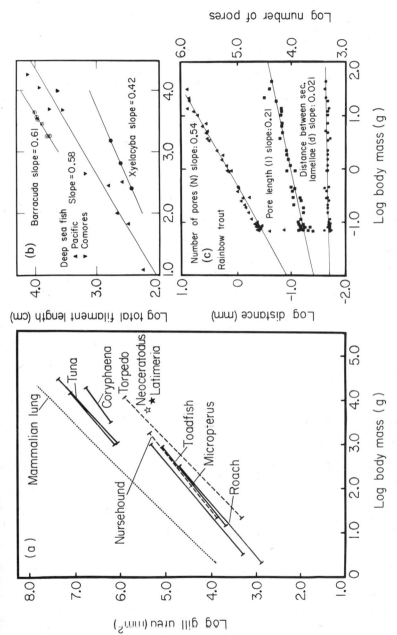

Fig. 2. Bi-logarithmic plots showing relationships between body mass and (a) total gill area for a number of fish species. The dotted line shows the relationship for the surface area of a mammalian lung with a slope of 1·0 based on the studies of Weibel[57]; (b) total filament length for gills of a number of different species of deep water fish collected during the British/French/American expedition to the Comores (▼ 5 species) and by a Japanese expedition to the Pacific Ocean (▲ 4 species). Regression lines are also given for 3 specimens (●) of *Xyelacyba myersi* from the Comores. (○) of the barracuda from the Comores (from Hughes[28]); and (c) dimensions of the gill sieve during development of rainbow trout based on measurements by Morgan.[38]

surfaces are relatively recent, already a greater range of species has been investigated than for mammals (see Fig. 2(a)). The gills are more exposed and methods for area measurements are now well established, though laborious (Hughes[25], Muir and Hughes[40]).

TABLE II. Values for the exponent b in the relationship $y = a.m^b$ for heart mass, resting oxygen consumption and gill area of a range fish species. Values given are mean values and the appropriate statistical data is given in the references. Ratios between oxygen consumption ($\dot{V}_{O_2}$) and gill area (A_g) and vice versa are also given. Numbers in square brackets are reference numbers.

	Heart Mass	$\dot{V}_{O_2}$	Gill area Ag	$\dfrac{\dot{V}_{O_2}}{Ag}$	$\dfrac{Ag}{\dot{V}_{O_2}}$
Trout	1·0†[38]	0·75[38]	0·95†[38]	−0·2	0·2
Tench		0·42†	0·7†	−0·28	0·28
			(0·52)	(−0·1)	(0·1)
Ray	0·93†	1·15†	0·97†	0·18	−0·18
Torpedo	1·01†	0·97†	0·94†	0·03	
Herring		0·82[15]	0·78[15]	0·042	−0·042
Plaice		0·65[14]	0·85[14]	−0·2	0·2
Skipjack		1·0‡	0·85[40]	0·15	−0·15
Mud-eel		Summer 0·72 ⎫[41] Winter 0·71 ⎭	0·73[41]	−0·011 −0·023	0·011 0·023

† G. M. Hughes, unpublished.
‡ E. Don Stevens, personal communication.

Determination of gill area requires the measurement of total filament length, frequency of the secondary lamellae and the average surface area of a secondary lamella. The results obtained for each of these dimensions tend to be complementary in that fish with large total gill areas usually have a large filament length, a high frequency of secondary lamellae, but not necessarily a large surface area of the individual lamellae (Hughes and Shelton[31]). Thus a great deal can be learnt from the simplest of these measurements, i.e. total filament length, in conditions which do not allow more detailed measurements. Such data is summarised in Fig. 2(b), for fish obtained during a British–French–American expedition to the Comores in the Indian Ocean and a Japanese deep water expedition to the Pacific. Total filament length for these deep water fish tends to lie on the same straight line of a log–log plot. Included in this data are two points for the coelacanth. *Latimeria*, indicating that the poor development of the gills of this fish is probably related to its deep water habit rather than to its particular taxonomic position (Hughes [28]).The figure also shows how the closeness of fit is much better for specimens of a single species such as *Xyelacyba* or the shallower-water barracuda. Relationships for body mass and total gill area are now available for many species as indicated in Fig. 2(a) and the exponents given in Table II range from 0·5 to 1·0. This is a rather different position from that envisaged during the first studies of this kind when an exponent of 0·85 for Tuna was thought to be close to that for the average exponent relating body mass to oxygen consumption (Muir and Hughes[40]). Figure 2(a) also shows that in all cases the respiratory area for fish is significantly less than that for a mammal of the same mass. The single line with a slope of 1·0 for mammals has already been subdivided into two types (Weibel[57]) according to whether they are "free-living" or "captive" species. With further investigations of mammalian lung surface areas it is possible that many different lines of varying slopes will be recognised, as shown by the fish studies. Among the latter, less active species have gill areas which are smaller than those of more active fish such as tuna.

Air-breathing Fish

Studies on the development of the respiratory organs of certain species indicate that there may be changes in slope during the life cycle. This has become especially apparent in some studies on the respiratory organs of Indian air-breathing fish. The exponents summarised in Table III show how the development of, for example, the suprabranchial chamber of *Anabas* shows a discontinuity at about 45g (Hughes *et al.*[29]). Appropriate tests (Fisher[17]) showed that this was statistically significant, and it later emerged that above this mass the fish depends far more upon its air-breathing organs than its gills for obtaining oxygen. The fact that the slope of the line for the air-breathing organs is greater than that for the gills is further evidence for the relatively greater development of these surfaces with greater body size. Partly as a result of these studies, experiments were carried out to test the possibility of "drowning" these fish. Such experiments showed, for example, that the lower weight groups of *Anabas* can survive submerged for a long time whereas larger fish die when not allowed to breathe atmospheric air. Specimens of *Channa punctatus*, however, can survive for a long time when not allowed to come to the surface whether they are small or large. Thus it can be concluded that the gills of *Channa* and *Saccobranchus* are quite capable of meeting the increase in oxygen demand with increasing body mass although the gill area per gram decreases comparatively rapidly in these species ($b = -0.35$). In consequence it has been suggested that the suprabranchial chamber may be capable of gaseous exchange even when it remains under water. In the gourami the suprabranchial cavity is ventilated with water during an air-breath (G. M. Hughes and H. M. Peters, unpublished).

These studies on air-breathing fish also emphasised the need to think of

TABLE III. Data obtained from least square regression analyses of bi-logarithmic plots in relation to body mass for some air-breathing fishes. Slopes of the lines for surface area of the gills and the air-breathing organs are given, (for detailed statistical data see the original reference); actual areas are given for a 100 g fish. Diffusing capacity per kg (D_{O_2}) is calculated for a 100 g fish from these measurements. Different slopes of the regression lines for the air-breathing organs at different body sizes are given.

	Slope	Gill area for 100 g fish (mm^2 g^{-1})	Aquatic D_{O_2} for 100 g fish (ml O_2 min^{-1} mm Hg^{-1} kg^{-1})		Slope	Area of air-breathing organ for 100 g fish (mm^2 g^{-1})	Air-breathing D_{O_2} for 100 g fish (ml O_2 min^{-1} mm Hg^{-1} kg^{-1})	References
Anabas testudineus	0·62	47	0·007	suprabranchial chamber	0·33 (0–45 g) 0·78 (45–120 g)	7·5	0·054	Hughes, Dube and Munshi[29]
				labyrinthine organ	0·80	32·0	0·23	
Saccobranchus fossilis	0·75	58	0·024	air sac	0·66	30·7	0·029	Hughes *et al.*[32]
				skin	0·69	200	0·003	
Amphipnous cuchia				air sac	0·71 (0–60 g) 0·20 (60–250 g)	4·8	0·017	Hughes *et al.*[33]
				skin	0·71			
Channa punctatus	0·59	72		suprabranchial chamber	0·70	39·2	0·075	Hakim, Munshi and Hughes[20]
Macrognathus aculeatum	0·73	62	0·052	skin	0·71			Ojha[41]

gas exchange surfaces in a more functional way by taking into account the barrier thickness through which oxygen transfer occurs. In the case of *Anabas*, for example, the thickness of the barrier in the air-breathing organs is very much less than that in the gills and consequently their diffusing capacity is greater (Hughes *et al.*[29]), although their surface area is less than that of the gills for the same specimens (Table III).

Dimensional Analysis

Measurements on the gills in relation to body mass also prove of value in the dimensional analysis of the basic mechanisms of gas exchange. Hills and Hughes[24] carried out analyses on the development of the gills of the small-mouthed bass (*Micropterus dolomieu*), using data obtained by Price[43]. In relating different gill dimensions to body mass a model was chosen which was similar to that given in Fig. 1 and enabled several hypotheses to be tested. Validity of the model was tested by use of the exponent values summarised in Table IV.

TABLE IV. Exponents for the relationships between body mass and different parameters relating to the gill dimensions of *Micropterus* and the rainbow trout. For details see text and Hills and Hughes[24]

	Micropterus	*S. gairdneri*
Distance between secondary lamellae	0·09	0·022
Number of pores	0·28	0·54
Pore length	0·24	0·21
Ventilation volume	0·73	0·93
Gill area	0·78	0·95

As a result of this analysis the following conclusions were reached: (a) water flow past the secondary lamellae is probably laminar; (b) there is a constant perfusion of the gill surfaces with blood at all body sizes; (c) counter-current flow between water and blood is more probable than co-current flow; (d) the resistance to overall oxygen transfer due to the water film is 5–10 times that due to the tissue barrier. This proportion remains almost constant at all body sizes, decreasing as $m^{-0.02}$ in the case of *Micropterus* and as $m^{-0.03}$ for the trout. Finally (e) the velocity of water between the secondary lamellae increases slightly with body size, in the case of *Micropterus* $\propto m^{0.10}$, and for the trout $\propto m^{0.17}$. Scheid and Piiper[46] have also emphasised the importance of the water component of the overall resistance to O_2 transfer. Nevertheless the tissue itself still remains a significant proportion of this resistance.

The analyses for trout were carried out using measurements made by Morgan[38] (Fig. 2c). She calculated the ventilation rate and obtained a slope of 0·93, whereas 0·73 had been obtained for *Micropterus*. For the bass the gill area increases as $m^{0·78}$, whereas the calculations (Hills and Hughes,[24]) suggested that for laminar flow this should be $\propto m^{0·75}$. In the trout, gill area is proportional to $m^{0·95}$ and dimensional analysis predicted $m^{0·89}$ for laminar flow. Thus despite certain differences in the detailed exponents (Table IV) relating different parameters to body mass, the final conclusions from this analysis for both *Micropterus* and the rainbow trout are almost identical. It is clear from such examples that the application of methods of this kind could have valuable results in other aspects of respiratory biology.

Dimensional considerations on the perfusion side of the exchange surface reveal several interesting features. Analysis of the flow of blood through the gills has been made by Muir and Brown.[39] Their equation may be rearranged as follows:

$$\frac{\dot{Q}}{A_g} = \frac{\Delta p \cdot \pi \cdot c^3}{256 \cdot \mu \cdot l_1^2}$$

where c is the diameter of the blood vessels, l_1 is their length, and A_g is the gill surface area. Δp is the pressure drop, $\dot{Q}$ the cardiac output and μ is blood viscosity.

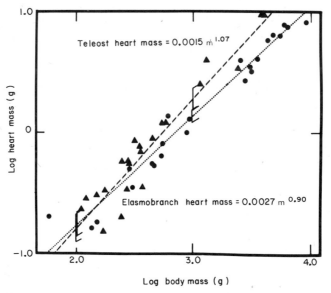

Fig. 3. Bi-logarithmic plot of heart mass against body mass for a number of teleost (excluding trout) and elasmobranch fish. 95% confidence limits are indicated for fish of 100 g and 1000 g.

Recent measurements of morphological characteristics enable us to estimate that cardiac output ($\dot{Q}$) increases linearly with body mass. This is suggested by the finding that heart mass $\propto m^{1 \cdot 0}$ (Fig. 3) and cardiac frequency is approximately $\propto m^{-0 \cdot 05}$. Similarly for rainbow trout gill area $\propto m^{0 \cdot 95}$ and consequently the cardiac perfusion per unit of gill surface ($\dot{Q}/A_g$) will remain roughly constant at all body sizes, a conclusion in agreement with that obtained using a different method of analysis by Hills and Hughes[24]. Thus from the equation of Muir and Brown it follows that $\Delta p c^3 / \mu l_1^2$ remains constant at all body sizes. Unfortunately there are no detailed measurements of the diameter and length of the blood channels in trout or rays of different body size, but from data given in the paper of Muir and Brown it can be deduced that c^3 / l_1^2 is more or less constant at all body sizes. Hence it follows that Δp/blood viscosity is also constant at all body sizes. Limited evidence is available suggesting that Δp increases with m for salmon and trout and hence, if all this data is confirmed and if Muir and Brown's theory is correct, it follows that blood viscosity must also increase with body mass.

Thus it appears that small fish, in order to conform with the hydrodynamic limitations of the Muir and Brown formula, may have lower blood viscosity than larger fish. It is likely that a lowered haematocrit and haemoglobin concentration would be associated with a decrease in size of the red blood cells which would reduce resistance to blood flow through the gills. This gives some explanation of the apparently paradoxical finding that smaller fish with a greater specific oxygen consumption have blood with smaller oxygen carrying capacities. Among higher vertebrates this does not seem to be true (Bartels[2]) but larger mammals are reported to have a greater oxygen affinity i.e. a lower oxygen tension at 50% saturation (Schmidt-Nielsen and Larimer[48]).

Gill Areas and Oxygen Consumption

A further aspect of fish respiratory systems is the relationship between gill dimensions and oxygen uptake (Hughes[26, 27]). This became of especial interest when I found that values for the ratio between oxygen uptake and gill area for fish of several different species were similar (Hughes[28]). This enabled me to suppose that for the coelacanth, *Latimeria*, this ratio would be about $200 \text{ cm}^3 \text{h}^{-1} \text{m}^{-2}$ ($1 \cdot 12 \text{ Js}^{-1} \text{m}^{-2}$). Surface area of the gills is extremely low ($0 \cdot 019 \text{ m}^2 \text{kg}^{-1}$) and hence leads to an estimate of specific oxygen consumption of less than $5 \text{ cm}^3 \text{kg}^{-1} \text{h}^{-1}$ ($0 \cdot 028 \text{ Js}^{-1} \text{kg}^{-1}$). Such a low value is not unreasonable for fish of such habits.†

Differences in this ratio for different body sizes may be considered by

† Note added in proof: measurements on fish kept at depths of 1230 m have shown such low levels of O_2-consumption (Smith, K. L. and Hessler, R. R. *Science*, **184**, 72–73, 1974).

means of the exponents relating body mass to metabolism and gill area. Some of the available data is summarised in Table II. It is apparent that there are some cases in which the exponent of this ratio $(A_g/\dot{V}_{O_2})$ is positive and others in which it is negative; in yet others there is little change with body size. Thus, there is now definite evidence that the three possibilities predicted by Hughes[26, 27] are all found among fish but their precise interpretation is not yet certain. In order to make such relationships more meaningful it would be useful to equate the slope of the gill area line with that for diffusing capacity as this takes into account barrier thickness. From the limited evidence available it appears that diffusing capacity has a slope slightly less than that for gill area (Hughes[27], Ojha[41]) but this may not be true for all species.

Morphometrically determined diffusing capacity indicates maximum possible oxygen uptake and hence gives an upper limit for active oxygen consumption. Consequently where the ratio of $Ag/\dot{V}_{O_2}$ is negative (Fig. 4c), as in the skipjack and ray, a point will be reached as body size increases where the oxygen demand will exceed the capacity of the gas exchanger to transfer oxygen, assuming a constant mean ΔP_{O_2} between the water and blood. Such considerations might be important in determining the upper size limit for a particular species but the possibility of other factors becoming limiting must not be forgotten. In those cases where the ratio is positive (Fig. 4(b)) the two lines diverge and consequently the capacity for taking up oxygen increases relative to the standard oxygen requirement. In such species one can suppose that increasing size would be accompanied by an increased scope for activity i.e. the difference between the standard and active rates of oxygen consumption (Fry[18]). The trout is in this category and, presuming a similar relationship for the salmon, these deductions fit in with the findings of Brett[6] and Brett and Glass[7] which show an increase in the scope for activity with body size. The third possibility is that the slopes are identical (Fig. 4(a)), suggesting that the scope for activity remains constant at all body sizes.

As will be indicated later (pp.74–5), corresponding investigations with reptiles have also shown species in which scope for activity increases with size and others where it remains constant (e.g. *Lacerta*). Furthermore among Amphibia there is evidence for a limitation of body size due to a reduction in oxygen exchange capacity relative to that of the oxygen requirement. Table V summarises the results of calculations from the regression lines for oxygen consumption and gill area for a number of species. They show how the ratio $\dot{V}_{O_2}/A$ is fairly constant for a 1 Kg fish and how this value may increase or decrease with body mass. Calculations for the air-breathing fish *Channa punctatus* are especially interesting as this species shows seasonal variations in oxygen consumption; all $\dot{V}_{O_2}/A$ values are very high for those fish which are forced to breathe water alone. These fish are active and hence cannot be compared with the "standard" figures used in the other calculations. When surfacing is allowed the air-breathing organs are also employed and the ratio of oxygen consumption to the total respiratory surface area has values comparable to other fish. The same is true for salamanders provided total respiratory area is used.

AMPHIBIAN RESPIRATORY SYSTEMS

As has been indicated, many species of fish supplement aquatic gas exchange through the gills by air-breathing. The Amphibia have respiratory mechanisms varying from those which are completely dependent upon aquatic gas exchange to those which live almost entirely by air-breathing. Respiratory surfaces are found mainly at the gills, skin and lungs. In some Amphibia the lungs are degenerate and they depend on the skin for gas exchange both in water and on land; others depend almost entirely on gills when they are submerged. During development there is often a change in the relative importance of different gas exchange surfaces. Furthermore the relative importance for carbon dioxide exchange may be different from that for oxygen exchange (Singh[49]). The release of carbon dioxide forms a vital part of the whole process and this is especially true in the transition from aquatic to air breathing (Hughes[25]). It is in this context that cutaneous respiration may be especially significant and there is now good evidence for its importance even

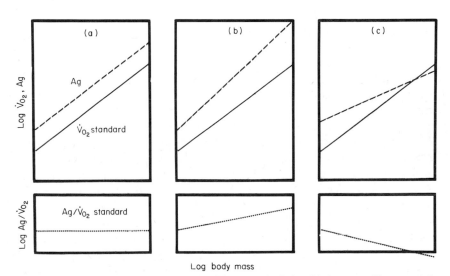

FIG. 4. Theoretical graphs to show three possible types of relationship between gill area and the standard oxygen consumption of fish when plotted against body mass on logarithmic coordinates. The plot of the ratio between area and oxygen consumption is shown in the graphs below. For description see text.

TABLE V. Numerical values for the ratio between resting oxygen consumption and respiratory surface area for fish and other vertebrates. In most cases, the particular values were calculated from the available regression analyses for animals of 10, 100 and 1000 g (1 cm^3 O$_2$ h^{-1} m^{-2} = 5·58 × 10^{-3} J s^{-1} m^{-2})

	Ratio $\dot{V}_{O_2}/Ag$ cm^3 O$_2$ h^{-1} m^{-2}				
	Ray	Torpedo	Skipjack	Trout	Tench
10 g	130	198	131	481	375
100 g	198	210	185	329	172
1000 g	298	226	263	240	162

Channa punctatus

	Surfacing prevented $\dot{V}_{O_2}/Ag$		Surfacing allowed $\dot{V}_{O_2}$/Total resp. S.	
	Winter	*Summer*	*Winter*	*Summer*
10 g	600	1001	195	372
100 g	632	1066	283	353
1000 g	685	1167	413	335

	Salamanders		*Lacerta*	Mammal	Bird
	Lunged	Lungless			
10 g	155	159	197	534	
100 g	178	203		312	255
1000 g	206	259		184	

in some reptiles where the integument is quite well-scaled (Standaert and Johansen,[50] Heatwole and Seymour[22]).

As with fish many studies have been at an interspecific level but have usually been limited to particular taxonomic subdivisions e.g. Anura (frogs and toads), Urodela (salamanders and newts); however, some recent intra-specific results have been most illuminating. Studies on the metabolic rate of anurans in relation to body mass have been summarised by Hutchison, Whitford and Kohl[35] who found an exponent of 0·71 to cover all the species investigated. However if one inspects their results when classified according to families, exponents are found which range from 0·52 among the Hylidae at 25°C to 0·94 for the Pelobatidae at 15°C. For the same animals skin surface area is related to body mass with an exponent of 0·58 for Anura. In this case the exponent for individual species ranges from 0·27 in *Bufo boreas* to 0·82 in *Hyla arborea*. Thus the relationship between the body surface area and metabolism is not a direct one; a similar relationship has been found for lung ventilation where the overall tidal volume is proportional to $m^{0·73}$. In their studies of individual species Hutchison *et al* found a positive correlation between lung tidal volume and the role of the pulmonary surfaces in respiration.

Salamanders have proved to be a most rewarding field for studies of this kind because within this one order there are lunged and lungless forms. Furthermore there have been extensive studies on the respiratory surface areas in relation to body size (Czopek,[12, 13] Szarski[51]). Whitford and Hutchison[58] found that for lungless salamanders (Fig. 5) the exponent for the relation between metabolism and body mass was 0·72, and for lunged forms it was 0·86 at 15°C. The overall relationship for the surface area of all

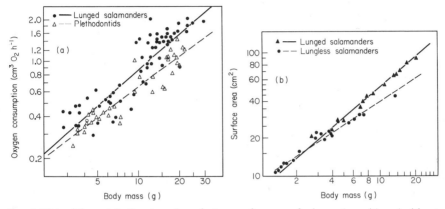

FIG. 5. Plots of the metabolic rate and respiratory surface area of salamanders with and without lungs against body mass ($1 \, cm^3 \, O_2 \, h^{-1} = 5\cdot8 \times 10^{-3} \, J \, s^{-1}$) (from Whitford and Hutchison[58]).

salamanders has an exponent of 0·69; for individual species it ranges from 0·61 to 0·72. As there is no significant difference between the *body* surface area of the lunged and lungless forms, the difference in metabolism cannot be related to this factor. However, when data for the total *respiratory* surface of the two groups was inspected a significant difference in slope was found (Fig. 5(b)). For the lunged species this was 0·79 and for the Plethodontids 0·61, a similar difference to that found for metabolism. Ultsch[56] recalculated respiratory surface relationships for salamaders as a whole using the more recent data of Czopek[13] on respiratory capillary length, and found that total capillary length of the lunged forms has an exponent of 0·86 and for the lungless forms, 0·73. These slopes are even closer to those found for metabolisms by Whitford and Hutchison, and reinforces their conclusion that the higher rate of metabolism of lunged urodeles is related to their greater ability to obtain oxygen because of their larger respiratory area.

For three species of *Sirenidae*, Ultsch[56] found that the exponent for respiratory surface area was close to 0·67, being 0·65 overall, and the slopes of the relationships between metabolism and body mass are almost identical.

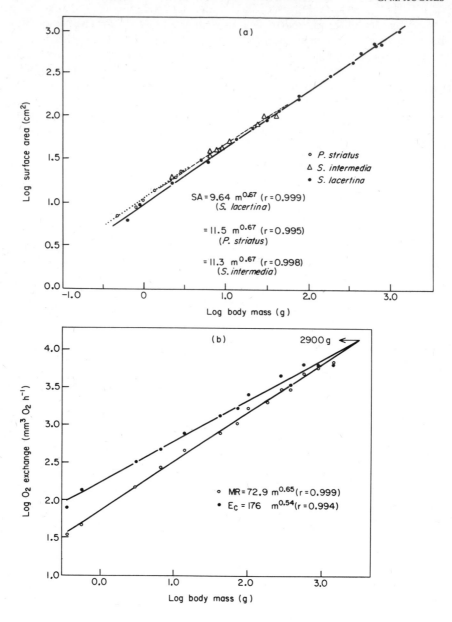

FIG. 6. Bi-logarithmic plots of (a) surface area against body mass for three species of *Siren* and (b) metabolic rate and body mass of submerged specimens of *Siren lacertina* together with changes in exchange capacity (after Ultsch[56]). (1 mm^3 O$_2$ h^{-1} = 5·58 × 10^{-6} J s^{-1}).

Larger forms of these animals die when prevented from surfacing whereas smaller specimens survive, although the ratio of metabolic rate to skin surface area is almost constant at all body sizes. Ultsch concludes that reduced survival with body size is due to changes in skin permeability. Analysing this he introduced the term oxygen exchange capacity (E_c) which is equal to the surface area times the permeability to oxygen for a given pressure difference across the exchange. For a 1 g animal this is about $1.74 \, cm^3 \, O_2 \, h^{-1}$ ($9.7 \times 10^{-3} \, J \, s^{-1}$) which is 74% greater than its resting metabolism and represents a sort of scope for activity. Ultsch finds that there is a reduction in permeability with body size and the slope of the line relating exchange capacity to body mass has a slope of 0.54 in *Siren lacertina* (Fig. 6). Consequently the lines for metabolism and E_c converge at 2.9 kg and suggests that such an animal might just survive when submerged at 25°C in water saturated at a P_{O_2} of 155 mmHg ($2.06 \times 10^4 \, Nm^{-2}$). As Ultsch points out, however, it would be expected that some reserve O_2 capacity would be maintained so that the animal had some scope for activity and consequently survival following submergence would only be expected for smaller animals. Experimentally it is known that death occurs at sizes of 0.6–0.8 kg and above this size *Siren lacertina* is obliged to breathe air. In fact this species grows to over 1.7 kg because it is able to use both water and air breathing. Another species. *Siren intermedia*, however, only reaches 45 g in the same type of habitat. Nevertheless a comparable analysis suggests that this species also would achieve a similar maximum size! Ultsch has suggested that the difference may be due to differences in feeding habit and thus exemplifies the way in which factors other than those being investigated in detail may become limiting.

This summary of scaling of some respiratory parameters among Amphibia illustrates the value of inter- and intra-specific approaches. It is not possible to discuss here the detailed physiological mechanisms involved, but this has been done in a number of cases (see, for example, Piiper, Gatz and Crawford[42]). Clearly studies such as those of Ultsch invite further investigations of, for example, the respiratory properties of the blood both *in vitro* and *in vivo*.

REPTILE RESPIRATORY SYSTEMS

Reptiles are a more diversified group than amphibians and reach much greater body sizes. Their respiratory system shows greater complexity with the elaboration of a trachea and bronchial system. There is an increase in folding and compartmentalism of the lung which far exceeds that of amphibians. In general reptiles are much more active and live in a greater variety of habitats, it being unnecessary for them to be near water. Nevertheless in some aquatic forms cutaneous respiration has been shown to play an important part, but gill respiration is absent.

One type of reptile investigated in relation to body size and activity is the giant tortoise (Hughes, Gaymer, Moore and Woakes[30]). It was found that the standard oxygen consumption is proportional to $m^{0.82}$. The animals were kept in a respirometer and some indication of their activity recorded under conditions in which the animals were relatively inactive. When the tortoises were more active a steeper slope (0·79) was found. Thus in larger animals the difference between these two levels increases suggesting that they have a greater scope for activity. The numbers of animals available and, of course, the conditions for measuring activity in such reptiles were not ideal. However, as far as I know no previous study had related activity to body size in such animals. Dr H. Prange tells me that he has made some recent measurements with marine turtles which show a similar increase in scope with body size.

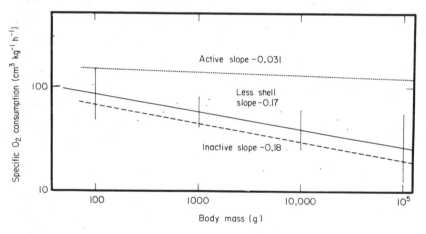

FIG. 7. Log–log plot of change in oxygen consumption per unit body mass with increasing body size in *Testudo gigantea*. Lines are shown for animals at rest and when moving about in the respirometer (after Hughes *et al.*[30]). The standard oxygen consumption has also been calculated in relation to the total body mass minus the weight of the carapace and skeleton and is plotted as a full line, together with 95% confidence limits at different masses (1 cm^3 O$_2$ kg^{-1} h^{-1} = 5·58 × 10^{-3} J kg^{-1} s^{-1}).

Another general question which arises in investigations of animals with such exoskeletons is the extent to which the "active" mass, i.e. body mass less the skeleton and shell, should be used (Benedict, [4]). Some authors have suggested for mammals that, as the mass of the skeleton increases disproportionately with body size, it is possible that the divergence in the relationship between metabolism and body mass from direct proportionality is due to a relative increase in low metabolic rate tissues such as skeletal tissue or fat. Measurements available for tortoises indicate that the increase in mass of shell plus skeleton is almost directly proportional to body mass. Consequently data replotted in Fig. 7 showed scarcely any difference in slope when the mass of the shell plus skeleton was subtracted from the total body mass. The increased scope for activity of larger animals together with the lightness of the carapace and large size of the lungs in these giant tortoises may have survival value. Thus large specimens have considerable buoyancy and would be readily dispersed by ocean winds and currents; their relatively low metabolic rate would increase the time for which they could survive such journeys.

As with other groups interspecific comparisons of respiratory variables with respect to body mass have been carried out. Tenney and Tenney[54] found that the lung surface area was proportional to $m^{0.75}$ in reptiles, but to $m^{1.0}$ in amphibians. In the Amphibia the skin respiratory surface has a capillary density proportional to $m^{0.2}$ and hence the true respiratory area varies as $m^{0.8}$, virtually the same as in reptiles. As in amphibians and mammals, lung mass in reptiles was directly proportional to body mass, but over a range of four log scales they found that the lung volume was proportional to $m^{0.75}$. It has recently been shown that this was probably due to the conditions under which volumes were determined, using excised lungs. Inflation of lungs *in situ* to a physiological pressure is obviously preferable. Over-inflation probably had most effect on the lungs of smaller reptiles and consequently resulted in a reduction in the slope.

Recent studies at the intrageneric level with lizards (Cragg[11]) has provided what I believe to be a model study involving the application of both physiological and morphometric techniques. In order to obtain a sufficient mass range it was necessary to study four species of *Lacerta* and Cragg was able to show that oxygen consumption is proportional to $m^{0.75}$ both at standard and at maximum active rates of metabolism. Exercise of these lizards was carried out on a treadmill and a more or less linear relationship between speed and $\dot{V}_{O_2}$ was obtained (such experiments are extremely difficult with lizards). From these studies she was able to determine the metabolic minimum cost of exercise in accordance with the method of Taylor, Schmidt-Nielsen and Raab;[53] cost being determined as the slope for the linear part of the $\dot{V}_{O_2}$/speed relationship. Minimum cost of exercise was proportional to $m^{0.53}$. This data is plotted on Fig. 8 together with that for mammals and the points for *Varanus* given by Taylor[52]. These results on *Lacerta* thus confirm his general conclusion that lizards expend less energy to run than do mammals of the same size.

Studies on the ventilation mechanism of *Lacerta* showed that both lung and tidal volume are directly proportional to body mass. As ventilatory frequency is independent of size it was found that the so called 'wasted'

D

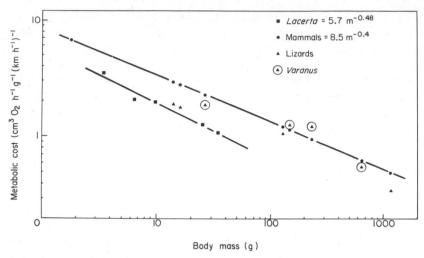

FIG. 8. Change in metabolic cost of locomotion in *Lacerta* at different body masses compared with studies on mammals and other reptiles (from Cragg[11]).($1 \, cm^3 O_2 \, h^{-1} \, g^{-1} \, (km \, h^{-1})^{-1} = 20 \cdot 1 \, J \, kg^{-1} \, m^{-1}$).

ventilation or ventilation requirement ($\dot{V}_E/\dot{V}_{O_2}$) increases with body size, whereas it is constant for mammals. Figure 9 shows the difference in slope between the ventilation and oxygen consumption lines which is $\propto m^{0 \cdot 25}$.

Detailed study of lung morphometry was only possible on five specimens. I would like to reemphasise the laborious nature of such work and it is for this reason that relatively few investigations have been carried out (Hughes

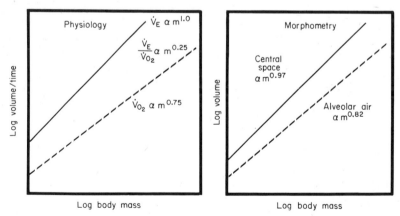

FIG. 9. Log–log plots showing changes in oxygen consumption and lung ventilation in specimens of different body size. Corresponding plots for volumes of the lung occupied by the central space and alveolar regions are shown (based on data of Cragg[11]).

and Weibel[34]). Results, however, are most rewarding and in *Lacerta* can be illustrated by comparing the volume of the central space of the lung and that of the alveolar space as shown in Fig. 9. The ratio between these two increases as $m^{0.15}$. This gives a possible anatomical basis for the reduction in ventilation requirement comparable to that determined physiologically. Thus in larger lizards a greater proportion of the inspired air can only reach the alveoli by diffusion from the central air space with a consequent increase in ventilation requirement.

CONCLUSIONS

One of the main conclusions I wish to emphasise from this study is the way morphological features of the respiratory system can limit the overall oxygen consumption of the animal. This is particularly true for active metabolism and unfortunately very little data is available. In fish, amphibians and reptiles comparison of oxygen uptake data with available gas exchange surface area has provided examples of this kind (Table V). It must also be emphasised that such a limitation is not valid in all cases, even when it is possible theoretically. There are other limiting factors involved which I have not had time to discuss, notably features of the cardiovascular system. It would seem that just as Lighthill (in this volume) defines limiting conditions in relation to hovering flight by means of a polygon (in the length-frequency diagram), so in dealing with respiratory, cardiovascular and other organ systems, if sufficient data were available we should be able to define a polygon showing the size range over which an animal could function. (In terms of Fig. 4, if regression lines were drawn for other systems, e.g. heart size or aortic diameter, then those which intersected the active oxygen requirement relationship at smaller body sizes than the intersection with the respiratory surface line would be more limiting factors.) For example in some cases the supply of available substrates through the feeding and digestive mechanisms could be limiting.

Thus although studies of locomotory mechanisms inevitably emphasise the skeletal and muscular systems, it is the functioning of the whole organism which governs the supply of oxygen and substrates to provide energy for the activity of such muscles, and these factors may be the most important at a given stage of evolution. It is only by such thinking that we can begin to understand the complexity of the relationship between metabolism and body mass which has been plotted in the mouse/elephant line, and to appreciate that it summarises a remarkable range of equilibria which can be different for each of the individual species. In fact it is not inconceivable that the overall slope (0·75) may be the statistical result of individual lines which have slopes ranging randomly from 1·0 to 0·5.

In this survey I have tried to emphasise the value of combining studies of the dimensions of respiratory systems with physiological investigation. At the moment one of the main barriers to further generalisation is the lack of detailed studies involving both types of approach on the same animals. The combination of a physiological approach to morphological data with the analytical techniques of engineers is not available in many laboratories. Symposia such as this, by encouraging scientists specialising in these different disciplines to discuss similar problems, clearly make a valuable contribution to such syntheses. Unlike Francis Bacon, we can longer take the whole of knowledge as our field of interest, and I believe that the best results are now obtained by the collaboration of individual specialists rather than by each scientist attempting to acquire the necessary know-how of the other's techniques. An important requirement, however, is a sympathy for the other scientist's type of approach, an understanding of his methods and perhaps some knowledge of the nature and meaning of the particular numerical values with which he deals.

REFERENCES

1. Barlow, G. W. Intra- and interspecific difference in rate of oxygen consumption in gobiid fishes of the genus *Gillichthys*. *Biol. Bull. mar. biol. lab. Woods. Hole*, **121**, 209–229 (1961).
2. Bartels, H. Comparative physiology of oxygen transport in mammals. *The Lancet* (Sept.) 599–604 (1964).
3. Beamish, F. W. H. Respiration of fishes with special emphasis on standard oxygen consumption II. Influence of weight and temperature on respiration of several species. *Can. J. Zool.* **42**, 177–188 (1964).
4. Benedict, F. G. The physiology of large reptiles. *Publs. Carnegie Instn.* **425** (1932).
5. Benedict, F. G. Vital energetics, a study of comparative basal metabolism. *Publs. Canegie Instn.* **503** (1938).
6. Brett, J. R. The metabolic demand for oxygen in fish, particularly salmonids, and a comparison with other vertebrates. *Respir. Physiol.* **14**, 151–170 (1972).
7. Brett, J. R. and Glass, N. R. Metabolic rates and critical swimming speeds of sockeye salmon (*Oncorhynchus nerka*) in relation to size and temperature. *J. Fish Res. Bd. Canada*, **30**, 379–387 (1973).
8. Brody, S. *In* "Bioenergetics and Growth," Reinhold, New York (1945).
9. Carey, F. G., Teal, J. M., Kanwisher, J. W., Lawson, K. D. and Beckett, J. S. Warm-bodied fish. *Am. Zoologist*, **11**, 137–145 (1971).
10. Carter, G. S. "Animal Evolution." Sidgwick and Jackson Limited, London (1951).
11. Cragg, P. A. "Respiration and Body Weight in the Reptilian Genus *Lacerta*: A Physiological, Anatomical and Morphometric Study." Ph.D. thesis, Bristol University (1975).
12. Czopek, J. Vascularization of respiratory surfaces in some Plethodontidae. *Zool. Polon.* **11**, 131–148 (1961).
13. Czopek, J. Quantitative studies on the morphology of respiratory surfaces in amphibians. *Acta. anat.* **62**, 296–323 (1965).

14. de Silva, C. Development of the respiratory system in herring and plaice larvae, *In* "The Early Life History of Fish." (J. H. S. Blaxter, Ed.), Springer-Verlag, Berlin, Heidelberg, New York (1974).
15. de Silva, C. D. and Tytler, P. The influence of reduced environmental oxygen on the metabolism and survival of herring and plaice larvae. *Netherlands J. Sea Res.* **7**, 345–362 (1973).
16. Dube, S. C. and Datta Munshi, J. S. A quantitative study of the erythrocytes and haemoglobin in the blood of an air-breathing fish, *Anabas testudineus* (Bloch), in relation to its body size. *Folia Haematol.* **100** (4), 436–446 (1973).
17. Fisher, R. A. "Statistical Methods for Research Workers." Edinburgh, Oliver and Boyd (1948).
18. Fry, F. E. J. The aquatic respiration of fish. *In* "Physiology of Fishes," Vol. 1, pp. 1–63. (M. E. Brown, Ed.), Academic Press, London and New York (1957).
19. Graham, J. B. Body temperature of the sea snake *Pelamis platuras. Copeia*, 531–533 (1974).
20. Hakim, A., Munshi, J. S. D. and Hughes, G. M. Morphometrics of the respiratory organs of the Indian green snake-headed fish *Channa punctatus* (1976) (in preparation).
21. Haldane, J. B. S. "Possible Worlds." Chatto and Windus, London (1925).
22. Heatwole, H. and Seymour, R. Respiration of marine snakes. *In* "Respiration of Amphibious Vertebrates" (G. M. Hughes, Ed.), Academic Press, London and New York, pp. 375–387 (1975).
23. Hemmingsen, A. M. Energy metabolism as related to body size and respiratory surfaces and its evolution. Reports of the Steno Mem. Hosp. and the Nordisk Insulinlab. IX (1960).
24. Hills, B. A. and Hughes, G. M. A dimensional analysis of oxygen transfer in the fish gill. *Respir. Physiol.* **9**, 126–140 (1970).
25. Hughes, G. M. The dimensions of fish gills in relation to their function. *J. exp. Biol.* **45**, 177–195 (1966).
26. Hughes, G. M. Gill dimensions in relation to other respiratory parameters. *In* "Proc. IX. Internat. Congr. Anat. Leningrad" (1971).
27. Hughes. G. M. Morphometrics of fish gills. *Respir. Physiol*, **14**, 1–25 (1972).
28. Hughes, G. M. On the respiration of *Latimeria chalumnae. J. Linn. Soc. (Zool.)* **59**, 195–208 (1976).
29. Hughes, G. M., Dube, S. C. and Munshi, J. S. Datta. Surface area of the respiratory organs of the climbing perch, *Anabas testudineus* (Pisces: Anabantidae). *J. Zool. Lond.* **170**, 227–243 (1973).
30. Hughes, G. M., Gaymer, R., Moore, M. and Woakes, A. J. Respiratory exchange and body size in the Aldabra giant tortoise. *J. exp. Biol.* **55**, 651–665 (1971).
31. Hughes, G. M. and Shelton, G. Respiratory mechanisms and their nervous control in fish. *In* "Advances in Comparative Physiology and Biochemistry", (O. Lowenstein, Ed.) Academic Press, London and New York, **1**; 275–364 (1962).
32. Hughes, G. M., Singh, B. R., Guha, G., Dube, S. C. and Munshi, J. S. Datta. Respiratory surface areas of an air-breathing siluroid fish *Saccobranchus* (≡ *Heteropneustes*) *fossilis* in relation to body size. *J. Zool. Lond.* **172**, 215–232 (1974).
33. Hughes, G. M., Singh, B. R., Thakur, R. N. and Munshi, J. S. Datta. Areas of the air-breathing surfaces of *Amphipnous cuchia* (Ham). *Proceedings of the Indian National Science Academy*, **40**(B), 4, 379–392 (1974).
34. Hughes, G. M. and Weibel, E. R. Morphometry of fish lungs. *In* "Respiration of

Amphibious Vertebrates". (G. M. Hughes, Ed.), Academic Press, London and New York, pp. 213–231 (1976).

35. Hutchison, V. H., Whitford, W. G. and Kohl, M. Relations of body size and surface area to gas exchange in anurans. *Physiol. Zool.* **41** (1), 65–85 (1968).
36. Huxley, J. S. "Problems of Relative Growth". Methuen, London (1932).
36a Joysey, K. A. The fossil species in space and time: some problems of evolutionary interpretation among pleistocene mammals. *In* "Studies in Vertebrate Evolution." (K. A. Joysey and T. S. Kemp, Eds.), Oliver and Boyd, Edinburgh, pp. 267–280 (1972).
37. Kleiber, M. Body size and metabolism. *Hilgardia,* **6**, 316–353 (1932).
38. Morgan, M. "Gill Development, Growth and Respiration in the Trout, *Salmo gardneri* (Richardson)". Ph.D. thesis. University of Bristol (1971).
39. Muir, B. S. and Brown, C. E. Effects of blood pathway on the blood pressure drop in fish gills with special reference to tunas. *J. Fish. Res. Bd. Canada,* **28**, 947–955 (1971).
40. Muir, B. S. and Hughes, G. M. Gill dimensions for three species of tunny. *J. exp. Biol.* **51**, 271–285 (1969).
41. Ojha, J. "Studies on the Structure and Function of the Respiratory Organs and Enzyme Histochemistry of the Respiratory Muscles of a Freshwater Mud-eel, *Macrognathus aculeatum* (Bloch)". Ph.D. thesis Bhagalpur University, India (1974).
42. Piiper, J., Gatz, R. N. and Crawford, E. C. Gas transport characteristics in an exclusively skin-breathing salamander, *Desmognathus fuscus* (Plethodontidae). *In* "Respiration of Amphibious Vertebrates" (G. M. Hughes, Ed.), Academic Press, London and New York, pp. 341–356 (1976).
43. Price, J. W. Growth and gill development in the small-mouthed black bass, *Micropterus dolomieu,* Lacépède. *Studies, State Univ. Ohio* (1931).
44. Rao, G. M. M. Oxygen consumption of rainbow trout (*Salmo gairdneri*) in relation to activity and salinity. *Can. J. Zool.* **46**, 781–786 (1968).
45. Romer, A. S. Notes on branchiosaurs. *Am. J. Sci.* **237**, 748–761 (1968).
46. Scheid, P. and Piiper, J. Quantitative functional analysis of branchial gas transfer: theory and application to *Scyliorhinus stellaris* (Elasmobranchii). *In* "Respiration of Amphibious Vertebrates" (G. M. Hughes, Ed.), Academic Press, London and New York, pp. 17–37 (1976).
47. Schmidt-Nielsen, K. Energy metabolism, body size and problems of scaling. *Fedn. Proc.* **29**, 1524–1532 (1970).
48. Schmidt-Nielsen, K. and Larimer, J. L. Oxygen dissociation curves of mammalian blood in relation to body size. *Am. J. Physiol.* **195**, 424–428 (1958).
49. Singh, B. N. Balance between aquatic and aerial respiration. *In* "Respiration of Amphibious Vertebrates", (G. M. Hughes, Ed.), Academic Press, London and New York, pp. 125–160 (1976).
50. Standaert, T. and Johansen, K. Cutaneous gas exchange in snakes. *J. comp. Physiol.* **89**, 313–320 (1974).
51. Szarski, H. The structure of respiratory organs in relation to body size in Amphibia. *Evolution,* **18**, 118–126 (1964).
52. Taylor, C. R. Energy cost of animal locomotion. *In* "Comparative Physiology" (L. Bolis, K. Schmidt-Nielsen and S. H. P. Maddrell, Eds.), North-Holland Publishing Company (1973).
53. Taylor, C. R., Schmidt-Nielsen, K. and Raab, J. L. Scaling of energetic cost of running to body size in mammals. *Amer. J. Physiol.* **219**, 1104–1107 (1970).

54. Tenney, S. M. and Tenney, J. B. Quantitative morphology of cold-blooded lungs: Amphibia and Reptilia. *Respir. Physiol.* **9**, 197–215 (1970).
55. Thompson, D'Arcy, W. "On Growth and Form". Cambridge University Press (1942).
56. Ultsch, G. R. Eco-physiological studies of some metabolic and respiratory adaptations of sirenid salamanders. *In* "Respiration of Amphibious Vertebrates", (G. M. Hughes, Ed.), Academic Press, London and New York, pp. 287–311 (1976).
57. Weibel, E. R. Morphological basis of alveolar-capillary gas exchange. *Physiol. Rev.* **53** (2), 419–495 (1973).
58. Whitford, W. G. and Hutchison, V. H. Body size and metabolic rate in salamanders. *Physiol. Zool.* **40**, 127–133 (1967).
59. Winberg, G. G. (1) Rate of metabolism and food requirements of fishes, (2) New information on metabolic rate in fishes. *Fish Res. Bd. Can.* (translation series 194 & 362) (1956).
60. Zeisberger, E. Die Abhängigkeit des Standardmetabolismus vom Gewicht der Fische, *Zeitschrift für Fischerei*, **10**, 1–3 (1961).
61. Zeuthen, E. Body size and metabolic rate in the animal kingdom. *C.R. Lab. Carlsberg, Ser. chim.* **26**, 15–161 (1947).

Editorial note. In this paper, as elsewhere in the book, oxygen consumption values are quoted together with their energy equivalent in S.I. units, on the assumption that 1 cm^3 of oxygen (STP) when used in oxidative metabolism can be equated with 20·1 J of metabolic energy. This is the value quoted, for example, by Taylor (p. 128 below), for terrestrial animals.

5. The Biological Basis for Development of Scaling Factors in Animal Locomotion

ARTHUR B. DUBOIS

John B. Pierce Foundation Laboratory and Yale University, New Haven, Conn., U.S.A.

ABSTRACT

The intrinsic properties of the materials of which the body is composed place a limit on the stresses that the tissues can tolerate. These limits cannot be exceeded safely during locomotion, or under the stress of gravity. One such limit is imposed by the intrinsic weakness of the cell membrane. It can withstand surface tension forces of approximately 10^{-3} N m^{-1}. Hence, the transmural pressure of cells must be limited to about 100 N m^{-2} (1 cm H$_2$O). Another limit is imposed by the amount of transmural filtration pressure withstood by the capillaries before they filter too much fluid and cause tissue edema, or before reabsorption of too much fluid produces tissue dehydration. Appreciable fluid shifts are produced by changes of filtration pressure of only about 100 N m^{-2}. To protect the tissues against excessive forces, extracellular supporting tissues are laid down in an amount and location which redistribute the external and internal forces in such a way that transcellular and transcapillary pressure gradients of even large animals are small in amount. Some mechanisms favouring this redistribution are described, and examples are drawn from experiments with dogs immersed in water and bluefish supported in air.

Since amoebae and jellyfish do not have to go very far very fast, they do not require much structure to harden the body. However, if one were to push or drag such a soft animal through the water at a high speed, it would be deformed away from its normal shape. The body of fast animals is forced to acquire internal structure as a defense against the strain resulting from external stress.

The cell is an assembly of proteins surrounded by a semipermeable membrane. This membrane is rather weak, and it would burst as a result of osmotic pressure (attraction of proteins for water) except for its ability to pump out sodium or other ions, and thereby reduce the osmotic pressure (Table I).

TABLE I

---

1. Membrane surface tension 5×10^{-4} N m^{-1}
2. Hence transmural pressure $= 100$ N m^{-2}
3. Protein attracts water
4. Salt must be pumped out
5. Water follows salt
6. Pressure gradient also would displace water, slowly
7. If interstitial fluid becomes depleted, perhaps cellular fluid will follow

---

The strength or weakness of the cell membrane is determined by measuring the force required to distort (flatten or elongate) the cell, and is expressed in units of surface tension. This surface tension is approximately 5×10^{-4} N m^{-1} (Harvey[6, 7], Cole[1]). Thus a cell with a diameter of 20 μm can withstand an internal pressure of only 100 N m^{-2} (1 cm H$_2$O) before becoming distorted and bursting. Now the pressure encountered by the front end of a fish is approximately equal to $\frac{1}{2}\rho V^2$, where ρ is the density of water and V is the speed. This is equal to 100 N m^{-2} when $V = 0.45$ m s^{-1} (1mph), so cellular distortion or rupture could occur at speeds greater than 0.45 m s^{-1}. Similarly, a land animal more than 1 cm high could have compressional distortion of the cells near the bottom. Protection of the individual cells becomes a problem in aquatic or terrestrial animals of any size or speed.

Although diffusion carries oxygen and nutrients across short distances, such as half a millimeter, reasonably fast, animals larger than this require bulk flow for the transport of substances to and from their cells. Such animals therefore require a circulatory system.

Small solutes pass to the interstitial space and cells by diffusion through the membranes of the capillary wall, but fluids and larger solutes are filtered under the influences of pressure gradients from the capillary into the interstitial space, and are drawn back into the downstream end of the capillary by osmotic attraction and a lessened outward pressure (Fig. 1). In this process, a pressure reduction of only 100 N m^{-2} in the venous end of the capillary will attract appreciable amounts of fluid from the interstitial space into the venules, and thereby reduce the extravascular fluid volume. Since pressure changes exceeding 100 N m^{-2} occur during the mechanical stress present during swimming or under the pull of gravity, the body requires mechanisms to protect the tissues against either excessive outward filtration or too much reabsorption of interstitial fluid.

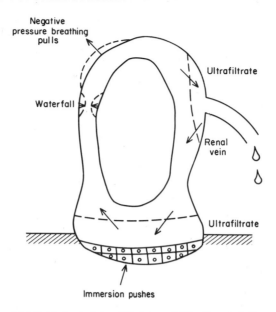

FIG. 1. The interstitial fluid, formed by diffusion and filtration, is influenced by transmural filtration pressure, which in turn derives from capillary pressure, tissue pressure, and external pressure, versus osmotic pressure. Filtration in the kidneys readjusts fluid volume. Collapsibility of the veins protects the tissues of the body against excessive changes of filtration pressure.

The weakness of the cell membrane and the vulnerability of capillary filtration are intrinsic properties that apply equally to large as well as small animals. The general principle emphasized in this paper is that all animals must reduce the pressure distorting their cells and the transmural pressure changes across the capillary wall to values less than 100 Nm^{-2}. This must be accomplished in the face of hydrodynamic pressures and hydrostatic pressures that otherwise would limit an animal's speed through water to 0·45 $m s^{-1}$ (1 m.p.h.), or its height on land to 1 cm of vertical distance. How can this be done?

A somewhat similar problem, weakness of the basic structural material, was encountered by ship builders in the days of sail. Larger ships required larger areas of canvas, and therefore the sails weighed more. But heavy sails would tear under their own weight, or under wind pressure. The ship builders solved the problem of the intrinsic weakness of cloth by dividing the canvas into smaller units and suspending the units from yard arms. Square riggers were designed because of the weakness of canvas rather than the desirability of yard arms. The animal body uses many devices to reduce the transmural pressure on the cell wall and capillary wall to safe limits. Instead of sails suspended from yard arms, we have flesh hung from our bones.

The circulation has its own built-in safety features. Successful animals such as man have a closed brain case which is filled with cerebrospinal fluid. This column of fluid is connected to the veins in the abdomen, whose pressure is taken as the reference pressure. Without this system, the brain would shrink when the head is up, and swell when the head is down. Father William discovered, while still a youth, that it was safe to stand on his head. However, if he had had an incomplete skull, like that of a dinosaur, this would have been safe only if he had had a small brain, as did the dinosaur. The cerebrospinal fluid acts as an hydraulic counterpoise to prevent filtration and reabsorption of the brain's interstitial fluid over the range of positions assumed by the body.

The hands are not protected by an enclosing case. When a student raises his hand, the fluid might drain out, leaving him with shriveled fingers. However, there is a protection against this, and it comes under the heading of "sluice", or "waterfall effect" (Fig. 1). The basic idea is that the veins collapse if subjected to a compressive transmural pressure of more than $100 \, \mathrm{N \, m^{-2}}$. Once this limit is reached, further negative pressure inside or positive pressure outside the veins aspirates no additional blood or tissue fluid. As the hand is raised, the venous column of blood creates a negative pressure inside the veins, but this collapses the veins, allowing no more than $-100 \, \mathrm{N \, m^{-2}}$ of pressure to be transmitted to the inside of the capillaries.

A firm handshake is a sign of vigor. But would it not squeeze the tissue juices out of the recipient's right hand, making him unable to grasp his sword, should the need arise? No! Once again the waterfall effect comes to the rescue! As the pressure outside the veins is raised by more than $100 \, \mathrm{N \, m^{-2}}$ in excess of that inside they collapse, and the increment of net reabsorption pressure in the capillaries therefore is limited to $100 \, \mathrm{N \, m^{-2}}$, no matter how much the pressure outside the skin is raised.

Too rapid an infusion of blood can dilate the heart. It also causes pulmonary edema. The thoracic muscles, which are strong enough for breathing and weight lifting, can produce a large negative pressure. If a person closes his larynx and attempts to breathe in, he creates a negative pressure around the heart of about $10^4 \, \mathrm{N \, m^{-2}}$. This would dilate the heart if blood could be aspirated into the veins in the thorax. But instead the veins collapse at the point they enter the thorax (Fig. 1). Negative pressure breathing creates a negative pressure in the veins of only $100 \, \mathrm{N \, m^{-2}}$, but no more; as at a waterfall, the pressure and flow upstream are isolated from the depth of the gorge or pressure downstream.

By now, I may have convinced you that heights greater than 1 cm are dangerous. A mouse is safe, but a man is not. But with all of these safety devices, perhaps a "hydrostatic equivalent man" (one sketched out in terms of the vertical variation of transmural pressures, rather than geometric height alone) may be no taller than a mouse (i.e. may have a transmural pressure variation of only $100 \, N \, m^{-2}$). Let us begin with the abdomen (Fig. 2). Since it contains fluid, the transmural pressure across the blood vessels is independent of height in the abdomen. Let us call the pressure at the top of the abdomen "zero". Next, consider the thorax. The heart is 5 cm above the diaphragm and therefore has a height of -5 cm, but intrapleural pressure around the heart is -5 cm H_2O ($-500 \, N \, m^{-2}$), therefore the

Transmural pressure within the abdomen is independent of height

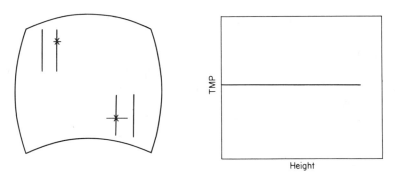

FIG. 2. Transmural filtration pressures in the abdomen are independent of height. The different lengths of the arrows at the top and bottom of the abdomen indicate the different pressure magnitudes there, although the pressure inside a blood vessel is equal to that outside.

hydrostatic component of transmural pressure of the heart is equal to that in the abdomen, and both are zero. The abdomen and heart are in "flatland", and are no "higher" than those of a mouse! The brain is surrounded by cerebrospinal fluid which filters into the abdominal veins and takes its pressure from the abdomen (Fig. 3). Therefore, the change of filtration pressures with height for the abdomen, heart, and brain are all zero. The major muscle groups are hung from bones. The pressure under the skin of a dependent muscle, hung like a sack (Fig. 4), can be deduced to be zero when the concavity of the skin near the top inflects into a convexity of the skin near the bottom. Thus, the pressure outside muscle groups can be "read" from their shape. The principle of the square-rigger applies to the arm and leg muscles, and it can be observed that the pressure is close to zero.

Obviously, there are parts of the body where transmural filtration pressures are not zero. The blood in the arteries and veins is subject to hydrostatic pressure differences. In the blood vessels of the feet, the pressure in the capillaries is high, and serum will ooze from a cut in the tissue of the toes. To take

Filtration pressures of brain and abdomen

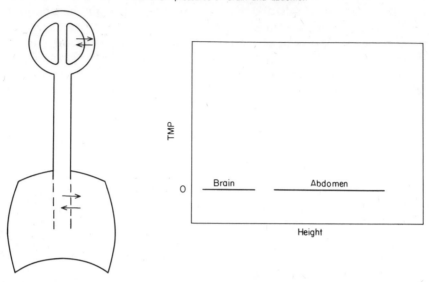

FIG. 3. Transmural filtration pressures in the brain are independent of height.

Skeletal suspension of soft tissues

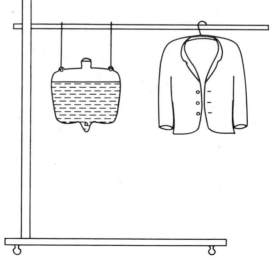

FIG. 4. Since the soft tissues are suspended from skeletal structures, fluid tends to stay inside (left), and the tissues tend to remain unwrinkled (right).

care of these gravitational differences, the body maintains an elevated plasma protein concentration which attracts fluid into the blood stream. But, the body simultaneously maintains a high blood pressure to force fluid out of the blood stream. The balance between these two large forces can be controlled as a function of height. The capillary pressure is varied by regulation of the arteriolar resistance. In the topmost regions, the arteriolar resistance may be low, causing a high capillary pressure which counteracts the reabsorption that otherwise would occur due to the osmotic pressure of the proteins. In the lower regions, the arteriolar resistance is high, lessening the capillary pressure, and therefore favoring reabsorption of fluid in the face of a large hydrostatic pressure in the arteries and veins. If the animal lies down, the arterioles can change their tone, readjusting capillary pressure appropriately. Other mechanisms such as pumping action of the muscles, valves in the veins, and ability of the subcutaneous tissues to withstand pressure also contribute to return of fluid to the core.

An animal moving at high speed through water encounters a large transmural pressure on the front of the head and a large negative transmural pressure on the shoulder. These regions must be hardened at the surface by extracellular dermal plates of cartilage and bone. The surface structures are supported by skeletal bone. Since the thrust of a fast fish comes from its tail, and the inertial and drag forces occur in and on the body, the tail and body must be connected by compression struts. This necessitates a strong vertebral column. Tension forces are sustained by the basement membrane, by ligaments, tendons, fascia and muscles. Such tissues have intrinsic length-tension relationships and breaking strengths. The body structure must make the best use of the intrinsic properties of the materials available. These extracellular supporting tissues redistribute the mechanical stresses before the are delivered to the cells and circulation, protecting them against the stress and strain that otherwise would rupture the cell membrane or cause displacement of intracellular fluid.

Experimental work with animals provides glimpses of the mechanisms that protect the cells and their circulation against the forces of gravity, locomotion, and hydrodynamics.

Example 1.

The death of the monkey "Bonnie" in Biosatellite III suggested that we knew very little at that time about the process of body fluid redistribution during weightlessness. Of course, it set off many experiments in many laboratories. In our laboratory, Dr J. T. Davis and I[2] immersed anesthetized dogs in warm water, with counterpressure on the breathing tube, to simulate a weightless circulatory system. We attributed the resultant diuresis which occurred during immersion to a shift of fluid from the tissues into the blood stream, and thence out through the kidneys. The evidence for this view is that the hematocrit drops initially during immersion indicating hemodilution, and that the diuresis occurs independently of any up or down change of left atrial transmural pressure. The fluid shifts were compatible with the known amounts of filtration that take place when the filtration pressures are changed by the amount observed during immersion.

Example 2.
The fluid shifts seen during simulated weightlessness suggested a new problem, namely that fish, which have a circulatory system supported by water, might never have evolved any circulatory protection against gravity. How would amphibians have been able to crawl out on the land? In pursuit of this question (DuBois, Cavagna and Fox[3, 4]) it was found that fish are subjected to two other kinds of forces, namely inertial forces during acceleration of the body, and transmural pressures due to hydrodynamic forces during swimming, and that the magnitude and direction of these forces are not unlike those due to hydrostatic pressures in land animals of a size equivalent to that of fish (Fig. 5).

Example 3.
Because the stresses on the circulatory system of fish are not very dissimilar from those due to gravity, DuBois *et al.,*[5] studied the response of the blood pressure of bluefish to gravitational stress induced by tilting the fish head up in air. A bluefish tolerated a 30° head up position (sin 30 = 0·5, therefore a 30° inclination is equivalent to a gravitational acceleration of $\frac{1}{2}g$) for at least 35 minutes, with no significant drop in blood pressure. Afterward, he swam vigorously back to sea. During other tilts, of shorter duration, we

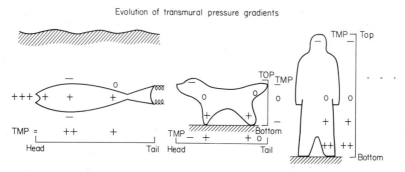

FIG. 5. This scheme compares the transmural pressures on a fish swimming with the hydrostatic pressures in the blood vessels of a quadruped and biped. Body structures seem to have evolved to withstand these pressure gradients and the forces of locomotion.

elicited a tachycardia and initial overshoot of blood pressure not unlike those found during exercise. This was reduced by atropine. In exercising bluefish, we found a tachycardia and an increase of blood pressure. If atropine was administered, these persisted during the rest periods, abolishing the change with exercise. Apparently, the bluefish has a resting vagal tone that slows the heart and drops the diastolic pressure. During exercise, this tone is diminished and the heart released from vagal inhibition. Atropine abolishes the parasympathetic vagal inhibition of the heart found at rest in bluefish. The tilting experiment suggested that bluefish, which never before have encountered gravitational stress on their circulation, are able to tolerate at least 0·5 g for long periods and hence could crawl up a 30° slope if only they had legs. Perhaps this tolerance for gravity evolved as a response to the need for tolerance to the stresses of swimming. This is speculative, because there is no way to prove it.

Since isolated examples can only illustrate and not prove a more general view, it would be impossible to conclude that forces involved in locomotion favored the development of body shape and structure. There must have been many other influences on body form, including the need for protection against predators. However, the bodily forms of modern animals appear to be so well equipped to deal with external and internal pressures encountered while swimming in water or moving on the land that it seems almost self evident that forces of locomotion played a major role in the evolution of body shape. My own view is that one of the goals in the evolution of body structure was to protect the cell membrane and the interstitial fluid against forces in these regions exceeding $100 \, \mathrm{N} \, \mathrm{m}^{-2}$ (1 cm H_2O) of transmural pressure. The extracellular supportive tissues appear to be laid down in the right amount and the right place to counterbalance forces encountered at rest and during locomotion. There is a suggestion, but, of course, no proof, that the adaptation of the body required for swimming may have been a prelude to the body's tolerance for g forces on the land's surface.

REFERENCES

1. Cole, K. S. Surface forces of the Arbacia egg. *J. Cell. & Comparat. Physiol.* **1**, 1 (1932).
2. Davis, J. T. and DuBois, A. B. Mechanism of diuresis in dogs during body immersion in water (abstract). *Proc. Internat. Union of Physiol. Sci.* **9**, 133 (1971).
3. DuBois, A. B., Cavagna, G. A. and Fox, R. S. Pressure distribution on the body surface of swimming fish, *J. Exp. Biol.* **60**, 581–591 (1974).
4. DuBois, A. B., Cavagna, G. A. and Fox, R. S. Locomotion of bluefish. *J. Exp. Zool.* **195**, 223–235 (1976).
5. DuBois, A. B., Carniol, P. and Morris, R. D. B. Blood pressure and heart rate in bluefish: the effects of exercise, gravity, and atropine (abstract), *Federation Proc.* **35**, 796 (1976).

6. Harvey, E. N. A determination of the tension at the surface of eggs of the annelid
   Chaetopterus. Biol. Bull. **60**, 67–71 (1931).
7. Harvey, E. N. Methods of measuring surface forces of living cells. *Tr. of the Faraday
   Soc.* **33**, 943–946 (1937).

PART II

TERRESTRIAL LOCOMOTION

6. Mechanics and Scaling of Terrestrial Locomotion

R. McN. ALEXANDER

Department of Zoology, The University of Edinburgh, Scotland.

ABSTRACT

The forces which mammals and birds exert on the ground change direction in the course of a step in such a way as to keep more or less in line with the most proximal joint of the limb or (in bipeds) with the centre of mass of the body. It is shown theoretically that this requires less energy than if the force were kept vertical.

Data on the relationships between stride length, speed and gait conform quite well to simple general rules.

The mechanics of various gaits is considered and equations are derived relating the energy cost of locomotion to speed, stride length and step length. Costs calculated from measurements of oxygen consumption agree satisfactorily with the predictions of the equations for large mammals but not for small ones.

INTRODUCTION

The main aim of this paper is to calculate from mechanical considerations the energy requirements of locomotion on legs. Simple models of the principal bipedal and quadrupedal gaits will be formulated and equations showing their energy requirements will be derived. Conclusions will be drawn as to how animals should relate their gait to their size and speed, to travel most economically. The conclusions will be discussed in the light of the actual performance of mammals and running birds. The analysis is taken further than in previous discussions of the models.[2, 3]

As an animal runs its parts accelerate and decelerate, so that its kinetic energy fluctuates. Its centre of mass rises and falls so that its gravitational potential energy fluctuates. Tendons are stretched, storing elastic strain energy, and recoil.[1] Each of these forms of energy goes through a cycle of

93

changes in every stride. Whenever their total increases, work must be done by the muscles. When it decreases, the muscles must dissipate the excess energy, degrading it to heat: it is sometimes convenient to describe this process by saying that the muscles do negative work.[21] When an animal is travelling over level ground with the same mean velocity in successive strides, there is no net change of potential, kinetic or elastic energy over a complete stride: the positive and negative work performed by the muscles for purposes considered so far cancel each other out. Net positive work is done against the drag which the air exerts on the body but Pugh[25] has shown that this work is relatively small for a human running in still air, and a more general argument shows that it is unlikely to represent a large proportion of the energy cost of terrestrial locomotion in still air for any animal.[2] Drag is therefore ignored in the remainder of this paper.

Metabolic energy is used both when muscles perform positive work and when they perform negative work. Experiments by Margaria[21] on man indicate that performance of 1 J positive work requires expenditure of 4 J metabolic energy, and performance of 1 J negative work requires 0·8 J metabolic energy.

DEFINITIONS AND SYMBOLS

The following terms are widely used but may not be familiar to some readers:

Stride length is the distance travelled by an animal's centre of mass in a complete cycle of limb movements.

Step length is the distance travelled by the trunk, between the instant when a foot is set down and the instant when the same foot is lifted. Note that step length is not necessarily the same for all the feet.

Net cost of transport is the metabolic energy, in excess of resting requirements, needed to transport unit mass of animal unit distance.

The following symbols are used:-

b half the step length (see definition above).

d the length of a limb segment.

g the acceleration of free fall.

h the height of the hip joint from the ground in normal standing. When the fore limbs of a quadruped are being considered, this height will be measured to the axis of rotation of the scapula relative to the trunk.

m the mass of the body (for a biped) or of the fore or hind half of the body (for a quadruped).

t time. Zero time occurs when the hip or shoulder is vertically over the supporting foot.

u	the horizontal component of the velocity of the trunk or, in the case of galloping, of the fore or hind quarters.
$\bar{u}$	the mean velocity of the animal over a complete stride.
x	the distance by which the hip is in front of a vertical line through a hind footprint, or the shoulder is in front of a vertical line through a fore footprint, at time t.
y	the height of the hip or shoulder from the ground at time t. The height of the shoulder is measured to the axis of rotation of the scapula.
C	the compliance of a leg to vertical forces applied through the hip or shoulder.
E	a metabolic energy cost.
F	the vertical component of the force exerted by a foot on the ground.
Q	the value of an integral which occurs in Eqn. (2).
T	net cost of transport (see definition above).
W	work, mechanical energy cost.
α, β, θ	angles defined in Figs 2 and 4.
η	an efficiency such that $1/\eta$ units of metabolic energy are required to provide 1 unit of positive work plus 1 unit of negative work.
λ	stride length (see definition above).

It will sometimes be convenient to express quantities in dimensionless form, as follows:

$$\hat{b} = b/h, \qquad \hat{u} = \bar{u}(gh)^{-\frac{1}{2}},$$
$$\hat{C} = mgC/h \qquad \hat{T} = T/g,$$
$$\hat{\lambda} = \lambda/h.$$

The following subscripts will be used:-

$0, \pm b, \pm \lambda/4$ the value of a quantity when $x = 0$, $x = \pm b$ or $x = \pm \lambda/4$.

h, r, w   relating to hopping, running and walking, respectively.

E, K, P   relating to elastic strain energy, kinetic energy and gravitational potential energy, respectively.

The mass of the legs will be ignored in most sections of this paper, but account will be taken of it eventually. Bipeds will be assumed to have their centres of mass located at the hip joint (the centre of mass of a standing man is about 5 cm above the hip joints.[13] Quadrupeds are treated as two bipeds in tandem, with centres of mass located at the hip joint and the axis of rotation of the scapula.

THE DIRECTION OF THE GROUND FORCE

Records have been made of forces exerted on the ground in locomotion by men,[7, 8, 19] cats,[20] dogs,[4, 19 and unpublished observations] horses,[5] a wallaby[3] and quail.[11] In all cases the force on the ground has a forward component when the foot is first placed on the ground and a backward component later in the step. It seems to be a general rule for man and quadrupeds that the force on the ground changes direction in the course of a step, so as always to be more or less in line with the hip joint or the dorsal part of the scapula (Fig. 1). Wallabies and quail are bipeds with their hip joints posterior to the centre of mass. If they kept the ground forces in line with their hip joints, they would fall on their faces. Instead, they keep them more or less in line with the centre of mass of the body.

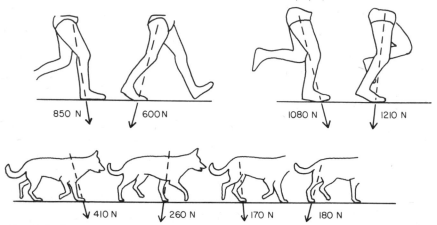

Fig. 1. Outlines traced from selected frames of films of a 68 kg man walking and running, and a 36 kg dog trotting, over a force platform. The magnitude and direction of the force recorded at the same instant is indicated in each case. From unpublished data obtained by the author and Miss A. Vernon, using the method of Ref. 3.

The horizontal components of the ground force seem unnecessary, and they are responsible for fluctuations of the kinetic energy of the body and so for a large part of the energy cost of locomotion. It would seem at first sight that energy could be saved by eliminating them, but it will be shown that this is not so.

Consider a leg consisting of two segments (a thigh and a shank) of equal length d (Fig. 2(a)). Let the hip travel horizontally at a height h from the ground. The foot is set down a distance b in front of the hip and is lifted again when the hip has passed over it and is b in front of it: the step length is thus $2b$. While the foot is on the ground it exerts a vertical component of force F which we will suppose for the present to be constant.

Suppose first that the force on the ground has no horizontal component (Fig. 2(b)). Consider the instant when the hip is a distance x in front the foot

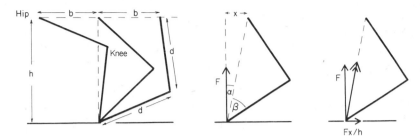

FIG. 2. Diagrams illustrating the discussion of the direction of the ground force.

so that the ground reaction exerts a moment Fx about the hip. The work δW done by the hip muscles as the animal advances a small distance δx is given by

$$\delta W = Fx(\delta\beta - \delta\alpha)$$

where α, β are as shown in Fig. 2(b). It is assumed that the orientation of the trunk relative to the ground is constant. Note that

$$\alpha = \arctan(x/h),$$

and

$$\beta = \arccos[(x^2 + h^2)^{\frac{1}{2}}/2d],$$

whence

$$\frac{dW}{dx} = -Fx\left[\frac{x}{(x^2 + h^2)^{\frac{1}{2}}(4d^2 - x^2 - h^2)^{\frac{1}{2}}} + \frac{h}{(x^2 + h^2)}\right]. \quad (1)$$

To determine the metabolic energy cost of a step we must obtain separate values for positive and negative work. dW/dx is always negative while x is positive. It is positive when x is negative except when $x < -(2dh - h^2)^{\frac{1}{2}}$, which only occurs if $d < (b^2 + h^2)/2h$. Hence the net work $W_{-b,0}$ which is done by the hip muscles as x increases from $-b$ to 0 may be the algebraic sum of positive and negative work done at different times, while the work $W_{0,b}$ done as x increases from 0 to b consists solely of negative work. We will take $W_{-b,0}$ as representing the total positive work done by the hip muscles during a step and $W_{0,b}$ as representing the negative work, and so obtain minimum values for the positive and negative work. From Eqn. 1.

$$W_{-b,0} = -\int_{-b}^{0} \frac{Fx^2 \, dx}{(x^2 + h^2)^{\frac{1}{2}}(4d^2 - x^2 - h^2)^{\frac{1}{2}}} - \int_{-b}^{0} \frac{Fhx \, dx}{(x^2 + h^2)}$$

$$= -Q + \tfrac{1}{2}Fh \ln[1 + (b/h)^2] \quad (2)$$

(The symbol $-Q$ is introduced for the first integral, for which no analytical expression will be required).

Similarly

$$W_{0,b} = -Q - \tfrac{1}{2}Fh \ln[1 + (b/h)^2]. \tag{2a}$$

Since the hip is moving at right angles to the ground force positive work done by the hip muscles must always be matched by negative work done by the knee muscles, and vice versa. (It is assumed that separate muscles operate the two joints, see Ref. 2). Hence positive work is done by the hip muscles as x increases from $-b$ to 0 and by the knee muscles as x increases from 0 to b and totals at least

$$W_{-b,0} - W_{0,b} = Fh \ln[1 + (b/h)^2].$$

Similarly negative work totalling at least $Fh \ln[1 + (b/h)^2]$ is done in the course of a step and the metabolic cost of the step is

$$\begin{aligned} E_1 &\geqslant (Fh/\eta) \ln[1 + (b/h)^2], \\ &\geqslant (Fb^2/h\eta)\left[1 - (b^2/2h^2)\right]. \end{aligned} \tag{3}$$

with approximate equality when b/h is small.

Now consider the situation in which the ground force is kept in line with the hip, by a horizontal component $-Fx/h$ (Fig. 2(c)). Since the ground force has no moment about the hip only the knee muscles do work. The animal decelerates as x increases from $-b$ to 0 and the knee muscles do (negative) work

$$\int_{-b}^{0} (Fx/h)\,dx = -Fb^2/2h.$$

A corresponding amount of positive work is done as x increases from 0 to b and the metabolic energy cost of the step is

$$E_2 = Fb^2/2h\eta. \tag{4}$$

If $b < h$ (as it is for all mammals and birds known to me), $E_2 < E_1$ and the incorporation of the horizontal component in the ground force reduces the energy cost of locomotion. The argument has been presented formally only for the case in which F is constant while the foot is on the ground. It is easily extended to show that the conclusion is unaltered in the more general case in which F is a function of x such that $F(x) = F(-x)$.

It has been assumed that the orientation of the trunk relative to the ground is constant. This assumption is reasonable for most quadrupedal gaits but would be false for a biped which ran or hopped in the manner of Fig. 2(b). However the angles through which the trunk would pitch and the fluctuations of kinetic energy of rotation of the trunk would be large only at low speeds.[2]

The models of gaits presented in subsequent sections of this paper are based on the assumption that the ground force is kept in line with the proximal joint of the limb.

SPEED, GAIT AND STRIDE LENGTH

As animals increase their speed they generally lengthen their stride. Data on the relationship between speed and stride length have been gathered from Muybridge[23] and other sources and are presented in Fig. 3. So that animals

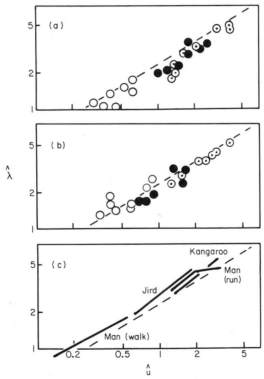

FIG. 3. Graphs plotted on logarithmic coordinates showing the relationship between stride length, speed and gait. The dimensionless quantities $\hat{\lambda}$ (representing stride length) and $\hat{u}$ (representing speed) are defined on p. 95. The broken line on each graph shows $\lambda = 2 \cdot 3\hat{u}^{0 \cdot 6}$. (a) Data from Muybridge's[23] photographs of horses, mostly with riders. (Photographs of horses drawing carts have not been used). (b) Data from Muybridge's[23] photographs of other mammals and an ostrich. (c) Data for men walking[28] and running[10], for a kangaroo hopping[3, 12] and for a jird (*Meriones unguiculatus*) walking and galloping (unpublished observations by Mr S. N. G. Frodsham and the author). O, walk; ●, run, trot or rack; ⊙, canter or gallop.

of different size can be compared, stride length and speed have been expressed in their non-dimensional forms $\hat{\lambda}(=\lambda/h)$ and $\hat{u}(=\hat{u}(gh)^{-\frac{1}{2}})$. It appears that for a wide range of animals and speeds $\hat{\lambda} \simeq 2\cdot3\hat{u}^{0\cdot6}$.

The following major groups of gaits can be distinguished (see also Ref. (15)).

(i) *Walking.*

The feet of a pair move alternately and are never off the ground simultaneously. Two types of walk are distinguished in this paper:

(a) *The stiff walk.* Each leg is kept extended while its foot is on the ground, so that the hip (or shoulder) tends to describe a circular arc about the foot and is highest as it passes over the supporting foot. The total force exerted on the ground by a pair of feet is greatest during the double support phase, i.e. while both are on the ground. This is the normal slow walk of man[7, 8] cats[20], dogs[4, 19], horses[5] and probably of the larger mammals in general (see photographs in [23]). It is also used by the domestic duck (unpublished force platform records by Mr D. J. Letten).

(b) *The compliant walk.* The knee (or elbow) is allowed to flex while the foot is on the ground so that the hip (or shoulder) is lowest as it passes over the supporting foot. The force on the ground is greater at this time than during the double support phase. This is the normal walk of quail[11]. It seems possible that it may also be used by the (generally) small mammals described by Jenkins[18] as "non-cursorial", but this possibility has not been investigated.

(ii) *Running, trotting and racking.*

The feet of a pair are placed on the ground in turn at equal intervals and are off the ground simultaneously for part of the stride.

(iii) *Bipedal hopping and symmetrical galloping.*

The feet of a pair are placed on the ground simultaneously.

(iv) *Cantering and asymmetrical galloping.*

The feet of a pair are placed on the ground asynchronously but not at equal intervals. They are off the ground simultaneously for part of the stride.

The photographs of Muybridge (Ref. 23; Fig. 3(a), (b)) show that horses and various other mammals normally make the transition from walking to trotting or racking at speeds corresponding to values of $\hat{u}$ between about 0·7 and 0·9. Men break into a run at about $2\cdot5\,\mathrm{m\,s}^{-1}$ ($\hat{u} = 0\cdot8$,[8]). An 18 kg kangaroo changes from pentapedal locomotion to hopping at $1\cdot8\,\mathrm{m\,s}^{-1}$ ($\hat{u} = 0\cdot8$,[12]). 2·5 kg cats change from walking to trotting at roughly $0\cdot9\,\mathrm{m\,s}^{-1}$ ($\hat{u} \simeq 0\cdot6$,[16]).

The data of Fig. 3 show that the change from trotting or racking to cantering or galloping tends to occur when $\hat{u} \simeq 1\cdot5$. However, horses can be made to canter with $\hat{u}$ as low as 1·3 and to trot and rack with $\hat{u}$ as high as 1·8 and 2·6 respectively. Heglund et al.[17] found that riderless racehorses changed from a trot to a gallop at about $5\,\mathrm{m\,s}^{-1}$ ($\hat{u} = 1\cdot4$) and that mice changed at $0\cdot8\,\mathrm{m\,s}^{-1}$ ($\hat{u} = 1\cdot5$). Goslow et al.[16] report that 2·5 kg cats change from a trot to a gallop at about $2\cdot7\,\mathrm{m\,s}^{-1}$ ($\hat{u} = 1\cdot8$). However 90 g jirds gallop at speeds down to $0\cdot55\,\mathrm{m\,s}^{-1}$ ($\hat{u} = 0\cdot9$).

Muybridge's[23] photographs and Gambaryan's[15] films show that for a wide range of mammals and speeds, step length is roughly equal to hip height ($\hat{b} = 0\cdot5$), but McMahon (this volume) finds $\hat{b}$ proportional to (body mass)$^{-0\cdot10}$.

STIFF WALKING

The model represents a biped but a quadruped can be represented as two bipeds in tandem. Each foot is lifted as the other is set down so the double support phase is infinitesimally brief (this assumption is convenient but unrealistic). Each leg is kept straight while its foot is on the ground so the centre of mass of the body moves in a series of circular arcs of radius h (Fig. 4(a)). At time t a line from the supporting foot to the hip joint is at an

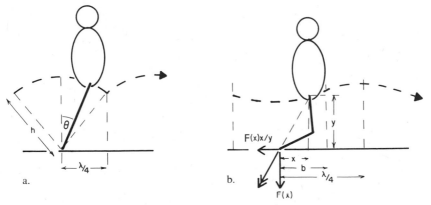

FIG. 4. Diagrams representing the models of (a) walking and (b) running.

angle θ to the vertical. The velocity of the centre of mass is u_0 at the instant when $t = 0$ and $\theta = 0$, and averages $\bar{u}$ over a complete stride.

When $t = 0$, $d\theta/dt = u_0/h$. The weight mg of the animal has a moment $mgh \sin \theta$ about the supporting foot so

$$d^2\theta/dt^2 = g \sin \theta/h,$$

whence

$$d\theta/dt = [(u_0^2/h^2) + (2g/h)(1 - \cos \theta)]^{\frac{1}{2}}. \tag{5}$$

θ will always be reasonably small: its maximum value is $\hat{\lambda}/4$ which is unlikely to exceed 0·5 at walking speeds (Fig. 3). Hence $\cos \theta \simeq 1 - \frac{1}{2}\theta^2$ and

$$d\theta/dt \simeq [(u_0^2/h^2) + (g\theta^2/h)]^{\frac{1}{2}}. \tag{5a}$$

By integration

$$\operatorname{arcsinh}\left[\theta(gh)^{\frac{1}{2}}/u_0\right] \simeq (g/h)^{\frac{1}{2}}t.$$

At the end of a step $t = \lambda/4\bar{u}$ and $\theta = \arcsin(\lambda/4h) \simeq \lambda/4h$, therefore

$$\operatorname{arcsinh}[(\lambda/4u_0)(g/h)^{\frac{1}{2}}] \simeq (\lambda/4\bar{u})(g/h)^{\frac{1}{2}},$$
$$u_0^2 \simeq (\lambda^2 g/16h)\operatorname{cosech}^2[(\lambda/4\bar{u})(g/h)^{\frac{1}{2}}]. \tag{6}$$

Inserting this value of u_0^2 and $\theta \simeq \lambda/4h$ in 5a

$$(d\theta/dt)_{\lambda/4}^2 \simeq (\lambda_g^2/16h^3)\coth^2[(\lambda/4\bar{u})(g/h)^{\frac{1}{2}}]. \tag{7}$$

The velocity of the animal at the end of the step is $h(d\theta/dt)_{\lambda/4}$ and its vertical component is $-(\lambda/4)(d\theta/dt)_{\lambda/4}$. The kinetic energy associated with this vertical component is $(m\lambda^2/32)(d\theta/dt)_{\lambda/4}^2$. The muscles must do negative work dissipating this kinetic energy, followed by a matching amount of positive work as the animal is given the upward component of velocity required for the start of the next step. The animal travels a distance $\lambda/2$ every time these quantities of positive and negative work are performed so the cost of transport is given by

$$T_w = (\lambda/16\eta)(d\theta/dt)_{\lambda/4}^2$$

and from (7)

$$T_w \simeq (\lambda_g^3/256h^3\eta)\coth^2[(\lambda/4\bar{u})(g/h)^{\frac{1}{2}}]. \tag{8}$$

This can be expressed in terms of the dimensionless quantities defined on p. 95.

$$\hat{T}_w \simeq (\hat{\lambda}^3/256\eta)\coth^2(\hat{\lambda}/4\hat{u}). \tag{8a}$$

The gait which has been described is impossible at high speeds, at which it would involve falling with acceleration greater than g. The centre of mass falls from its maximum height h to its minimum height $[h^2 - (\lambda/4)^2]^{\frac{1}{2}} \simeq h$ $[1 - (\lambda^2/32h^2)]$ in time $\lambda/4\bar{u}$, whence it follows that $\bar{u} \ll (gh)^{\frac{1}{2}}$ (i.e. that $\hat{u} \ll 1$). Men normally walk at speeds up to about $2\cdot5\,\mathrm{m\,s^{-1}}$ ($\hat{u} \simeq 0\cdot8$) but athletes in walking races exceed $3\cdot7\,\mathrm{m\,s^{-1}}$ ($\hat{u} \simeq 1\cdot2$). Such speeds are made possible by a peculiar hip action (see Ref. 13) which reduces the vertical excursion of the centre of mass below the value predicted by the model.

Discussion of compliant walking is deferred until after the discussion of running.

RUNNING

The model presented in this section describes bipedal running but can be applied to quadrupedal trotting and racking, if the quadruped is treated as two bipeds in tandem.

The model is depicted in Fig. 4(b). The step length ($2b$) is less than half the stride length (i.e. less than $\lambda/2$) so there are periods when neither foot is on the ground. While a foot is on the ground it exerts a force with vertical and horizontal components $F(x)$, $F(x)x/y$ respectively where

$$F(x) = (\pi mg\lambda/8b) \cos(\pi x/2b). \tag{9}$$

$F(x)$ is made proportional to $\cos(\pi x/2b)$ to imitate force platform records of men running and a dog trotting (unpublished observations). The factor $(\pi mg\lambda/8b)$ is introduced to make the average value of $F(x)$ over a half stride equal mg (see Ref. 3).

It will be shown that velocity fluctuations within a stride are likely to be small so that $x \simeq \bar{u}t$ and while a foot is on the ground

$$d^2y/dt^2 \simeq (\pi g\lambda/8b) \cos(\pi\bar{u}t/2b) - g.$$

By integration, the increase in height which occurs as x increases from 0 to b is

$$y_b - y_0 \simeq (gb/2\pi\bar{u}^2)(\lambda - \pi b). \tag{10}$$

While no foot is on the ground $d^2y/dt^2 = -g$ and the further increase in height which occurs as x increases to $\lambda/4$ is

$$y_{\lambda/4} - y_b = g(\lambda - 4b)^2/32\bar{u}^2. \tag{11}$$

Hence the change in gravitational potential energy which occurs as the centre of mass rises from its lowest point to its highest is

$$W_P = mg(y_{\lambda/4} - y_0)$$
$$\simeq (mg^2\lambda/32\bar{u}^2)(\lambda - 2\cdot9b). \tag{12}$$

The horizontal component of the ground reaction $F(x)x/y$ decelerates the animal as x increases from $-b$ to 0 and accelerates it as x increases further to b. It can be shown by putting realistic values in Eqn (10) that the changes in y which occur while the foot is on the ground are likely to be small. (Realistic values of λ and $\bar{u}$ for running and trotting can be obtained from Fig. 3, and b/h is about 0·5 for most mammals). Hence $F(x)x/y \simeq F(x)x/h$ and the change of kinetic energy which occurs as x increases from 0 to b is

$$W_K \simeq \int_0^b [F(x)\,x/h]\,\mathrm{d}x$$

$$\simeq 0·09\ mgb\lambda/h. \tag{13}$$

The difference between the velocities of the body when $x = b$ and when $x = 0$ is given by

$$u_b - u_0 \simeq W_K/m\bar{u}$$
$$\simeq 0·09\ gb\lambda/\bar{u}h. \tag{14}$$

It can be shown by putting realistic values in (14) that $(u_b - u_0)$ is likely to be $0·14\bar{u}$ or less, throughout the range of speeds at which running and trotting occur. This is the justification for the earlier assumption that $x \simeq \bar{u}t$.

The leg shortens and extends again during each step, by flexion and extension of its joints. These changes in its length are partly due to changes in length of muscles and partly to extension and recoil of elastic structures, principally tendons. The latter part only is considered in defining the elastic compliance C of the leg. As the force on a leg of elastic compliance C increases from zero to $F(0)$ elastic strain energy W_E is stored where

$$W_E = \tfrac{1}{2}[F(0)]^2 C$$
$$= (\pi mg\lambda/b)^2(C/128). \tag{15}$$

When $x = 0$ kinetic and gravitational potential energy are both minimal but elastic strain energy is at its maximum. Hence the positive work which the muscles must supply as x increases from 0 to b is $(W_P + W_K - W_E)$. This happens twice in each stride so the cost of transport is

$$T_r = 2(W_P + W_K - W_E)/\lambda m\eta.$$

Inserting W_P, W_K and W_E from equations (12), (13) and (15)

$$T \simeq \frac{g}{16h\eta}\left\{\frac{gh}{\bar{u}^2}(\lambda - 2·9b) + 2·9b - \frac{2·5\ mgC\lambda h}{b^2}\right\} \tag{16}$$

or in terms of the dimensionless quantities defined on p. 95.

$$\hat{T}_r \simeq \frac{1}{16\eta}\left\{\frac{(\hat{\lambda} - 2·9\hat{b})}{\hat{u}^2} + 2·9\hat{b} - \frac{2·5\hat{C}\hat{\lambda}}{\hat{b}^2}\right\}. \tag{16a}$$

COMPLIANT WALKING

The model of compliant walking is identical with the model of running, except that $2b$ is greater than $\lambda/2$ so that there is a double support phase. The changes of kinetic energy which occur in each step are less than the value of W_K given by Eqn (13) because the horizontal components of the forces exerted by the feet during the double support phase oppose each other, but one leg does work against the other and the totals of positive and negative work associated with the horizontal component of the ground force are as predicted by Eqn (13). The expression for W_E which was derived from running (Eqn 15) also applies. However, the expression for W_P does not apply. The potential energy changes which occur during each half stride have been evaluated by numerical methods, for particular values of $\hat{b}$, $\hat{\lambda}$ and $\hat{u}$.

At speeds so low that double support phases occupied over 50% of the time, the vertical component of the ground force would be greatest during the double support phase. This model of compliant walking has not been applied to such low speeds.

BIPEDAL HOPPING

The cost of transport can be worked out for bipedal hopping in precisely the same way as for running, if we note that only one step occurs in each stride but that it involves both feet. The compliance of both legs, acting together, is $\frac{1}{2}C$. Hence Eqn (16a) can be altered to apply to bipedal hopping, by changing $\hat{\lambda}$ to $2\hat{\lambda}$ and $\hat{C}$ to $\frac{1}{2}\hat{C}$ throughout.

$$\hat{T}_h \simeq \frac{1}{16\eta}\left\{\frac{(2\hat{\lambda} - 2\cdot9\hat{b})}{\hat{u}^2} + 2\cdot9\hat{b} - \frac{2\cdot5\hat{C}\hat{\lambda}}{\hat{b}^2}\right\}. \tag{17}$$

The fluctuations of velocity which occur in bipedal hopping can similarly be obtained by substituting 2λ for λ in Eqn (14)

$$(u_b - u_0)_h \simeq 0\cdot18gb\lambda/\bar{u}h. \tag{18}$$

GALLOPING

In the symmetrical gallop (the bound of Gambaryan[15]) the feet of each pair move in synchrony. The gait would be equivalent to two bipeds hopping in tandem if no forces were transmitted between the fore quarters and the hind quarters. In such a case differences in velocity up to $0\cdot18gb\lambda/\bar{u}h$ (from

E

Eqn (18)) would occur between the fore and hind quarters. During the period of time $2b/\bar{u}$ when the fore feet were on the ground the mean velocity of the fore quarters would be about $0.09\ gb\lambda/\bar{u}h$ less than the velocity of the hind quarters, so the hind quarters would catch up by $0.18gb^2\lambda/\bar{u}^2h$. They would fall behind again by the same amount while the hind feet were on the ground. Insertion of the realistic values $\hat{b} = 0.5$ and (from Fig. 3) $\hat{\lambda} = 2.3\hat{u}^{0.6}$ shows that the difference between the maximum and minimum length of the body would be about $0.1h$ when $\hat{u} = 1$ and less at higher speeds. Much larger fluctuations of body length commonly occur, both in symmetrical and in asymmetrical galloping (for examples see Ref. 15). Even symmetrical galloping is not at all precisely represented by two bipeds hopping in tandem.

KINETIC ENERGY OF THE LEGS

The kinetic energy of a system of particles can be expressed as the sum of the external kinetic energy (due to movement of the centre of mass of the system) and the internal kinetic energy (due to movement of parts of the system relative to the centre of mass). So far only the external kinetic energy of a biped, or of the fore or hind quarters of a quadruped, has been considered. Internal kinetic energy is associated with movement of the legs relative to the trunk. A rough calculation based on sparse data has led to the conclusion that this is likely to contribute about $0.03\ \hat{u}^2g/\hat{\lambda}\eta$ to the cost of transport.[2]

DISCUSSION

Equations (8a), (16a) and (17) can be used to predict relationships between cost of transport and speed, if values can be assigned to $\hat{b}$, $\hat{\lambda}$ and $\hat{C}$. It has been shown that for a wide range of animals and speeds $\hat{b} \simeq 0.5$ and $\hat{\lambda} \simeq 2.3\hat{u}^{0.6}$. It is not clear why these values of $\hat{b}$ and $\hat{\lambda}$ should be particularly advantageous, but they will be used. Cavagna[9] found that the elastic stiffness of the human leg was 19 to 28 $\mathrm{kN\,m^{-1}}$, depending on the exercise used to test it. The value which will be used is 25 $\mathrm{kN\,m^{-1}}$, which was obtained when the subjects landed from a jump on one foot: it corresponds to $\hat{C} = 0.03$. The results of calculations in which $\hat{C}$ was taken as zero will also be considered.

Theoretical relationships between cost of transport and speed, obtained in this way, are shown in Fig. 5(a). The continuous lines were obtained from Eqns (8a), (16a) and (17) which take no account of internal kinetic energy due to movement of limbs relative to the trunk. In each case the additional cost of internal kinetic energy changes must be added to obtain the total cost of transport. The rough estimate of this additional cost, which is the same for all gaits, is shown by the broken line, and the addition has been made in Fig. 5(b). Values for compliant walking are shown in Fig. 5 as extensions of the curves for running: the transition occurs when $\hat{u} = 0.8$, at which speed the assumptions regarding $\hat{\lambda}$ and $\hat{b}$ make stride length and step length equal.

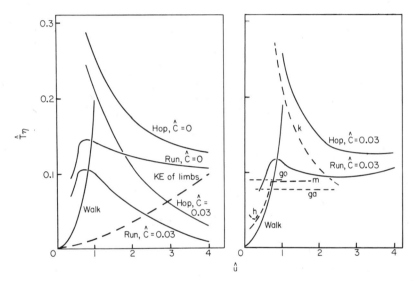

FIG. 5. Graphs showing relationships between cost of transport and speed. The quantities $\hat{T}$, η and $\hat{u}$ are defined on p. 95. (a) Values of $\hat{T}\eta$ calculated as explained in the text for stiff walking (Eqn (8a)); running (or trotting or racking, Eqn (16a); values for $\hat{u} < 0.8$ refer to compliant walking). These values exclude the part of the cost associated with fluctuations of internal kinetic energy due to leg movement, which are shown separately (broken line). (b) (Continuous lines) Theoretical values of $\hat{T}\eta$, including the part associated with fluctuations of internal kinetic energy. (Broken lines) Values of $\hat{T}\eta$ calculated from measurements of oxygen consumption of m, men[8, 24]; k, kangaroos[12]; go, goats[27]; ga, gazelles[27] and h, horses.[6]

The following tentative conclusions can be drawn from Fig. 5. An elastic compliance equivalent to $\hat{C} = 0.03$, which is believed on sparse evidence to be realistic, greatly reduces the cost of running and hopping at high speeds. Bipedal hopping incurs a much higher cost than running (with the same values of $\hat{\lambda}$, $\hat{b}$ and $\hat{C}$) at low speeds but only a slightly higher cost at high speeds. Trotting is equivalent to two bipeds running in tandem. Galloping is more like hopping than running and walking, though it has been shown that two bipeds hopping independently in tandem is not an accurate model of galloping. In so far as galloping resembles hopping it seems likely to be much more costly than trotting at low speeds, but only a little more costly than trotting at high speeds. It may even be less costly than trotting at high speeds if kinetic energy taken from the limbs can be stored as elastic strain energy in the back. The curves for running and compliant walking intersect the curve for stiff walking at about $\hat{u} = 0.8$, that is a little below the maximum speed at which stiff walking is possible. To save energy animals should use the stiff walk at speeds below $\hat{u} = 0.8$ and other gaits at higher speeds and this is, in general, what cursorial mammals do.

Fig. 5(b) shows that if the values used in compiling Fig. 5(a) are realistic the total cost of transport (including the component associated with internal kinetic energy) should increase with increasing speed in stiff walking, be more or less independent of speed in running and decrease with increasing speed in hopping. The fairly constant value for running is the sum of an increasing cost for internal kinetic energy and a decreasing cost for other purposes.

Some data calculated from measurements of oxygen consumption are included in Fig. 5(b). In calculating them it was assumed that metabolism involving 1 cm^3 oxygen uses 20 J, and that the efficiency η is 0.25. (The latter is slightly above the value of 0.21 which follows from the observations of Margaria[21] referred to in the introductory section of this paper). The predicted costs of transport agree quite well with the measured values for human walking and running and for kangaroo hopping: it would be bold to hope for better agreement, since the same values of $\hat{b}$, $\hat{\lambda}$ and $\hat{C}$ are being made to serve for all species in the models. The observed values for gazelle and goat are not too far below the level predicted for running but do not show the expected dip at the low speeds at which the animals presumably walked. Similar data are available for cheetahs[27] and dogs.[26] The values for horses refer to walking; according to Taylor (discussion) the values for running horses are significantly below the predicted curve.

All the examples cited so far are fairly large animals, of body mass over 15 kg. Costs of transport are higher for smaller animals. Values calculated from measurements of oxygen consumption correspond to $\hat{T}\eta = 1.4$ for 20 g mice[26] and $\hat{T}\eta = 0.6$ for 40 g quail.[14] These are far higher than the values predicted in Fig. 5.

It was assumed in calculating the theoretical curves that $\hat{b} = 0.5$, irrespective of body size. Gambaryan[15] reports that $\hat{b}$ ranges from 0.3 to 0.75 in mammals, and tends to be lowest in large mammals. McMahon (this volume) finds $\hat{b} \propto m^{-0.10}$. It was assumed that $\hat{\lambda} = 2.3\hat{u}^{0.6}$, irrespective of size. Figure 3 suggests that $\hat{\lambda}$ may generally be a little less than this for large mammals such as horses and a little more than this for small mammals such as jirds. It was assumed that $\hat{C}$ was independent of size as it would be for geometrically similar animals built of the same materials. No values of $\hat{C}$ are available for small mammals. The expression used to calculate fluctuations of internal kinetic energy is based on the assumption that the moments of inertia of legs are proportional to mh^2, which will be true if the legs are the same proportion of body mass in animals of different size and if their radius of gyration is a constant proportion of h. The scaling rule proposed by McMahon[22] does not imply any need to modify the expression for differences of body size.

These considerations suggest that the values of $\hat{b}$ and $\hat{\lambda}$ assumed in calculating theoretical costs of transport for Fig. 5 may be too low for small mammals and that the values of $\hat{C}$ may not be appropriate. However, the most extreme suppositions which seem reasonable would not predict values of $\hat{T}\eta$ above about 0·3, for running with $\hat{u} < 4$. They cannot explain the observed value of about 1·4 for 20 g mice at low speeds (0·1 to 0·3 m s^{-1}, probably equivalent to a range of $\hat{u}$ from about 0·2 to 0·6). It is of course possible that η may depend on body size but I know of no evidence that it is lower for small animals than for large ones. It is hard to escape the conclusion that there is some component of the cost of transport which has been overlooked, which has a very strong inverse relationship with body size.

REFERENCES

1. Alexander, R. McN. Mechanics of jumping by a dog. *J. Zool., Lond.* **173**, 549–573 (1974).
2. Alexander, R.McN. Mechanics of bipedal locomotion, *In* P. Spencer Davies, "Perspectives in Animal Biology". 493–504 Pergamon, Oxford (1976).
3. Alexander, R. McN. and Vernon, A. The mechanics of hopping by kangaroos (Macropodidae). *J. Zool. Lond.* **177** 265–303 (1975).
4. Barclay, O. R. Some aspects of the mechanics of mammalian locomotion. *J. Exp. Biol.* **30**, 116–20 (1953).
5. Björck, G. Studies on the draught force of horses. *Acta Agric. Scand.* Suppl. 4 (1958).
6. Brody, S. "Bioenergetics and Growth". Reinhold, New York (1945).
7. Carlsöo, S. "How Man Moves". Heinemann, London (1972).
8. Cavagna, G. A. Travail mécanique dans la marche et la course, *J. Physiol., Paris*, **61**, suppl. 1, 3–42 (1969).
9. Cavagna, G. A. Elastic bounce of the body. *J. Appl. Physiol.* **29**, 279–282 (1970).
10. Cavagna, G. A., Margaria, R. and Arcelli, E. A high speed motion picture analysis of the work performed in sprint running. *Research Film*, **5**, 309–319 (1965).
11. Clark, J. and Alexander, R. McN. Mechanics of running by quail (*Coturnix*). *J. Zool., Lond.* **176**, 87–113 (1975),
12. Dawson, T. J. and Taylor, C. R. Energy cost of locomotion in kangaroos. *Nature, Lond.* **246**, 313–314 (1973).
13. Dyson, G. H. G. "The Mechanics of Athletics", 6th Ed. University of London Press, London (1973).

14. Fedak, M. A., Pinshow, B. and Schmidt-Nielsen, K. Energy cost of bipedal running. *Am. J. Physiol.* **227**, 1038–1044 (1974).
15. Gambaryan, P. P. "How Mammals Run". Wiley, New York (1974).
16. Goslow, G. E., Reinking, R. M. and Stuart, D. G. The cat step cycle: hind limb joint angles and muscle lengths during unrestrained locomotion, *J. Morph.* **141**, 1–42, (1973).
17. Heglund, N. C., Taylor, C. R. and McMahon, T. A. Scaling stride frequency and gait to animal size: mice to horses. *Science, N,Y.* **186**, 1112–1113 (1974).
18. Jenkins, F. A. Limb posture and locomotion in the Virginia opossum (*Didelphis marsupialis*) and in other noncursorial mammals, *J. Zool., Lond.* **165**, 303–315 (1971).
19. Kimura, T. and Endo, B. Comparison of force of foot between quadrupedal walking of dog and bipedal walking of man. *J. Fac. Sci., Univ. Tokyo* (V) **4** (2), 119–130 (1972).
20. Manter, J. T. The dynamics of quadrupedal walking. *J. exp. Biol.* **15**, 522–540 (1938).
21. Margaria, R. Positive and negative work performances and their efficiencies in human locomotion. *Int. Z. angew. Physiol.* **25**, 339–351 (1968).
22. McMahon, T. Size and shape in biology. *Science N.Y.* **179**, 1201–1204 (1973).
23. Muybridge, E. "Animals in Motion", 2nd Ed. Dover, New York (1957).
24. Passmore, R. and Durnin, J. V. G. A. Human energy expenditure. *Physiol. Rev.* **35**, 801–840 (1955).
25. Pugh, L. G. C. E. The influence of wind resistance in running and walking and the mechanical efficiency of work against horizontal or vertical forces. *J. Physiol,., Lond.*, **213**, 255–276 (1971).
26. Taylor, C. R., Schmidt-Nielsen, K. and Raab, J. L. Scaling of energetic cost of running to body size in mammals. *Am. J. Physiol.* **219**, 1104–1107 (1970).
27. Taylor, C. R., Shkolnik, A., Dmi'el, R., Baharav, D. and Borut, A. Running in cheetahs, gazelles and goats: energy cost and limb configuration. *Am. J. Physiol.* **227**, 848–850 (1974).
28. Zarrugh, M. Y., Todd, F. N. and Ralston, H. J. Optimization of energy expenditure during level walking. *Europ. J. appl. Physiol.* **33**, 293–306 (1974).

7. Walking, Running and Galloping: Mechanical Similarities between Different Animals

GIOVANNI A. CAVAGNA, NORMAN C. HEGLUND AND
C. RICHARD TAYLOR

Istituto di Fisiologia Umana dell' Universita di Milano and Centro di Studio per la Fisiologia del Lavoro Muscolare del C.N.R., Milano, Italy, and Museum of Comparative Zoology, Harvard University, Cambridge, Massachusetts 02138, U.S.A.

ABSTRACT

By measuring the forward speed changes and the vertical displacement of the centre of mass of the body during a stride, one observes that the corresponding changes in kinetic energy of forward motion and potential energy of vertical displacement are almost completely out of phase (as in a pendulum) during the walking of man, dog and monkey, whereas they are practically in phase (as in a bouncing ball) during the running of man, the trotting of dog and monkey and the hopping of kangaroo. In galloping a substantial shift of potential energy into kinetic energy and vice versa takes place at low and average speeds; at high speeds this shift is small and the mechanics of galloping become similar to those of running.

The mechanical work necessary to move the centre of gravity W_{tot} (external work) is mainly spent against gravity (W_v) and to sustain the forward speed changes (W_f).

In the walking of man, dog and monkey, both W_v and W_f increase with speed and, since some energy is recovered $W_{tot} < W_v + W_f$. The power output expressed per kg of body mass, is similar in dog and in man ($\dot{W}_{tot}$/mass $\sim 0\cdot1$–$0\cdot4\,\mathrm{J\,kg^{-1}\,s^{-1}}$ (2–6 cal/kg min) at $0\cdot5$–$1\cdot7\,\mathrm{m\,s^{-1}}$ (2–6 km/h)).

In trotting and galloping (dog and monkey), in hopping (kangaroo) and in running (man), the mechanical power spent against gravity ($\dot{W}_v$) changes little with the speed of locomotion; $\dot{W}_v$ is maximal for kangaroo ($3\cdot5$–$4\cdot9\,\mathrm{J\,kg^{-1}\,s^{-1}}$), intermediate in man ($1\cdot5$–$2\cdot9\,\mathrm{J\,kg^{-1}\,s^{-1}}$) and minimal in dog and monkey ($0\cdot6$–$1\cdot3\,\mathrm{J\,kg^{-1}\,s^{-1}}$). $\dot{W}_f$ increases with the speed $\overline{V}_f$ in all the species.

$\dot{W}_{tot}$ increases linearly with the speed in all the species: this function has a y-intercept which depends on $\dot{W}_v$. For a given speed $\dot{W}_{tot}$/mass is maximum for the kangaroo, intermediate for man and monkey and minimum for dog.

The ratio $\dot{W}_{tot}$/total rate of energy expenditure for a hopping kangaroo was $0\cdot23$ at $2\cdot8\,\mathrm{m\,s^{-1}}$ and $0\cdot62$ at $7\cdot8\,\mathrm{m\,s^{-1}}$; these high values indicate a substantial recovery of "elastic" energy which becomes more efficient with increasing speed.

111

INTRODUCTION

When terrestrial locomotion is carried out by means of limbs, forward speed changes and vertical displacements of the centre of mass of the body inevitably take place within each step. A change of the forward speed V_f involves a change of the kinetic energy of forward motion $E_{kf} = \frac{1}{2}mV_f^2$ (m being the mass of the whole body); while a vertical displacement S_v involves a change of the potential (positional) energy $E_p = mgS_v$.

The instantaneous kinetic and potential energies of the centre of mass of the body have been measured in human locomotion: during each step of walking the variations of E_{kf} and E_p were found to be out of phase as in a pendulum (Cavagna, Saibene and Margaria[4]), whereas during each step of running they are almost exactly in phase (Fenn,[8] Cavagna et al.[5]). The pendulum-like mechanism of walking is economical since the shift of positional energy into kinetic energy and vice versa minimizes the (costly) oscillations of total mechanical energy $(E_p + E_{kf})$. The bouncing-like mechanism of running allows the recovery of an appreciable amount of energy through the peculiar "elastic" properties of contracted muscles (Cavagna et al.[5] Dawson and Taylor,[7] and Cavagna and Citterio[2]).

The aim of this study was to find out whether the mechanisms described for human walking and running are also found in other species.

MECHANISM OF WALKING

Experimental records obtained using the force plate technique described by Cavagna[1] are given in Fig. 1. The curves were recorded during walking in man, dog, and monkey. They show:

1. The vertical component F_v of the force exerted by the feet on the platform;

2. The time-integral of the difference between F_v and body weight as a function of time, i.e. the area between the F_v tracing and a horizontal line at a value of F_v equal to the subject's body weight. This integral is proportional to the vertical component of the velocity of the centre of mass plus an integration constant;

3. The forward component F_f of the force exerted by the feet on the platform. Negative values of F_f indicate that the force exerted by the platform on the body is directed backwards, causing a forward deceleration (braking); positive values indicate a forward push;

4. The time-integral of the function F_f which is proportional to the forward component V_f of the velocity of the centre of mass plus an integration constant.

The tracings in Fig. 1 indicate striking similarities between the different animals in spite of large size differences (from 3·7 to 60 kg) and in spite of the fact that we are comparing bipedal with quadrupedal walking. The mechanics of walking is essentially the same: the maximal forward speed (V_f), attained as a consequence of the forward push (positive F_f) takes place at the end of the period of downward velocity (V_v), i.e. when the centre of mass is at its lowest point. On the contrary, the minimal forward velocity takes

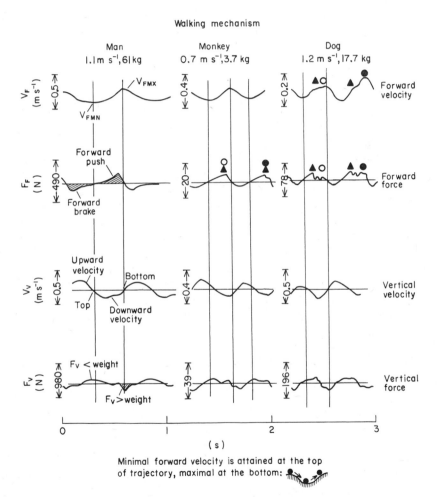

FIG. 1. Exact copy of records of vertical force, F_v, vertical velocity changes, V_v, forward force, F_f, and forward velocity changes, V_f, during walking of man, monkey and dog. The symbols over the curve refer to the foot-fall pattern as follows: O rear left; ● rear right; △ front left; ▲ front right.

place at the end of the period of upward velocity, i.e. when the centre of mass is at its highest. These tracings therefore indicate that the pendulum-like mechanism of walking also applies to quadrupedal locomotion: the body, by falling forward (downward velocity), increases the kinetic energy of its forward motion (E_{kf}) and decreases its potential or positional energy E_p. The maximal E_{kf} is attained when E_p is minimal, and vice versa.

MECHANISM OF HOPPING, TROTTING, AND RUNNING

The mechanisms of hopping and trotting are mechanically very similar to that of running. This is shown by the experimental records of Fig. 2 obtained in kangaroo, dog, monkey and man.

In hopping and trotting, as in running, a "flight period" takes place during each step. After this flight period the body is subjected to a forward decelera-tion which is immediately followed by a forward acceleration. It can be seen that the end of the forward deceleration, when the minimal forward speed is attained, coincides with the end of the downward velocity, i.e. when the centre of mass has reached its lowest point.

Correspondingly the maximal forward speed lasts for the entire "flight period" (neglecting air resistance) during which the centre of mass attains the highest point of its trajectory. As will be shown below this mechanism implies that very little or no transfer of potential energy E_p into kinetic energy E_k can take place in running, hopping and trotting. In other words the muscles are responsible for simultaneous increments of both E_p and E_{kf}.

The only difference between the mechanism of hopping, running and trotting is the duration of the flight period, which is greatest in hopping, intermediate in running and shortest in trotting. Actually at very low speeds of trotting the flight period can be nil (Fig. 4; dog's trot). In general the mechanism of run, hop and trot consists of a series of bounces with stretching and immediate shortening of the muscles of the lower limbs.

POTENTIAL AND KINETIC ENERGY CHANGES OF THE CENTRE OF MASS DURING THE STRIDE

The V_v and V_f tracings of Figs 1 and 2 allow the determination of potential energy E_p and the kinetic energies E_{kf} and E_{kv} associated with the forward and vertical components of the velocity of the centre of mass.

Curves of E_{kf}, $E_p + E_{kv}$ and E_{tot} ($= E_{kf} + E_p + E_{kv}$), plotted against time, are shown in Fig. 3 for the case of *walking* (of man, dog and monkey). It can be seen that in general the changes in E_{kf} and E_p (E_{kv} is neglible) are out of

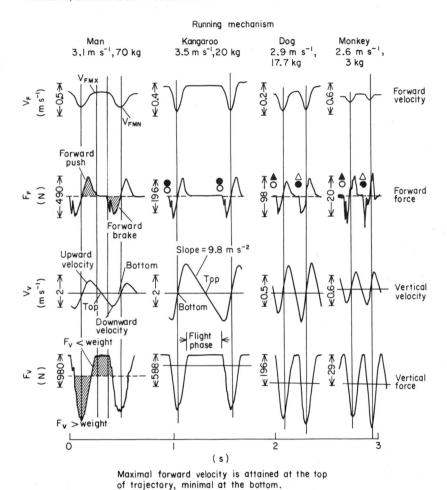

Maximal forward velocity is attained at the top
of trajectory, minimal at the bottom.

FIG. 2. Records obtained during running in man, hopping in kangaroos and trotting in dogs and monkeys. A filter was used to eliminate vibrations from the F_v tracing of the monkey: this however rounded-off the tracing. During the "flight period" the body decelerates upwards and accelerates downwards with $1 g = 9·8 \, \mathrm{m \, s^{-2}}$.

phase. This means that in one stride the sum of the positive work needed to accelerate the body forwards (discounting energy lost during deceleration), which we denote by W_f, and the positive work done against gravity in raising the centre of mass (discounting energy lost as the centre of mass falls), W_v, is greater than the positive external work actually done, W_{tot}. (W_f is defined as the sum of the increments of the curve E_{kf} between each minimum and the succeeding maximum (see Fig. 3), and is equal to the amplitude

Kinetic, potential and total mechanical energy of the centre of mass during walking.

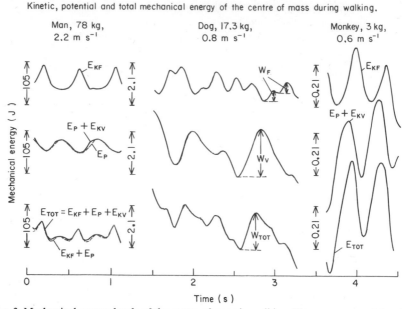

FIG. 3. Mechanical energy levels of the centre of mass in walking. The curves were determined from experimental records of velocity such as those in Fig. 1, by means of a computer. The increments of the curve $E_{kf} = \frac{1}{2}mV_f^2$ indicate the positive work W_f necessary to sustain the forward speed changes. The increments of the curve $E_p + E_{kv}$ indicate the positive work done against gravity, W_v. E_{kv} is negligible in the dog and the monkey. The increments of the curve E_{tot} indicate the positive external work done, W_{tot}: this is smaller than $W_f + W_v$ since the curves E_{kf} and E_p are almost completely out of phase as in a pendulum (curves for man from Cavagna, et al. [6]).

if the curve has only one maximum per stride; W_v and W_{tot} are defined similarly). The difference $(W_v + W_f) - W_{tot}$ indicates the mechanical energy which is recovered. The amount of this recovery can be expressed as a percentage of the maximum possible work, ie:

$$\% \text{ recovery} = \frac{(W_v + W_f) - W_{tot}}{W_v + W_f} \times 100. \qquad (1)$$

A 100% recovery would require the $E_p + E_{kv}$ and E_{kf} curves to be exactly out of phase and of equal amplitude; a 0% recovery would require the curves to be perfectly in phase. In walking the recovery attains a maximum of 60–70% in man (Cavagna et al.[6]), of 50–60% in the bigger dog (17·5 kg), and of 28% in the smaller dog (5 kg) and in the monkey.

The mechanical energy of the centre of mass during *running* (man), hopping (kangaroo) and trotting (dog and monkey) is given in Fig. 4. The horizontal tracts indicate the flight periods: in these periods E_{kf}, $E_p + E_{kv}$ and E_{tot}

remain constant because no external force other than gravity (air resistance being neglected) acts on the body. After take-off the kinetic energy of vertical motion, $E_{kv} = \frac{1}{2}mV_v^2$, is transformed into potential energy and the centre of mass is lifted until the highest point of its trajectory is attained; after this point the downward displacement begins, E_p decreases and E_{kv} increases by the same amount. At the instant of contact E_{kf}, $E_p + E_{kv}$ and their sum E_{tot}, all decrease, and reach a minimum when the centre of mass is at its lowest point. The energy is either dissipated as heat, mostly in the muscles doing negative work, or stored as elastic energy in the tendons and in the cross bridges of muscles. Immediately after the elastic storage and negative work phase, the body is propelled simultaneously upwards and forwards by release of the stored elastic energy and by active shortening (positive work) of the muscles. The transfer of mechanical energy between its kinetic and (gravitational) potential forms is very small, (on the average 2%) and about equal in trotting, hopping and running.

The changes of potential and kinetic energy of the centre of mass taking place during galloping are shown in Fig. 5. At low and average speeds (4·9; 5·4 and 5·9 m s^{-1}), after "landing" on the back legs, the potential energy

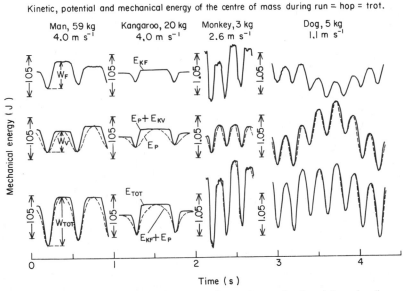

Kinetic, potential and mechanical energy of the centre of mass during run = hop = trot.

FIG. 4. Mechanical energy levels of the centre of mass in running (man), hopping (kangaroo) and trotting (monkey and dog). The curves indicate that the mechanism of these three types of gait is essentially the same. In the trotting of dog at the low speed of 1·1 m s^{-1} no aerial phase takes place.

Kinetic, potential and total mechanical energy of the centre of mass of dog galloping.

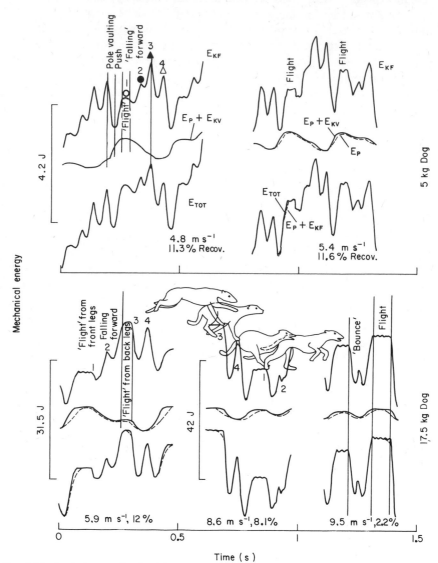

FIG. 5. Mechanical energy levels of the centre of mass in a gallop (dog). After landing on the back legs, the body "falls forward" (a decrease in potential energy and an increase in kinetic energy). "Pole vaulting" and a push directed upward and forward take place immediately before take off from a front leg. The % recovery (Eqn (1)) decreases with increasing speed, indicating that at high speeds the mechanism of galloping becomes similar to that of running. Only one take off per stride, from a front leg, takes place at low speeds, whereas an additional take off from the back legs takes place at high speeds (8·6 and 9·5 m s^{-1}). The symbols for the foot-fall pattern are the same as in Fig. 1.

E_p usually decreases and, simultaneously, the average level of kinetic energy E_{kf} increases. This indicates that, after landing, the body "falls forwards". After these events, a second flight period (usually shorter) may or may not take place. The forward fall is arrested by the contact of one front leg followed by the contact of the second front leg from which take off for the main flight period begins. Pole vaulting over this second leg is clearly indicated by a decrease of E_{kf} with a simultaneous increase of E_p, bringing the height of the centre of mass back to its initial level. Just before "take off", after pole vaulting, E_p and E_{kf} increase together indicating a muscular push upward and forward by the leg which is about to leave the ground. During the phases in which the body falls forward and pole vaults over the front leg, a substantial shift between E_p and E_{kf} takes place.

At high speeds ($8 \cdot 6$ and $9 \cdot 5 \, \mathrm{m \, s}^{-1}$ in Fig. 5) two flight phases are always clearly detectable; between them E_{kf} and E_p decrease more or less simultaneously as in running; the body mainly "bounces", first over the front legs and then over the back legs, storing and releasing elastic energy while the feet are on the ground. The interchange between potential and kinetic energy (Eqn 1) decreases with the velocity of gallop, attaining values similar to those characteristic of running.

WORK DONE AGAINST GRAVITY, TO SUSTAIN FORWARD VELOCITY CHANGES AND EXTERNAL WORK ACTUALLY DONE BY THE MUSCLES AT DIFFERENT SPEEDS OF LOCOMOTION

Walking

As mentioned above, gravity is used to accelerate the body forward in walking. Since the kinetic energy change, W_f, increases with the speed of locomotion (even if the forward speed change ΔV_f is maintained constant), the potential energy change, W_v, should show a corresponding increase with speed: in fact in a pendulum, W_f and W_v are equal and increase together with the amplitude of the oscillation.

This is about what happens with man walking: both W_v and W_f increase with walking speed. However, W_v exceeds W_f at walking speeds below $1 \cdot 1 - 1 \cdot 7 \, \mathrm{m \, s}^{-1}$ whereas W_f exceeds W_v at higher speeds: at $1 \cdot 1 - 1 \cdot 7 \, \mathrm{m \, s}^{-1}$, which happens to be the most economical speed of walking (Margaria[9]), W_v equals W_f and the percentage recovery, given by Eqn (1), is maximum (Cavagna et al.[6]).

The preliminary data hitherto obtained on quadrupedal walking (dog and monkey) indicate:

1. The order of magnitude of the work per unit mass done in walking for unit time at a given speed is the same in man and dog. $\dot{W}_{tot}/\mathrm{mass} \sim 0 \cdot 1 - 0 \cdot 4$

J kg^{-1}s^{-1} (2–6 cal/kg min) at 0·5–1·7 m s^{-1} (2–6 km/h). The work done against gravity is appreciably greater in the monkey.

2. Both W_f and W_v increase with the walking speed: up to 1·1 m s^{-1}, $W_v > W_f$ as in man.

3. In both dog and monkey, as in man, the total external work done in walking, W_{tot}, is smaller than the sum of W_f and W_v. As described above, this depends on the fact that the curves E_{kf} and $E_p + E_{kv}$ (Fig. 3) are almost completely out of phase. Actually the minimal forward speed V_f (and hence the minimum value of E_{kf}) is often attained somewhat before the maximal height of the trajectory of the centre of mass is reached. This is particularly evident in the case of the monkey: a muscular action, directed upward and forward, is necessary to complete the lift of the centre of mass.

The range of speeds (0·5–1·2 m s^{-1}) that can be covered by the walking mechanism in our dogs and monkeys is rather small. In man, speeds up to 2·2 m s^{-1} can be attained by walking without appreciably modifying the pendulum-like mechanism: this is probably because man's structure can afford greater potential energy changes and these in turn, can account for kinetic energy changes corresponding to a greater forward speed.

Running, Hopping, Trotting and Galloping

The work done against gravity at each step is given by

$$W_v = mg \cdot S_v \tag{2}$$

where mg is the body weight and S_v the vertical displacement of the centre of mass; W_v also equals the amplitude of the E_p curves in Figs 3 and 4. If f is the step frequency, the work done in unit time will be:

$$\dot{W}_v = mg \cdot S_v \cdot f. \tag{3}$$

This quantity, divided by the body mass m to compare animals of different size, is given as a function of the speed of hopping (kangaroo), or trotting and galloping (dog and monkey) in Fig. 6. The trend of the data obtained on man running (Cavagna et al.[6]) is also given for comparison.

It appears that $\dot{W}_v$ differs appreciably in the different species, being maximum for kangaroo, intermediate for man, and minimum for dog and monkey. In addition $\dot{W}_v$ does not change much with the speed of locomotion: from a rough inspection of the experimental data, $\dot{W}_v$ can be considered about constant, independent of the average forward speed.

In man, S_v decreases with increasing speed, but this is more or less balanced by an increase of f. Cavagna et al.[6] found that the decrease of S_v is mainly due to the decrease of the vertical displacement of the centre of mass taking place when the body is in contact with the ground, whereas the vertical displacement during the "flight" period is much less affected by the speed; correspondingly they found that the time t_c spent in contact with the ground during each step decreases appreciably with speed whereas the "flight" period increases or remains about constant. According to these authors the interpretation of the above findings is as follows: the reduction of the time of contact is a necessary consequence of the fact that the distance travelled by the centre of mass while a foot is in contact with the ground (the "length of contact"),

$$L_c = V_f \cdot t_c, \qquad (4)$$

cannot increase in proportion to the speed $\overline{V}_f$ because of anatomical limitations. Actually the experimental data on man indicate that L_c tends to remain constant above a given speed. A reduction of the time t_c during which

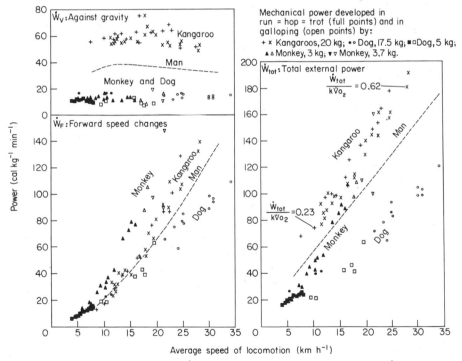

FIG. 6. Power spent against gravity, $\dot{W}_v$, to sustain the forward velocity changes, $\dot{W}_f$, and total external power, $\dot{W}_{tot}$, in monkey trotting and galloping, dog trotting and galloping, and kangaroo hopping. The interrupted lines were traced through the data obtained on man by Cavagna et al.[6] The two values of minimal or partial efficiency on the right hand graph were calculated from the oxygen consumption data reported by Dawson and Taylor[7] assuming 5 kcal (21 kJ) per litre of oxygen (1 cal/kg min = 0·07J kg^{-1}s^{-1}; 1 km/h = 0·278 m s^{-1}).

the body "bounces" over the lower leg implies a quicker bounce, i.e. a stiffening of the muscles (a stiffer "spring"). This tends to decrease the amount of vertical displacement during the time of contact, i.e. the work done against gravity at each step, but it increases the step frequency in such a way that $\dot{W}_v$ is maintained about constant.

Dawson and Taylor[7] found that in kangaroos the stride frequency also increases slightly with increasing speed: this is confirmed by the present study indicating that in spite of the fact that S_v decreases, $\dot{W}_v$ remains about constant. In kangaroos, as in man, the reduction of S_v with speed is mainly due to a reduction of the vertical displacement while the body is on the ground. The same argument may also apply to dog and monkey trotting and galloping.

Work against gravity can thus be viewed as due to a sequence of "bounces" of a ball which becomes stiffer as the speed of locomotion is increased. The height of these bounces seems rather arbitrary: the present data indicate that it does not depend on body mass alone, nor on the overall dimensions of the body. It seems more likely that it depends on the peculiar characteristics of the elastic system involved. In the case of the kangaroo, for example, this is represented mainly by the elastic elements of the gastrocnemii and by the long Achilles tendons: when the kangaroo decelerates downwards and forwards, his foot is flexed so that these elastic structures are stretched, storing elastic energy. In order to release this elastic energy an extension of the foot is necessary, but this implies an appreciable upward displacement of the body, of the order of magnitude of the dimensions of the foot of the kangaroo. The vertical displacements that we measure are in fact of this order of magnitude. In other words: in order to utilize the elastic energy stored by his elastic system, the kangaroo must "jump over his feet".

The work done during each step due to forward velocity changes is given by:

$$W_f = \tfrac{1}{2}m(V^2_{f_{max}} - V^2_{f_{min}}) \qquad (5)$$

where $V_{f_{max}}$ and $V_{f_{min}}$ are the maximal and minimal forward speeds of the centre of mass. The average power, i.e. the positive work done per unit time, is given by:

$$\dot{W}_f = W_f \cdot f. \qquad (6)$$

This quantity, divided by body mass, is also plotted in Fig. 6 against $\overline{V}_f$, the average forward speed. It can be seen that in contrast to $\dot{W}_v$, $\dot{W}_f$ is an increasing function of $\overline{V}_f$ which seems to start from the origin. Quadrupeds as well as hoppers and man (interrupted line) follow about the same trend in spite of large differences of size and in spite of the fact that apparently different mechanisms of locomotion are involved (running, hopping, trotting and galloping). Also the data on $\dot{W}_f$ obtained in walking follow almost the same curves, making it even clearer that $\dot{W}_f$ goes to zero when $\overline{V}_f$ goes to zero.

Cavagna et al.[6] found that the experimental data obtained in man running are fitted by the equation:

$$\frac{\dot{W}_f}{m} = k' \frac{\overline{V}_f^2}{1 + (t_v/L_c)\overline{V}_f}, \tag{7}$$

where t_v is the "flight period" at each step (0.11 s), L_c is the "length of contact" (Eqn 4, $L_c \approx 1$ m) and k' ($0.25\,\mathrm{s}^{-1}$) is a constant which is greater the more rigid is the connection between body and ground during deceleration. The ratio t_v/L_c is roughly constant in man (about $0.113\,\mathrm{s\,m}^{-1}$ at $2.8\,\mathrm{m\,s}^{-1}$, to about $0.123\,\mathrm{s\,m}^{-1}$ at $8.4\,\mathrm{m\,s}^{-1}$, with considerable scatter).

Equation (7) has been derived essentially from the observation that the forward deceleration of the body at each step increases linearly with speed: this observation was made both in man (Cavagna et al[6]) and in kangaroo (present work). In fact Eqn (7) also fits the data of kangaroo's hopping well. Further studies are required to check if Eqn (7) can be reasonably extended to other kinds of locomotion.

The total external work W_{tot} was measured by summing the increments of the curve E_{tot} (Fig. 4) during each step. The average power, $\dot{W}_{tot} = W_{tot} \cdot f$, is given in Fig. 6 as a function of the average speed of locomotion.

In man as well as in kangaroo, dog and monkey, this function is of the form:

$$\dot{W}_{tot} = a + b \cdot \overline{V}_f. \tag{8}$$

Since the function $\dot{W}_f = \mathrm{fn}(\overline{V}_f)$ is not linear (particularly at low speeds, see Eqn (7)), $\dot{W}_v$ cannot be constant, as a rough inspection of the data may suggest. The constant a of Eqn (8) therefore cannot be identified with $\dot{W}_v$ even if it clearly depends on that quantity. For monkey and dog the y-intercept is negative, while for man and kangaroo it is slightly positive.

COMPARISON OF MECHANICAL WORK WITH ENERGY EXPENDITURE

It is well known that for a variety of species, including man, the oxygen consumption (which is an expression of the chemical energy used) increases linearly with the speed of running, trotting and galloping, $\overline{V}_f$, but that the y-intercept of this linear relationship is above the resting energy consumption (Taylor et al.[11]). This energy expenditure is necessary to account for the mechanical work output which consists of: (a) the external work necessary to move the centre of mass ($\dot{W}_{tot}$ in Fig. 6); and (b) the internal mechanical work which is mainly used to sustain the velocity changes of the limbs relative to the centre of mass. The "internal work" in man walking and running has been measured (Cavagna and Curadini[3]): it increases approximately as $\overline{V}_f^2$ and follows a relationship which starts from the origin. Since the curves relating both $\dot{W}_f$ and $\dot{W}_{int}$ to $\overline{V}_f$ start from the origin, the only output of mechanical energy which can explain the observed y-intercept of the linear relationship between energy expenditure and running speed it the work done against gravity, W_v. As shown in Fig. 6 this work has to be performed at a high rate at low as well as high speeds, and this rate of working is required just to "start" the mechanism of running.

Efficiency

Since the mechanical work determined here is only a fraction of total mechanical work done, a minimum value of efficiency can be defined from the ratio of external work to total energy expenditure. This is shown for the kangaroo at two different speeds in Fig. 6: it can be seen that at high speeds this "minimal efficiency" is appreciably greater than the maximal efficiency of the transformation of chemical energy into mechanical work by the contractile component of muscles (0·25). This apparent contradiction can be easily understood by recalling that we have not taken into account the elastic energy stored during negative work and released during positive work. In other words not all positive work is sustained by the transformation of chemical energy. The same finding was obtained by Cavagna et al.[5] for a running man. The fact that in kangaroos the efficiency increases with speed suggests that the role played by elasticity at high speeds (when the muscles and tendons are more tense for the reasons described above) is relatively more important. This is in agreement with the finding of Dawson and Taylor[7] that in kangaroos the energy expenditure per unit time decreases with increasing speed.

In addition it should be pointed out that an increase in efficiency with speed would result in a decrease in the slope and an increase in the y-intercept of the linear relationship found between energy expenditure and the speed of locomotion.

REFERENCES

1. Cavagna, G. A. Force platforms as argometers. *J. Appl. Physiol.* **39**, 174–179 (1975).
2. Cavagna, G. A. and Citterio, G. Effect of stretching on the elastic characteristics and the contractile component of frog striated muscle. *J, Physiol. (London)* **239**, 1–14 (1974).
3. Cavagna, G. A. and Curadini, L. Internal work in locomotion: effect of walking and running speed on kinetic energy of upper and lower limbs. *IRCS Medical Science: Biomedical Technology: Physiology; Social and Occupational Medicine*, **3**, 294 (1975).

4. Cavagna, G. A., Saibene, F. and Margaria, R. External work in walking. *J. Appl. Physiol.* **18**, 1–9 (1963).
5. Cavagna, G. A., Saibene, F. and Margaria, R. Mechanical work in running, *J. Appl. Physiol.* **19**, 249–256 (1964).
6. Cavagna, G. A., Zamboni, A. and Thys, H. The sources of external work in level walking and running. *J. Physiol.* (*London*) (in press).
7. Dawson, T. J. and Taylor, C. R. Energetic cost of locomotion in Kangaroos. *Nature*, **246**, 313–314 (1973).
8. Fenn, W. O. Work against gravity and work due to velocity changes in running. *Am. J. Physiol.* **93**, 433–462 (1930).
9. Margaria, R. Sulla fisiologia e specialmente sul consumo energetico della marcia e della corsa a varie velocità ed inclinazioni del terreno. *Atti dei Lincei*, **7**, 299–368 (1938).
10. Muybridge, E. "Animals in motion" Lewis S. Brown Ed., Dover Publications Inc., New York (1957).
11. Taylor, C. R., Schmidt-Nielsen, K. and Raab, J. L. Scaling of energetic cost of running to body size in mammals. *Am. J, Physiol.*, **219**, 1104–1107 (1970).

8. The Energetics of Terrestrial Locomotion and Body Size in Vertebrates

C. RICHARD TAYLOR

Museum of Comparative Zoology and Biological Laboratories, Harvard University, Cambridge, Massachussetts 02138, U.S.A.

ABSTRACT

Since the support members (calcified bone) and the machines for generating force (striated muscle) are essentially the same in all vertebrates, one would expect to find generalities underlying the diversity in vertebrate locomotion. Studies of the energetics and mechanics of terrestrial locomotion have demonstrated a number of general relationships:

(1) metabolic power increases linearly with speed in most cases;
(2) energetic cost of moving unit mass through unit distance varies in a regular manner with body size;
(3) there appears to be a postural cost for locomotion which is independent of speed;
(4) simple empirically derived equations can be used to predict metabolic power of a running animal from body mass and speed;
(5) the transition points from one gait to another change in a regular manner with body size and probably reflect equivalent speeds for different sized animals.

Much work remains before these empirical findings can be explained in simple mechanistic terms; however, some progress has been made.

INTRODUCTION

Vertebrates move along the ground in a variety of ways. This rich diversity has captured the imagination of biologists, and the differences between various modes of vertebrate locomotion have recieved a great deal of attention. Yet the locomotion of the slithering snake, the hopping kangaroo, the sprinting cheetah, the galloping mouse and the running man have many features in common. Both the support members of vertebrate locomotory systems (calcified bone) and the machines for generating force (striated muscle) are essentially the same in all vertebrates. It is the search for general principles, underlying the mechanics of vertebrate locomotion, and resulting from the similarity in building blocks, which is the object of this paper.

Studying how locomotory parameters change with body size provides a powerful tool for uncovering these general principles. Unlike the engineer who builds a mechanical device for moving along the ground, the biologist does not know the design constraints under which the locomotory systems of vertebrates have been built. Yet the mechanical properties of bone and muscle change in a regular manner with their dimensions and, consequently, so must the locomotory parameters of vertebrates. The million-fold range in body mass found in vertebrates provides a natural experiment for studying these parameters.

There are potential pitfalls to be avoided in the search for generalities. The biologist is at a point in this search where he or she must rely on the tools and expertise of the engineer, the physicist and the mathematician. Yet the practitioners of these disciplines must deal with the complexity and reality of biological systems. Simplifying assumptions which are not in accord with reality must be avoided, and the models and predictions must continually be checked against the real animal. Thus, it is not possible for the biologist simply to turn over these problems to scientists in other disciplines; he or she must work in conjunction with them, checking their assumptions and the predictions of their models against reality. We must not be content with a simple, elegant, logically consistent scheme when it does not explain how animals work.

In this paper I should like to discuss some of the generalities which have emerged from studies of the energetics of vertebrate locomotion; to focus on some of the unanswered questions; and to point to some directions in which I feel the answers to these questions may lie.

SOME GENERAL RELATIONSHIPS IN VERTEBRATE LOCOMOTION

Metabolic Power Increases Linearly with Speed

It is fairly easy to train an animal to wear a gas mask and to measure the rate at which it consumes oxygen while it moves at a steady speed on a treadmill (or some similar device), although it can involve some hazards and often requires a great deal of patience. Each litre of oxygen (STP) when used in oxidative metabolism can be equated with about 20.1 kJ of metabolic energy. Measurements made in this manner have yielded a very general but

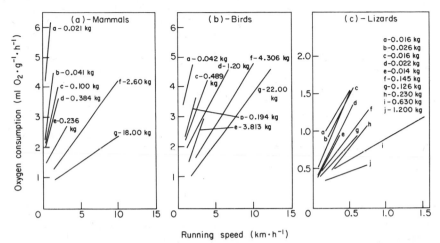

FIG. 1. Steady state oxygen consumption as a function of speed for quadrupedal mammals (a); bipedal birds (b); and quadrupedal lizards (c). 1 ml O_2 g^{-1} h^{-1} $\approx$ 5·6 J kg^{-1} s^{-1}; 1 km h^{-1} $\approx$ 0·28 m s^{-1}. Fig. 1(a) is redrawn from Taylor, Schmidt-Nielsen and Raab[27]: a—white mouse; b—kangaroo rat; c—kangaroo rat; d—ground squirrel; e—white rat; f—dog; and g—dog. Fig. 1(b) is redrawn from Fedak, Pinshow and Schmidt-Nielsen[12]: a—painted quail; b—bobwhite; c—chukar; d—guinea fowl; e—goose; f—wild turkey; and g—rhea. Fig. 1(c) is drawn from data presented in Bakker:[1] a—*Tupinambis*; b—*Varanus*; c—*Gerrhonotus*; d—*Varanus*; e—*Gerrhonotus*; f—*Varanus*; g—*Ctenosaura*; h—*Varanus*; i—*Varanus*; and j—*Tupinambis*.

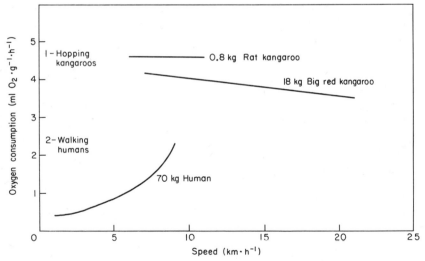

FIG. 2. Steady state oxygen consumption (V_{O_2}) as a function of speed for hopping kangaroos and walking humans. In both kangaroos (V_{O_2} either is unchanged or decreases slightly with increasing speed (data from rat kangaroo is by personal communication from Cook, Cooke and Taylor, data from big red kangaroo is from Dawson and Taylor[11]). V_{O_2} of the walking human is a curvilinear function of speed (data from Margaria et al.[19]). 1 ml O_2 g^{-1} h^{-1} $\approx$ 5·6 J kg^{-1} s^{-1}; 1 km h^{-1} $\approx$ 0·28 m s^{-1}.

surprising result: in most vertebrates metabolic power increases linearly with speed over a wide range of speeds. This linear increase has been observed in quadrupedal mammals (Fig. 1(a); Taylor *et al.*,[27] Taylor[25]); bipedal birds (Fig. 1(b); Fedak *et al.*[12]); bipedal mammals (Margaria *et al.*;[19] Taylor and Rowntree[26]); quadrupedal lizards (Fig. 1(c); Bakker[1]); and even in snakes using lateral undulatory motion (Chodrow and Taylor[7]). Recently in our laboratory (Parsons and Taylor, personal communication) we have found that this linear increase also occurs when spider monkeys or lorises move suspended beneath a support.

There are two situations where this linear relationship has not been found: human walking and kangaroo hopping (Fig. 2(a) and (b)). Margaria *et al.*[19] have shown that oxygen consumption is a curvilinear function of speed in walking man. There is one optimal speed at about $1 \cdot 1$–$1 \cdot 7$ m s^{-1} (4–6 km h^{-1}) where the energetic cost of traveling unit distance is minimal, and it is greater at slower or faster walking speeds. An exchange between kinetic energy and gravitational potential energy within each stride accounts for this curvilinear relationship (Cavagna *et al.*[5]). Both speed and height of the center of mass of an animal change within each stride, even when the speed averaged over a number of strides, or taken at an equivalent point in a sequence of strides, is constant. The kinetic-potential energy exchange is maximal at the optimal speed and is smaller at faster and at slower walking speeds. Cavagna and his colleagues[4] discuss these exchanges in detail in their paper in this volume.

Dawson and Taylor[11] found that once a big red kangaroo started to hop, the rate at which it consumed oxygen decreased slightly with increasing speed (Fig. 2). This independence of metabolic power and speed has also been found in the small (1 kg) rat kangaroo (Cook, Cooke and Taylor, personal communication). Cavagna *et al.*[4] have discussed the power required to account for the changes in speed and in the height of the centre of mass during kangaroo hopping. They have shown that a kinetic–elastic energy storage link must exist and becomes more important as the kangaroo hops faster.

Cost of Transport Changes in a Regular Manner with Body Size

A second generality is clear from Fig. 1(a), (b) and (c): metabolic power increases more rapidly with increasing speed in smaller animals. Since metabolic power increases linearly with speed, the rate of increase or "slope" of the relationship between the rate of oxygen consumption and speed is a constant for each animal. It has the units J kg^{-1} m^{-1} (the amount of energy required to move one unit mass one unit distance). The slope represents an "incremental cost of transport". This incremental cost is similar to "net cost of transport" which is frequently referred to in the literature. Net cost is usually defined as metabolic power per unit mass during running, minus metabolic power per unit mass during rest, divided by running speed (cf. Alexander's definition on p. 94 above). Its units are the same as those of the slope or incremental cost. Net cost and incremental cost would be identical if the y-intercept (metabolic power at the extrapolated zero running velocity) were equal to the resting metabolic power. The y-intercept however, turns out to be greater than the resting metabolic power. The magnitude of this difference and possible explanations for it will be dealt with in the section on postural cost of locomotion.

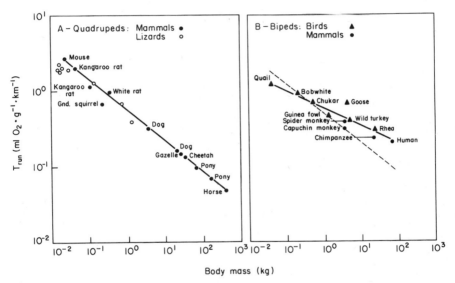

FIG. 3. Incremental cost of locomotion (T_{run}, or the slopes of the linear relationships between oxygen consumption and speed from Fig. 1(a), 1(b), and 1(c)) plotted as a function of body mass on logarithmic coordinates for quadrupeds (Fig. 1(a)) and bipeds (Fig. 1(b)). Data on mouse, kangaroo rats, white rat, ground squirrel, dog, and horse replotted from Taylor et al.;[27] data on gazelle and cheetah from Taylor et al.;[28] data from ponies is unpublished data from by laboratory; data from lizards from Bakker;[1] data from bipeds taken from Fedak et al.[12] $1 \, ml \, O_2 \, g^{-1} km^{-1} \approx 20 \cdot 1 \, J \, kg^{-1} m^{-1}$.

When the slopes or incremental costs for quadrupedal mammals ranging in size from mice to horses are plotted as a function of body mass on logarithmic coordinates, the points form a straight line (Fig. 3(a), Taylor et al.[27]). This enables us to formulate a simple equation for predicting the slope or incremental cost (T_{run}) from body mass (m):

$$T_{run-4} - 10 \cdot 7 \, m^{-0 \cdot 4} \tag{1}$$

where T_{run-4} has the units $J \, kg^{-1} m^{-1}$ and m is body mass in kg. It should be noted that, as has been pointed out by Tucker[29] and Fedak et al.,[12] if one plots weight specific power instead of mass specific power as a function of speed, then the slope becomes dimensionless, (also pointed out by Alexander on p. 95 above.) I have chosen to use mass specific power since I feel that cost of transport should be expressed in terms of the amount of energy required to move unit mass through unit distance; however the difference is immaterial as long as the gravitational acceleration, g, remains constant.

The slopes for quadrupedal lizards do not differ significantly from those predicted for quadrupedal mammals of the same weight (Fig. 3, Bakker[1]). This is somewhat surprising since the sprawling wide track of lizards is so different from the upright gait of mammals.

Fedak, Pinshow and Schmidt-Nielsen[12] measured the energy cost of bipedal running. They also found a regular relationship between incremental cost of running and body mass, but a different one than had been observed for quadrupeds (Fig. 3(b)):

$$T_{run-2} = 11 \cdot 5 \, m^{-0 \cdot 24} \tag{2}$$

where T_{run} has the units $J \, kg^{-1} \, m^{-1}$ and m is body mass in kg.

The relationships between log incremental cost of locomotion and log body mass for bipeds and for quadrupeds cross at a mass of about 1 kg: the cost is about $\frac{1}{2}$ as much for a 30–40 g bipedal bird as for a quadrupedal mammal of the same mass; while the cost is about twice that for a 70 kg bipedal ostrich or man as for a quadruped of the same mass.

Resting Metabolic Power: A Postural Cost of Locomotion

These simple relationships enable us to predict the slope of the relationship between metabolic power and speed reasonably well simply from body mass for both bipeds and quadrupeds. However, in order to calculate the total power input of an animal running at a given speed, both the slope and the y-intercept must be known. As mentioned earlier, the y-intercept is greater than resting metabolic power. Empirically it turns out that the difference between the y-intercept and resting metabolic power is a constant factor (1·7 for quadrupeds and 1·5 for bipeds). As discussed by Schmidt-Nielsen and Wilkie in their papers in this volume, the resting metabolic rate of both mammals and birds varies in a regular manner with body mass. Kleiber[17] has formulated a simple equation for predicting resting metabolic power of mammals from body mass:

$$W_{std-4} = 3 \cdot 5 \, m^{-0 \cdot 25} \tag{3}$$

where W_{std} has units $W \, kg^{-1}$ and m is body mass in kg. A very similar relationship has been formulated for nonpasserine birds by Lasiewski and Dawson:[18]

$$W_{std-2} = 3.7 \, m^{-0.28}. \qquad (4)$$

Predicting Power Used for Running from Body Mass and Speed

Using the mass dependent relationships for slope and resting metabolic power, it is now possible to formulate a simple empirically derived equation for calculating total power of a running quadruped or biped (W_{run}) simply from its mass and the speed at which it travels. Metabolic power (W_{run}) equals slope times speed plus the y-intercept. For quadrupeds this becomes:

$$W_{run-4} = (T_{run-4})(V) + 1.7 \, W_{std-4} \qquad (5)$$

where V is speed in ms^{-1}. Using Eqns (1) and (3), T_{run} and W_{std} in Eqn (5) can be expressed in terms of body mass so that

$$W_{run-4} = (10.7 \, m^{-0.4})(V) + 6m^{-0.25}. \qquad (6)$$

Similar equations can be written for bipedal birds:

$$W_{run-2} = (T_{run-2})(V) + 1.5 \, W_{std-2}. \qquad (7)$$

Substituting terms from Eqns (2) and (4) this becomes:
$$W_{run-2} = (11.5 \, m^{-0.24})(V) + 5.6 \, m^{-0.28}. \qquad (8)$$

It should be noted that the slope for bipedal mammals should be used with the y-intercept for mammals, and thus Eqn (8) is specific for bipedal birds.

These equations have been empirically derived using data from birds and mammals which regulate their body temperatures at a level about 40°C, whereas lizards can have quite a variable body temperature. We know that temperature has a significant effect on resting metabolic power, which generally *increases* by a factor (called Q_{10}) of between 2 and 3 for each 10°C increase in temperature. As they stand, these equations are not valid for animals whose body temperatures differ from 40°C, but it is possible to write equations which are valid. The incremental cost or slope (T_{run}) is independent of temperature and we have already seen in Fig. 3 that approximately the same relationship between T_{run} and mass is found in quadrupedal mammals and in quadrupedal lizards. Thus Eqn (1) can be used to predict T_{run} of quadrupedal lizards and mammals with low body temperatures, but the y-intercept will depend both on body temperature and on the phylogenetic position of the vertebrate (Dawson,[9] Dawson and Hulbert[10]). The resting metabolism of most mammals which have a low body temperature can be derived from Kleiber's equation using a Q_{10} of 2.5 to account for differences from the normal levels found in most mammals (Fig. 4). However, lizards and the primitive Madagascan insectivores, the Tenrec (*Tenrec ecaudatus*) and the Setifer (*Setifer setosus*), have a constant about $\frac{1}{4}$ that of mammals for the relationship between resting metabolic power and body mass, although the exponent is about the same (Hemmingsen,[14] Jagger et al.[16]). Thus both the Q_{10} effect and the different constant must be taken into account in calculating the *y*-intercept of the relationship between metabolic power and running speed for these animals. It should be noted that, contrary to intuition, the total metabolic power of a lizard is less than that of a mammal running at the same speed, because of the smaller *y*-intercept (Fig. 5).

Why is the *y*-intercept greater than resting metabolic power? Schmidt-Nielsen[23, 24] has proposed that this represents a postural cost of locomotion. Cavagna et al.[4] have shown that the power required to bring about the vertical displacements of the center of mass during each stride (allowing the feet to swing beneath the trunk) is more or less independent of speed and is of the proper magnitude to explain the difference between *y*-intercept and resting metabolic power, at least in dog, monkey and man.

Equivalent Speeds for Animals of Different Size

In order to build mechanical or mathematical models for scaling vertebrate locomotion to body size, we need to use equivalent speeds for different sized animals to compare locomotory parameters. A. V. Hill[15] developed a simple and fairly comprehensive model for scaling running parameters to body size for geometrically similar animals. He used top speed as an equivalent speed for animals of different size. This is a theoretically sound choice, but it is impractical, for how does one know that an animal is running at top speed? Furthermore, animals of different size running at the same speed are not in equivalent situations; for example, at a speed of $1\cdot7$ m s^{-1}, a 500 kg horse is walking slowly, a 10 kg dog is trotting, and 0·030 kg mouse is galloping.

Heglund et al.[13] have recently measured stride frequency as a function of running speed in quadrupeds of different size. They found that both the speed and the stride frequency at which animals changed from one gait to another changed in a regular manner with body size. The transitions from walk to trot, and from trot to gallop occurred at lower speeds and higher stride frequencies in smaller animals. Speed and stride frequency at the trot–gallop transition scaled with body mass (*m*) in the following manner:

$$\text{Speed}_{T-G} = 1\cdot53\,m^{0\cdot24} \tag{9}$$

$$\text{Stride frequency}_{T-G} = 4\cdot48\,m^{-0\cdot14} \tag{10}$$

where speed is in m s^{-1}, stride frequency in Hz, and *m* is in kg. These authors proposed that the speed at the trot–gallop transition is an equivalent speed for quadrupeds of different size. It has the advantage over top speed that it can be easily measured.

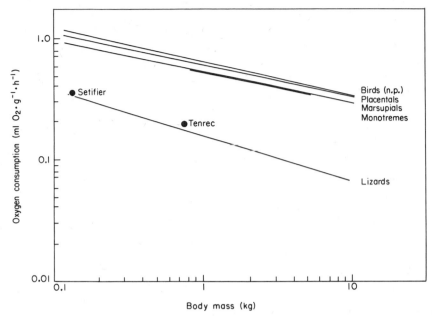

FIG. 4. Standard (resting) metabolism of vertebrates normalised to a body temperature of 38°C using a Q_{10} of 2·5. Equations taken from Dawson[9] and Dawson and Hulbert.[10] The Tenrec point is from Jagger, Taylor and Compton;[16] and the Setifer data point is unpublished data from my laboratory. $1 \text{ ml O}_2 \text{ g}^{-1} \text{ h}^{-1} \approx 5.6 \text{ J kg}^{-1} \text{ s}^{-1}$.

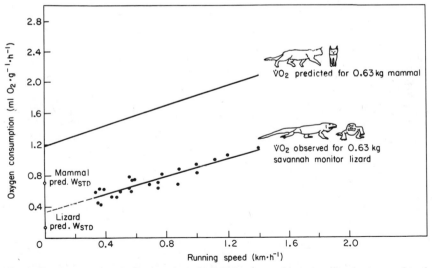

FIG. 5. Steady state oxygen consumption of a 0·63 kg savannah monitor lizard, compared to the oxygen consumption predicted for a 0·63 kg mammal, as a function of speed. Data for the lizard from Bakker.[1] $1 \text{ ml O}_2 \text{ g}^{-1} \text{ h}^{-1} \approx 5.6 \text{ J kg}^{-1} \text{ s}^{-1}$, $1 \text{ km h}^{-1} \approx 0.28 \text{ m s}^{-1}$.

Bipeds such as man do not exhibit a regular change in gait of this nature, and therefore it is more difficult to identify an equivalent speed for bipeds of different size. One possibility may be to use $(gh)^{\frac{1}{2}}$ where g is the gravitational acceleration and h is the height of the hip joint, as suggested by Alexander in his paper. This would be proportional to $m^{\frac{1}{6}}$.

SOME UNANSWERED QUESTIONS IN VERTEBRATE LOCOMOTION— AND SOME SUGGESTIONS ABOUT WHERE TO LOOK FOR THE ANSWERS

Why Does Metabolic Power Increase Linearly With Speed?

Hill[15] and others have assumed that once an animal attains a constant speed, most of the power required for running is used to accelerate and decelerate the limbs. Having made this assumption, we are able to predict that metabolic power should increase at a rate between the square and the cube of speed. An animal can increase speed by increasing either stride length or stride frequency. If we consider the extreme case where the increase in speed is achieved entirely by an increase in stride length, then the number of times the limbs are accelerated and decelerated per unit time remains the same as speed increases. The velocity of the foot relative to the ground will have to reach zero while the foot is on the ground and accelerate to a speed exceeding that of the trunk of the animal during the recovery stroke. Since kinetic energy equals $\frac{1}{2}MV^2$, the power in this case should increase as the square of running speed. Let us now consider the other extreme where an animal relies entirely on an increase in stride frequency to increase its speed. Velocity of the foot relative to the ground will still alternate between zero and a value exceeding trunk speed during each stride, but the number of times that this will occur per unit time will increase directly with the animal's speed. Thus the power required in this case will be proportional to $(\frac{1}{2}MV^2) \times (V)$ or the cube of running speed. Furthermore the work per stride needed to overcome air resistance would be proportional to V^3; however, even at the highest attainable speeds, this contributes only 10–15% of the total energy required (Cavagna et al.[6]). The power required for respiration etc. is built into the efficiency factor.

Neither of these predictions is in accord with the empirical observations. There are also many other observations which conflict with the assumption that a major part of the metabolic power input is required to move the limbs. Taylor et al.[28] have found that, despite large differences in moments of inertia of the limbs, cheetahs, gazelles and goats of about the same mass use the same metabolic power to run at the same speed over a wide range of speeds. Also the incremental cost of lizards and mammals is about the same despite very different limb configurations. Thus in models of vertebrate locomotion we cannot assume that most of the metabolic power input is being used to move the limbs.

Cavagna and Curadini[3] have recently measured the kinetic energy changes involved in acceleration and deceleration of the arms and legs in human running. Assuming no transfer of energy between the limbs, they found, as might be expected, that the power required to account for kinetic energy changes of the limbs increased at about the square of speed both for walking

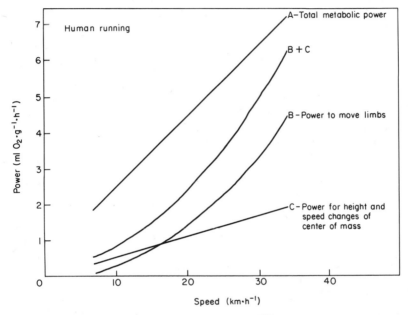

FIG. 6. Total metabolic power—(a) (from Margaria et al.[19]); power required to move the limbs relative to the center of mass—(b) (from Cavagna and Curadini[3]); and power required to account for changes in height and speed of the center of mass within a stride—(c) (from Cavagna et al.[6]). 1 ml O_2 g^{-1} h^{-1} ≈ 5·6 J kg^{-1} s^{-1}, 1 km h^{-1} ≈ 0·28 m s^{-1}.

and for running (but with different constants). At speeds above 5 m s^{-1}, they found that more power was required to account for these changes in kinetic energy than the total metabolic power input of the animal (assuming a muscular efficiency of 25% in converting chemical energy into muscular work). Clark and Alexander[8] found a different situation in a bipedal bird; the power required to account for kinetic energy changes of the limbs of a running quail was small in comparison to total power input.

It is clear that we need more information on how the power required to move limbs, relative to the centre of mass, changes as a function of speed in a

F

size range of quadrupeds and bipeds before we can understand the energetics and mechanics of vertebrate locomotion. It is equally obvious that this information, in itself, will not be sufficient. Cavagna et al.[4] have pointed out that significant amounts of power are also required to raise and reaccelerate the centre of mass within each stride, even when an animal is running at a steady speed. In human running the power required for this purpose is greater than the power required to move the limbs relative to the trunk at running speeds up to $5\,\mathrm{m\,s^{-1}}$ (Fig. 6). At this speed the power required to account for the measured energy changes is twice as great as the metabolic power input (assuming a muscular efficiency of 25%). This is obviously not possible. It simply points to the important role which elastic storage of energy plays within the stride. Cavagna[2] has pioneered the study of elastic energy storage in isolated muscles.

To understand the mechanics and energetics of terrestrial locomotion, we need to understand not only the absolute energy changes that take place during locomotion, but also the kinetic-gravitational and the kinetic-elastic storage links. As discussed by McMahon,[20] a plausible explanation for the linear increase in power with increasing speed is that once the running animal achieves a steady speed it becomes primarily a spring system. Muscles then contract nearly isometrically to maintain the proper spring tension, rather than shortening as they contract. The linear increase in metabolic power with speed is explained because metabolic power appears to increase linearly with the force generated by muscles (Ruegg[22]). To test this hypothesis we need information on the length changes of muscles and tendons, and on the forces developed by muscles during running as a function of speed.

Is it Valid to use the Gait Transition Points as Equivalent Speeds for Comparing Animals of Different Size?

In order to use speeds at the walk–trot transition or the trot–gallop transition as equivalent speeds for different sized animals, we need to know what happens to the active muscle fibres at these transitions. It has been suggested[20] that equivalent forces per cross sectional area of active muscles are developed at these transitions in animals of different size. If this is the case, we would indeed have made a sound choice of equivalent speeds. Measurements determining which muscles are active, together with the forces which they are generating at the transitions, would answer this question.

What is the Relation Between Resting Metabolic Power and Metabolic Power During Running?

Hemmingsen[14] and Wilkie,[30] as well as many other authors, have assumed that maximal power input is a constant multiple (10–20 times) of standard resting metabolic power. If this were the case then the scaling arguments used to predict maximal power during locomotion would also explain the regular changes in resting metabolic power with body size. Pasquis *et al.*[21] have found that the concept of a constant multiple is not in accord with empirical data. The observation that incremental cost of locomotion of lizards is the same as that of mammals, although resting metabolism may differ by 20-fold, also suggests that the relationship between the resting metabolic power and metabolic power used during muscular activity needs careful examination.

TABLE I. Ratio of predicted metabolic power (W_{run}) at the trot–gallop transition point to predicted resting metabolism (W_{std}) for quadrupeds of different size.

Animal	Body mass (kg)	Speed at trot–gallop transition (m s^{-1})	$\dfrac{W_{run\ TG}}{W_{std}}$
pygmy mouse	0·009	0·49	4·8
mouse	0·038	0·69	5·2
rat	0·362	1·20	6·0
dog	1·1	1·56	6·4
gazelle	27·3	3·38	8·0
horse	500·0	6·78	9·9

Speed at trot–gallop calculated using Eqn (9); W_{run} at this speed calculated using Eqn (6); and W_{std} calculated using Eqn (3). I wish to acknowledge my graduate student, Norman Heglund, who initially used these calculations to prove a point during a discussion of scaling problems.

If we assume that the speed at the trot–gallop transition is an equivalent speed, then it is interesting to calculate how metabolic power predicted for a quadruped running at this speed compares with its resting metabolic rate. Table I shows that metabolic rate at the trot–gallop transition is 4·4 times that predicted at rest for a 9 g pygmy mouse and that this factor increases with increasing size to 9·2 in a 500 kg horse. This may be equivalent to an increase with body size of "scope for activity", as discussed by Hughes in his paper. If these transition points are indeed equivalent speeds, then one would expect to see a similar increase with body size at another equivalent speed: the speed of maximum sustained power input. An increase in the ratio of maximal sustained power to resting power with increasing body size is in accord with the empirical findings of Pasquis *et al.*[21] It is clear that we must exert caution in extending scaling arguments which have been derived to explain power input of running animals to the resting metabolic power of these animals.

CONCLUSION

In summary, a number of generalities have emerged from studies of the energetics and mechanics of terrestrial locomotion in vertebrates:

(1) metabolic power increases linearly with speed in most cases;
(2) cost of moving a unit mass through unit distance varies in a regular manner with body size (although differently in quadrupeds and bipeds);
(3) there appears to be a postural cost for locomotion which is independent of speed;
(4) metabolic power of a running animal can be predicted fairly well simply from body mass and the speed of the animal;
(5) the transition points from one gait to another change in a regular manner with body size and are useful as a measure of equivalent speeds for different sized animals.

We have some insight into the explanation of these generalities, but a great deal of work remains before we can either verify or disprove our ideas.

REFERENCES

1. Bakker, R. T. Locomotory energetics of lizards and mammals compared. *The Physiologist*, **15** (1972).
2. Cavagna, G. A. and Citterio, G. Effects of stretching on the elastic characteristics and the contractile component of frog striated muscle. *J. Physiol.* **239**, 1–14 (1974).
3. Cavagna, G. A. and Curadini, L. Internal work in locomotion: effect of walking and running speed on kinetic energy of upper and lower limbs. *IRCS Medical Science: Biomedical Technology; Physiology; Social and Occupational Medicine* **3**, 294 (1975).
4. Cavagna, G. A., Heglund, N. C. and Taylor, C. R. Walking, running and galloping: mechanical similarities between different animals. (This volume).
5. Cavagna, G. A., Saibene, F. P. and Margaria, R. External work in walking. *J. Appl. Physiol.* **18**, 1–9 (1963).
6. Cavagna, G. A., Saibene, F. P. and Margaria, R. Mechanical work in running. *J. Appl. Physiol.* **19**, 249–256 (1964).
7. Chodrow, R. W. and Taylor, C. R. Energetic cost of limbless locomotion in snakes. *Fed. Proc.* **32** (1973).
8. Clark, J. and Alexander, R. McN. Mechanics of running by quail (*Coturnix*). *J. Zool. Lond.* **176**, 87–113 (1975).
9. Dawson, T. J. Primitive mammals. *In* "Comparative Physiology of Thermoregulation, Vol III, Special Aspects of Thermoregulation", (G. C. Whittow, Ed.) Academic Press, New York and London (1973).

10. Dawson, R. J. and Hulbert, A. J. Standard metabolism, body temperature and surface areas of Australian marsupials. *Am. J. Physiol.* **218**, 1233–1238 (1970).
11. Dawson, T. J. and Taylor, C. R. Energetic cost of locomotion in kangaroos. *Nature*, **246**, 313–314 (1973).
12. Fedak, M. A., Pinshow, B. and Schmidt-Nielsen, K. Energy cost of bipedal running. *Am. J. Physiol.* **227**, 1038–1044 (1974).
13. Heglund, N. C., Taylor, C. R. and McMahon, T. Scaling stride frequency and gait to animal size: mice to horses. *Science* **186**, 1112–1113 (1974).
14. Hemmingsen, A. M. Energy metabolism as related to body size and respiratory surfaces, and its evolution. *Reports of the Steno Memorial Hospital and the Nordisk Insulinlaboratorium*, **9** (2), 1–10 (1960).
15. Hill, A. V. The dimensions of animals and their muscular dynamics. *Science Progr.* **38**, 209 (1950).
16. Jagger, J. A., Taylor, C. R. and Crompton, A. W. The tenrec: a primitive mammal with reptilian energetics. *Fed. Proc.* **33** (1974).
17. Kleiber, M. "The Fire of Life. An introduction to animal energetics". John Wiley and Sons, New York (1961).
18. Lasiewiski, R. C. and Dawson, W. R. A re-examination of the relation between standard metabolic rate and body weight in birds. *Condor*, **69**, 13–23 (1967).
19. Margaria, R., Cerretelli, P., Aghemo, P. and Sassi, G. Energy cost of running. *J. Appl. Physiol.* **18**, 367–370 (1963).
20. McMahon, T. A. Scaling quadrupedal galloping: frequencies, stresses and joint angles (this volume).
21. Pasquis, P., Lacaisse, A. and Dejours, P. Maximum oxygen uptake in four species of small mammals. *Resp. Physiol.* **9**, 298–309 (1970).
22. Ruegg, J. C. Smooth muscle tone. *Physiol. Rev.* **51**, 201–248 (1971).
23. Schmidt-Nielsen, K. Locomotion: energy cost of swimming, flying and running. *Science*, **177**, 222–228 (1972).
24. Schmidt-Nielsen, K. "Animal Physiology, Adaptation and Environment." Cambridge University Press, London (1975).
25. Taylor, C. R. Energy cost of animal locomotion. *Comp. Physiol.* (L. Bolis, K. Schmidt-Nielsen, and S. H. P. Maddrell. Eds), North Holland, Amsterdam pp. 23–42 (1974).
26. Taylor, C. R. and Rowntree, V. J. Running on two or four legs: which consumes more energy? *Science* **179**, 186–187 (1973).
27. Taylor, C. R., Schmidt-Nielsen, K. and Raab, J. L. Scaling of energetic cost of running to body size in mammals. *Am. J. Physiol.* **219**, 1104–1107 (1970).
28. Taylor, C. R., Shkolnik, A., Dmi'el, R., Baharav, D. and Borut, A. Running in cheetahs, gazelles, and goats: energy cost and limb configuration. *Am. J. Physiol.* **227**, 848–850 (1974).
29. Tucker, V. A. Energetic cost of locomotion in animals. *Comp. Biochem. Physiol.* **34**, 841–846 (1970).
30. Wilkie, D. R. The work output of animals: flight by birds and man-power. *Nature*, **183**, 1515–1516 (1959).

9. Scaling Quadrupedal Galloping: Frequencies, Stresses, and Joint Angles

THOMAS A. McMAHON

Division of Engineering and Applied Physics, Harvard University, Cambridge, Massachussetts 02138, U.S.A.

ABSTRACT

A previously proposed model for elastic similarity in mammals may be used to predict the angular excursion of the limbs, the peak vertical acceleration, and the supported fraction of the stride in animals running at the lowest galloping speed (trot–gallop transition). Force plate, light photography, and X-ray photographic methods are used to test the predictions of the model and to make direct measurements of muscle force during running.

When A. V. Hill[4] published his important study of the scaling of animal muscular dynamics, he assumed that the animals compared were geometrically similar and that they were running at top speed, under conditions where the peak muscle stress and muscle work per stroke per gram body mass was the same in large and small animals. I have argued elsewhere that elastic similarity, rather than geometric similarity, is maintained between animals of dissimilar size, and this leads to a relative thickening of all body proportions as body size increases (McMahon[5, 6]). The elastically similar model may also be used to make predictions for the changes in running performance with body weight (McMahon[7]). In this paper I shall briefly review the scaling of elastically similar animals running at the transition speed between trotting and galloping, which is taken to be a physiologically equivalent speed. I shall then discuss several new observations and experiments designed to test the theory further. The overall objective in this work is to elucidate the organizing principle which causes larger animals to be more heavily built and capable of higher speeds than smaller ones. This understanding, which builds on the principles of Hill's early work, promises insight into the fundamental physics of gait, with applications to comparative zoology, ecology, and natural history.

METHODS

Theoretical

1. *Body dimensions.* As discussed in more detail elsewhere (McMahon[5, 6]), the elastically similar scaling model requires that the length of any bone, muscle, or tendon of the body is proportional to $m^{1/4}$, (m = body mass) whereas the diameter is proportional to $m^{3/8}$. This relative thickening of diameter compared to length maintains the static deflection of a structure under its own weight proportional to its length.

2. *Stride frequency.* The natural frequency of vibration of a system of masses, springs and levers representing an elastically similar animal is proportional to $m^{-1/8}$, or $l^{-1/2}$, where l is an appropriate length dimension, e.g. the length of a limb or a bone (McMahon[6]). Heglund et al.[3] found stride frequency at the trot–gallop transition proportional to $m^{-0.14}$, in agreement with the postulate that animals gallop at their natural frequency of lowest-mode vibration. Pennycuick[8] found that ungulates of different size gallop freely at a stride frequency proportional to $l^{-0.50}$, also in agreement with the present predictions. It should be noted that animals of any design would be expected to walk at a frequency proportional to $l^{-1/2}$ (pendulum frequency), but only animals maintaining elastic similarity vibrate (gallop) at a given multiple of their walking frequency.

3. *Height of rise of centre of mass.* During the aerial phase of galloping, the centre of mass falls through a height $h_1 = \frac{1}{2}gT_1^2$, where T_1 is the falling period. If T_1 is proportional to $m^{1/8}$, h_1 is proportional to $m^{1/4}$. During the period the feet are on the ground, the dynamic excursion of the centre of mass, h_2, is proportional to the static deflection of the body under its own weight (McMahon[5]). Since the definition of elastic similarity requires that this static deflection be proportional to l (e.g. the length of the vertebral column), the total vertical excursion of the centre of mass is $h = h_1 + h_2 \propto m^{1/4}$ at the trot–gallop transition.

4. *Vertical acceleration during supported period.* The force applied to the ground by an animal's foot in general has a vertical and horizontal (and small lateral) component. During the passage of each foot under the body, there is a moment when the horizontal force goes through zero and the vertical

force is near maximum. If the vertical displacement h is proportional to $m^{1/4}$, and the period available for acceleration upwards, T_2, is proportional to T_1 and thus to $m^{1/8}$, then the vertical acceleration is scale-invariant and the peak vertical force applied to the ground by each foot should be proportional to body weight. Both this result and that of paragraph 3 depend on the assumed scale invariance of the ratio of the supported period to the total stride period.

5. *Limb kinematics and running speed.* As developed in detail elsewhere (McMahon[7]) the elastic similarity model may be used to predict that the angular excursion of any joint of the skeleton should be proportional to l/d, where d is the diameter of the limb or bone, and thus to $m^{-1/8}$ under physiologically equivalent conditions, but that stride length should be associated with the d length scale and thus be proportional to $m^{3/8}$. These two results are not compatible unless the two hindlimbs and two forelimbs move asynchronously, in such a way as to add the steps of limbs on opposite sides of the body in series, with the degree of asynchrony increasing as body size increases.

Experimental

1. *Light photography.* To test the predictions of the model and guide its further development, we photographed mice, rats, dogs and horses at framing rates up to 200 frames s^{-1}. The animals ran at constant speed on a treadmill, except for the horses, which were allowed to run freely on a racetrack.

2. *Force plate.* In a separate series of experiments, ground squirrels, dogs and a hyrax were contained in a moving floorless cage and caused to run across a Kissler physiological force plate. The force plate system is linear to within 1% in the force range we employed it, with less than 3% cross-talk between channels, and has a loaded natural frequency greater than 150 Hz. The force plate system therefore had an adequate frequency response for our measurements, since no deflections of significant amplitude were observed at frequencies greater than 20 Hz.

3. *Direct tendon force measurement by X-ray cine.* An extremely important parameter of locomotion at the trot–gallop transition is muscle force and stress. In order to achieve a direct measurement of muscle force during running, a new technique was developed which effectively used the tendon as a force transducer. The skin of the dorsal ankle was opened and steel markers were placed within the Achilles tendon of seven ground squirrels (*Citellus tridecelineatus*) and one small dog (Sammy, a 4·2 kg Fox terrier). The proximal marker was placed close to the junction of tendon and muscle belly; the distal

marker was near the calcaneal insertion. The markers were 0·36 mm stainless steel wires in the ground squirrels and 1·57 mm steel ball bearings in the dog. In each case, the skin wound healed completely and the animals never showed any lameness, even immediately after the insertion operation.

The animals were photographed running on a treadmill with Siemans X-ray cine equipment about 1 week after the operation. A steel wire length standard or gnomen was taped over the animal's tibia during filming. Both the length between the two tendon markers and the length of the gnomen was read from each frame of the film using a Vanguard motion analyzer. After a sequence of running experiments, the animal was deeply anesthetized with halothane and 60 mg/kg nembutal, and a 2·29 mm steel wire (in the dog's case a light steel chain) attached rigidly to the tendon proximal and distal to the marked region. The two tied ends of the intact tendon were now held in a special jig and quickly stretched from the resting length to a series of longer lengths, and the force which resulted was recorded on a Harvard Apparatus HA373 isometric tension transducer. While the step length changes were carried out, the tendon and the gnomen were photographed on the X-ray cine equipment. Using this calibration procedure, the tendon force could be known as a function of the actual length between the tendon markers.

RESULTS

Comparative Kinematics

In Fig. 1 (top), the fraction of the total stride period during which either one of the two front feet supports the body at the trot–gallop transition is shown

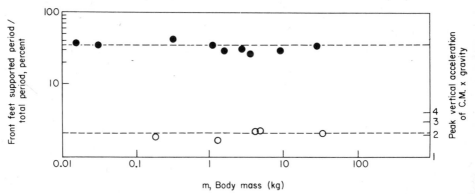

FIG. 1. Top: period of support by the forelimbs/total stride period, as a function of body size. Bottom: peak vertical acceleration vs. body size. Neither quantity depends on body size, as required in the elastic similarity model. All measurements are taken at the trot–gallop transition.

as a function of body mass. As predicted by the elastic similarity model, this fraction is independent of body mass.

Another test of the same feature of running at the trot–gallop transition is shown in Fig. 1 (bottom). Here the peak vertical force contributed by the two front limbs divided by body mass is shown for the several animals used in force-plate trials. The mean of this peak vertical acceleration is about 2·1 times the acceleration of gravity, and is uncorrelated with body size, as predicted.

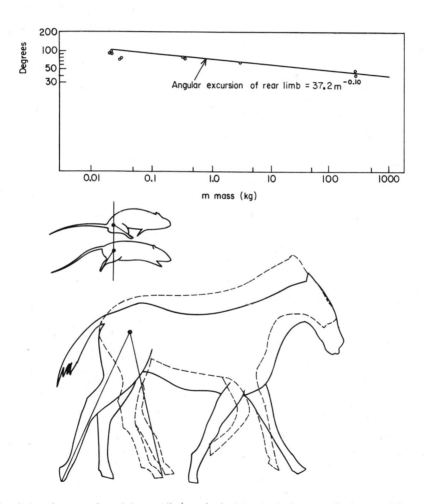

FIG. 2. Angular excursion of the rear limb vs. body mass. At the lowest galloping speed, larger animals move all the joints of the skeleton through smaller angles. This is demonstrated by the illustrations of mouse and horse, which were traced from cine studies at the trot–gallop transition.

In Fig. 2, which has been reproduced from an earlier paper (McMahon[7]) the maximum angular excursion of the hind limb is shown as a function of body mass. The angular excursion diminishes as body size increases, and is in fact proportional to $m^{-0.10}$, as compared to the model prediction of $m^{-1/8}$.

Tendon Forces

The animals whose tendons were given metallic markers performed on the

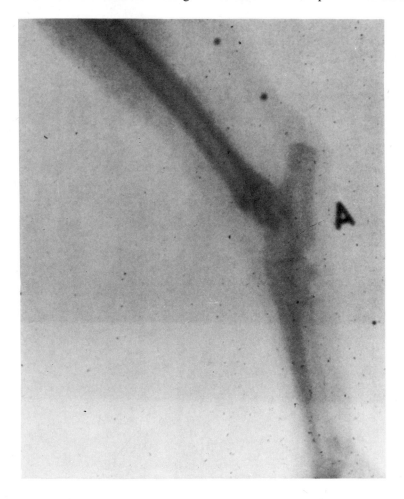

FIG. 3. (a) Ground squirrel, *Citellus tridecelineatus*, and (b) dog, *Canis familiaris* (Fox terrier), with steel markers in the tendons. Stretches of the marked segment of the tendon served to measure muscle force directly. The length standard is not present in the dog radiograph.

treadmill quite normally, encouraging us to believe that their tendons had suffered little damage by the procedure. A ground squirrel is shown in one frame of a galloping sequence in Fig. 3(a); the dog is shown in a trotting sequence in Fig. 3(b). We found that the gnomen appeared to change its length periodically by about 9%, while the distance between the tendon markers varied by more than 20% during a stride. Using the gnomen as a length standard, it was possible to know the actual distance between the tendon markers in each frame.

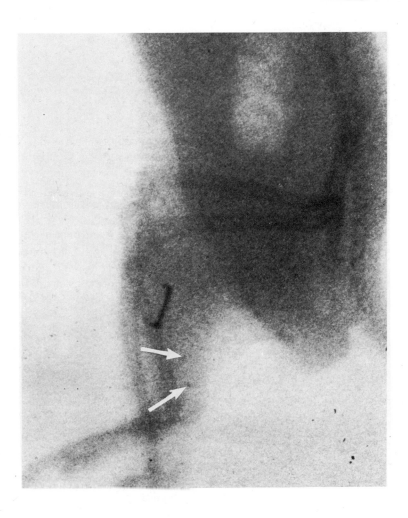

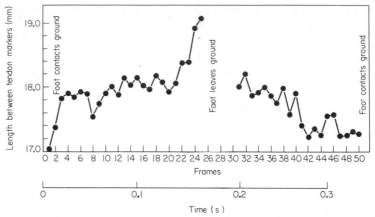

FIG. 4. Length of tendon markers vs. time for one trotting stride at a speed near the trot-gallop transition in the dog. Just after the foot left the ground, the animal's leg passed out of view of the camera for five frames.

The actual length between the tendon markers is shown in Fig. 4 as a function of time for one trotting stride of the dog at a speed close to the trot–gallop transition. The tendon stretch is greatest at the last moment of the contact phase, just before the foot leaves the ground.

As a preliminary comment, one might point out that tension development in the tendon and muscle would be necessary before landing if the process of impact were to store elastic energy reversibly. Several recently proposed conceptual models of running, including the elastically similar model discussed earlier, depend on the ability of tendons and muscles to store and return elastic energy to the body (Cavagna,[2] Alexander,[1] McMahon[7]).

CONCLUSION

Is the trot–gallop transition really a physiologically equivalent condition? What is the stimulus or principle which causes animals to change gait? Using the techniques for direct measurement of muscle force described above, we plan to take up the challenge of these and other questions. Although the elastically similar model is far from complete, its ability to predict many of the allometric rules of running make it useful for suggesting experiments and interpreting their results. Comparative questions are easier than absolute ones. While it may never be possible to understand the physical design of any one animal from first principles, a fairly subtle understanding of the physical scaling laws of the animal kingdom is a reasonable hope and may not be far away.

ACKNOWLEDGEMENTS

This work was supported in part by Grant#GB36588 from the National Science Foundation and the Division of Engineering and Applied Physics, Harvard University, Cambridge, Mass.

REFERENCES

1. Alexander, R. M. The mechanics of jumping by a dog (*Canis familiaris*) *J. Zool., Lond.* **173**, 549–573 (1975).
2. Cavagna, G., Heglund, N. and Taylor, C. R. Walking, running, and galloping: mechanical similarities between different animals. (In this volume) (1977).
3. Heglund, N., McMahon, T. A. and Taylor, C. R. Scaling stride frequency and gait to animal size: mice to horses. *Science*, **186**, 1112–1113 (1974).
4. Hill, A. V. The dimensions of animals and their muscular dynamics. *Sci. Progr. London*, **38**, 209–230 (1950).
5. McMahon, T. A. Size and shape in biology. *Science*, **179**, 1201–1204 (1973).
6. McMahon, T. A. Allometry and biomechanics: limb bones in adult ungulates, *Am. Nat.* **109**, 547–563 (1975).
7. McMahon, T. A. Using body size to understand the structural design of animals: quadrupedal locomotion. *J. Appl. Physiol.* **39**, 619–627 (1975).
8. Pennycuick, C. On the running of the gnu (*Connochaetes taurinus*) and other animals. *J. Exp. Biol.* **63**, 775–799 (1975).

10. Problems of Scaling in the Skeleton

J. D. CURREY

Department of Biology, University of York, England

ABSTRACT

Skeletons will fail if they are stressed too much or if they deflect too much. The usual concomitant of an increase in size of an animal, if stresses and deflections are to remain tolerable, is that the skeleton must increase in mass more rapidly than the mass of the animal. This is not always true; a rigid gas-filled buoyancy tank, designed to go to a particular depth before it implodes through elastic instability, can have effectively a constant mass relative to that of the animal whatever the animal's size. Burrowing molluscs may also have no scaling problems. However in general a larger animal does impose larger stresses and strains on its skeleton if geometrical similarity is maintained.

There are various ways round this problem:

1. The animal can accept a lower locomotory performance—in general the largest animals are not the fleetest.
2. The animal can change its shape so as to reduce stresses and deflections. This usually will produce a lowered locomotory performance.
3. The animal may accept a lower safety factor in its skeleton.
4. The mechanical efficiency of the material can be increased.

Possible examples of all these are given, but the facts are as yet so meagre that most are informed guesses. Particularly, we have little idea of what the skeleton is designed to optimise (deflection, impact resistance, static strength etc.) and we have little idea of the metabolic cost of skeletons. Cases are discussed in which the animal's skeleton alters markedly because of scaling effects. The question of dynamic loading is discussed, but not resolved.

INTRODUCTION

The problem of scaling in skeletons is one of peculiar difficulty, because it is very hard to design experiments, even armchair experiments, that will

153

determine what skeletons are designed *for*. As this symposium is concerned with locomotion, we can ignore to a large extent skeletal parts like skulls, that are protective only, except insofar as they increase the weight or mass of the animal.

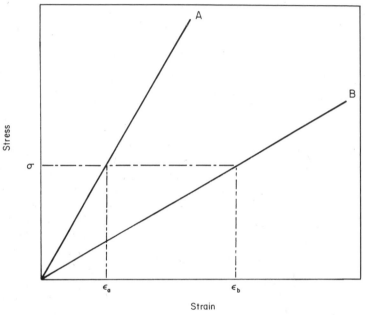

FIG. 1. Two linearly elastic materials, A and B have stress/strain curves A and B. At any arbitrary stress level σ, the energy absorbed by A, U_a is less than that absorbed by B, U_b. $U_a = \frac{1}{2}\sigma.\varepsilon_a$ $U_b = \frac{1}{2}\sigma.\varepsilon_b$.

In general one can suppose that the function of skeletal limb elements is to provide rigid levers through which muscles can act on the environment. If this is so, then their primary function is to be stiff. This stiffness needs to be considered per unit mass, because almost any structure can be made stiff by being made massive, but for any active animal a massive skeleton is a disadvantage. Greater stiffness of a structure can also be achieved by making the material of which the structure is composed stiffer per unit mass. Many animals take up this option by using calcium carbonate and other ceramics which have a high stiffness/mass ratio. Unfortunately stiffness must be paid for. Stiff materials necessarily absorb less energy for a given stress level than floppy material (Fig. 1). Furthermore, biological stiff materials tend to be of a composition (ceramic as opposed to protein, for example) that makes them brittle and weak.

However, although the primary function of a skeletal element is to be stiff, hard tissues, unlike say tendon, are able to withstand various types of load and usually we cannot say which is critical. Any skeletal element has a certain mass, which is determined by natural selection; that is: any slight increase in mass will produce advantages in energy absorption, stiffness, bending strength and so on, but these advantages will be outweighed by the disadvantages of greater mass, increased weight, increased moment of inertia, increased metabolic cost and so on. Any slight decrease in mass will be a disadvantage for the opposite reasons. In theory it might be possible to determine selective values for alterations of the various mechanical parameters and then ask the following question: If we reduce the mass by $x\%$, where x is small, what will be the selective disadvantage to the animal because (a) the element is less stiff (b) the element's energy absorption in torsion is reduced (c) the static bending strength is reduced, and so on? We might then find that although, say, the torsional stiffness was reduced more than the Euler buckling strength, yet it was, of all possible mechanical variables, the reduction of the Euler buckling strength that had the greatest selective disadvantage to the animal. In this analytically enviable case we could say that the element was designed against Euler buckling.

It is important to be clear that, from the point of view of natural selection, catastrophic failure and mild inefficiency may have the same effect, if the number of individuals affected is appropriate. For instance, suppose a mutation occurs in a population which prevents one animal in 100 from suffering a fatal fracture before it reaches reproductive age. However, suppose also it does this by making the skeleton slightly less stiff, so that the animals expend slightly greater energy getting around, and that thereby the energy available for producing offspring is reduced by 1%. In engineering terms the occasional failure by fracture would be obvious, but the lack of stiffness would be as serious a disability.

McMahon[8, 9] argues that it is the need for stiffness that determines the scaling effects we see in many skeletons, and produces evidence that these scaling effects are not markedly at variance with the hypothesis.

In view of this it is important to realise that fracture is by no means a rare event in wild populations. For instance Buikstra[1] shows that 40% of the adults of a complete population of macaques (*Macaca mulatta*) showed healed fractures of the long bones or clavicles. Presumably there were many individuals whose fractures did not heal and who therefore died. Race-horses often break their legs while galloping, even without stumbling, and many ballet dancers near retiring age show limbs with a number of healed fatigue fractures each of which probably immobilised them for a short while (Schneider *et al.*[11]). Stiffness may be important, but fracture is important too.

STRATEGIES FOR OVERCOMING SCALING EFFECTS

If we take it as given that larger size will produce a tendency for stresses and deflections to get larger, then there are at least four strategies that can be adopted by animals. (1) Locomotory performance can be reduced. (2) The shape of the skeleton can be altered. (3) The safety factor of the skeleton can be reduced. (4) The mechanical efficiency of the skeletal material can be increased or modified.

In what follows I discuss these four strategies, mostly very briefly, and then consider some cases in which the scaling factor does not work in the way taken as given.

Change in Locomotory Performance and Skeletal Shape

If one takes land quadrupeds of roughly similar build it is quite clear that the largest animals are not the fastest, and certainly not the most agile. It is notoriously difficult to get good data for maximum speeds, but certainly the elephant, the hippopotamus and the rhinoceros, though speedy, are not as fast as animals rather smaller than themselves (Hill[7]). There are, as shown in this symposium, a variety of reasons for this but obviously one is the necessity for large bodies to have graviportal limbs whose moments of inertia round the joints are very large compared with those of smaller animals. One cannot help feeling that these large animals have given up the struggle for speed; in the horse the limb muscles are mostly near the centre of rotation of the limb, thereby minimising the moment of inertia, but in the elephant the muscles have moved down again, as if the selective advantage of reducing the moment of inertia had become nugatory. Of course, it can be argued that large animals do not need to be fast; size is an advantage in itself, as it usually deters predators, so the need to flee disappears. There are no natural experiments to enable us to distinguish the lack of need and the lack of ability to travel fast. However, the fact remains that one cannot design very large quadrupeds without resorting to graviportal legs, and graviportal legs would make fast locomotion very expensive in muscle and in stresses at the joint, and in general would render it imprudent.

There are various cases in which the problems of scaling are such that the mode of locomotion is changed entirely during the animal's development. Of particular interest here because of its clarity is the case of the swimming bivalve molluscs such as *Pecten*, the scallop. These swim by jet propulsion, but as they grow larger swimming becomes increasingly difficult (Gould[5]) because the shell is so dense relative to water. There comes a stage in the life of many of these molluscs when they must become sedentary. They then often attach temselves firmly to the substrate by means of collagen byssus threads, or by cement. What is interesting about many of these species is the change in skeletal structure made possible by giving up swimming. The shells of the swimmers tend to be thin for their overall size. In sedentary forms, since lightness is no longer at a premium they can, and do, markedly thicken their shells, which in consequence become more effective at protecting the animal. This is well seen, for instance in *Hinnites* (Fig. 2) (Thayer[12]). We have here a change in the way of life, necessitated by scaling factors, which has a marked effect on skeletal proportions and structure.

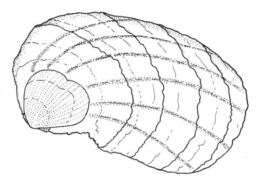

FIG. 2. *Hinnites gigantea.* The ontogenetic stage that was free-swimming is the small, scallop-like part at the left. The entirely sessile stage is oyster-like in general appearance.

In contrast to this, one can show cases where, as the animal grows, scaling effects require a considerable alteration in the structure of the skeleton, even though its locomotion is unaltered. For example snails have evolved their characteristic shell in order to carry around with them a protective housing whose centre of gravity is fairly near the base, yet whose aperture is rather small in relation to the total volume enclosed. It is essentially a shape adapted for locomotion (unlike the bivalve shape, which is in general more suitable for a sedentary life except in the case of some active burrowers like *Tellina*, and some occasionally active swimmers like scallops). Many snails show almost perfect gnomonic growth, in that the shape of the shell remains constant throughout growth. As they approach adult size, snails have a scaling problem that is not faced by vertebrates, who reconstruct their skeleton as they grow, or by arthropods, who shed theirs and start again. As the shell grows it becomes thicker, so the last grown parts are appropriate for the size of the animal. However the parts laid-down earlier remain thin-walled. Though there is no locomotory penalty for this, the shell is excessively

thin and therefore vulnerable at the tip. Of course, in the younger animals
the shell needs to be thin, and therefore vulnerable, but probably many snails
are like forest trees—life expectancy increases with age. A thin shell is acceptable in a young snail while it is putting out considerable metabolic efforts to
become larger, but would be the Achilles heel of a larger animal which can
expect to live for some time. There are various stratagems snails can adopt to
overcome this problem. (1) They have a shape such that the younger stages

Living chamber

FIG. 3. *Nautilus* shell split in two to show the series of chambers. The gas in these remains at
effectively the same pressure whatever the depth.

of the shell are enclosed by or strengthened by the later growing parts. This is
exemplified by *Nautilus* (Fig. 3) (not a snail, of course) and *Conus* (Fig. 4(a)).
(2) They can remove the youngest part of the shell, forming a new wall beneath
which the living body retreats. This method is not common though it is
adopted by, for instance, *Rumina decollata*. (3) The Shell can be filled in
with a crude rubble of calcium carbonate, which though not particularly
strong as a material, is sufficient to prevent the tip being easily broken. This
method is adopted for instance by *Trochus* (Fig. 4(b)), and also by *Conus*.

Reduction of Safety Factors

I have suggested that an animal may adopt the strategy of accepting a lower
safety factor as one way of dealing with scaling problems in the skeleton.
Unfortunately this suggestion, though obvious and intuitively correct, is

virtually impossible to back up with facts, The only animals for which we have much evidence concerning the incidence of fractures are ourselves and domesticated animals. In presenile humans the incidence of fractures goes *down* with age; this is presumably a product of the rash impetuosity of youth rather than the greater safety factor present in the adult skeleton. Domestic animals are bred with particular traits in mind. Racehorses probably suffer

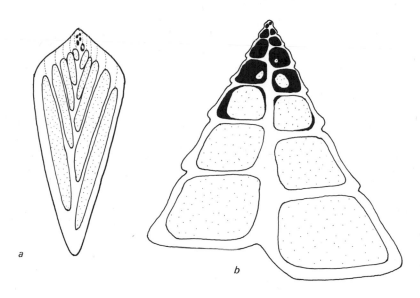

FIG. 4(a) Mid-section through a shell of *Conus*. Shell laid down initially left blank, living space stippled, secondary shell "rubble" black. The older, thinner parts of the shell are partially surrounded by younger, thicker material (b) Mid-section through a shell of *Trochus*.

no more injuries than greyhounds and they are both bred primarily for speed. But in general breeders will increase the speed of their animals until such time as the incidence of bad damage to the musculoskeletal system approaches the unacceptable, so it would say little about scaling effects in natural populations if greyhounds and racehorses had the same incidence. The ludicrously small trench which is capable of imprisoning the elephants at London Zoo shows that these large animals cannot move more violently than normal without danger, and presumably they are aware of it. Such evidence as this is hopelessly anecdotal, and it will be hard to improve.

Nevertheless a reduction in what the skeleton is able to bear over and above "normal" locomotory activity (which may itself be severely reduced in large animals), is a possible strategy for larger animals, one which they may well have employed.

Change in Skeletal Material

One way in which an animal can prevent its limbs from becoming too bulky is to increase the mechanical quality of the material of which the limbs are made. I give here two cases which *may* be examples of this. In insects showing the hemimetabolous life history there is rather little change in the shape of the larva from one instar, or stage, to the next, but after the last moult the animal becomes sexually mature and develops functional wings. The locust is such a hemimetabolous insect. In the locust, then, the final, adult stage is the largest, and so problems of scaling should apply to it more than to the smaller immature stages. It can also fly, which probably produces considerable additional problems of stressing. It can be argued (as by Wainwright *et al.*[14]) that in most insects fracture is a less important mode of failure than buckling. Therefore, for an insect, it is important to produce a stiff skeleton, and one way to do this is to produce a material with a high modulus of elasticity. Adult locusts do have skeletons with a higher modulus than that of the immature stages (Hepburn and Joffe[6]), brought about by a more thorough tanning of both exo- and endocuticle (the two layers of the cuticle). This would seem to be a fine example of what we are looking for, but unfortunately the situation is not so simple. There is probably not a progressive increase in stiffness after each moult but rather a difference only between the last juvenile stage and the adult. A reason for this is that once the cuticle has become properly tanned there is no easy way of digesting it, in order to get rid of it at the next moult. The juvenile stages remove part of their cuticle before moulting, which helps them to wriggle out of their exoskeletons, and they also use the products of the partial digestion of the cuticle in the building of the next, larger version of their exoskeleton. If the larval cuticle were heavily tanned both these processes would be much more difficult. As regards scaling it may simply be that the adult locust is released from the necessity of having a somewhat labile skeleton and so can increase its mechanical efficiency. Nevertheless it is fortunate for the locust that it is able to do so.

A more convincing case is the development of the mechanical properties of bone in man (Currey and Butler[3]). As a child grows to adulthood the stresses on its skeleton will, for reasons with which we are now familiar, tend to get larger. However, the proportions of the body alter, certainly as between the toddler and the adult, in a way opposite to that which would seem to be necessitated by a constancy of elastic stability as proposed by McMahon (Fig. 5); that is, the limb bones become relatively more slender as the child grows. There are no doubt good mechanical reasons for this, but it is obvious that if it were possible to improve the stiffness of the bone material, such an improvement would be valuable. This is indeed what happens, at least in the femur. The modulus of elasticity increases markedly towards the age of

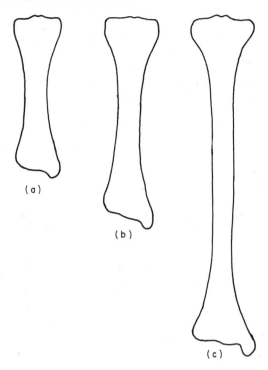

FIG. 5. Outline drawing of human tibias, not to scale. (a) Newborn, (b) Three year old, (c) Adult.

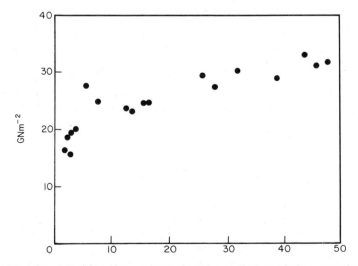

FIG. 6. Modulus of elasticity of human femoral specimens. Each spot is the mean value of at least four specimens. Ordinate: Modulus of elasticity. Abscissa: Age in years.

10 and more slowly, but surely, thereafter (Fig. 6). The static bending strength behaves similarly. The improvement is associated with an extremely modest increase in the mineral content of the bone so that the density of the bone material is increased only trivially. It might seem, then, that as the adult bone is stiffer than the juvenile bone, the problem of scaling has been partially overcome. As stiffness becomes more important, the bone becomes stiffer.

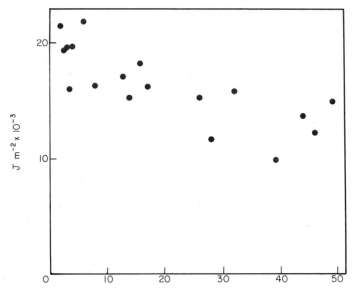

Fig. 7. Work absorbed by human femoral specimens. Each spot is the mean value of at least four specimens. Ordinate: Work absorbed p.u. cross-sectional area. Abscissa: Age in years.

Unfortunately, of course, there is something to be paid for this increase in stiffness, and in this case it is reduction in energy absorption (Fig. 7). This change in energy absorption is brought about in three ways. Compared with adult bone:

(a) Juvenile bone has a lower modulus of elasticity. This implies that for any particular stress level in the bone the energy absorbed will be greater (Fig. 1).

(b) Juvenile bone yields more; that is it shows more plastic deformation before fracturing (Fig. 8).

(c) When juvenile bone fractures it often does not break cleanly but splinters progressively. This produces the characteristic "greenstick" fracture of children's bones which is very rarely seen in adults. Such a fracture mode absorbs a great deal of energy.

These factors together allow toddlers' bones to absorb more energy per unit volume than adult bones before finally breaking in two (Currey and Butler[3]). We do here seem to have an example of the trade-off in quality:

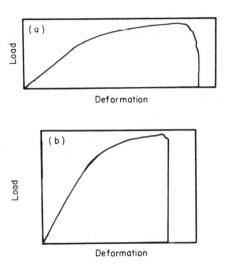

FIG. 8. Load deformation curves of human femoral specimens. (a) 2 year old, (b) 48 year old. Note how there is much more plastic deformation in the younger specimen.

in adventurous childhood when the stiffness of the skeleton is not tested in ordinary locomotion, but falls out of apple tress are common, the ability to absorb energy without breaking is important. In serious, heavy, adulthood when the skeleton must act as stiffly as possible, the bone may break more easily but it can withstand the loads placed on it in ordinary locomotion with suitable rigidity. I suspect that in childhood it is breaking that is the common mode of failure, but that as adulthood is approached stiffness becomes more and more important, and the question of which is the more important selectively becomes quite evenly balanced.

TWO EXAMPLES OF NO DISADVANTAGE IN LARGE SIZE: *NAUTILUS* AND *TELLINA*

Let us consider some cases where either there are no scaling effects or they may work to the advantage of larger animals. The first case is that of an animal, *Nautilus*, which uses its rigid skeleton as a buoyancy tank (Fig. 3). The animal (a cephalopod mollusc, rather like an octopus) lives in the last chamber, and the other chambers are filled with gas at slightly less than atmospheric pressure. For present purposes one can assume that the chambers have no connection to the outside. As the animal can descend to great depths the pressure differences between inside and outside can be considerable, and indeed the shells have been found capable of withstanding pressures of 50 atmospheres or more (Denton and Gilpin Brown[4]). The walls are thin in comparison to their overall diameters and we can apply the simple theory of thin pressure vessels. How do these shells fail under pressure? In my laboratory we have observed that the compressive strength of *Nautilus* shell material is of the order of 400 MNm^{-2}, and with such a strong material it is most probable that the shell would fail through buckling, not crushing. Now, the formulae for buckling of pressure vessels always have a term t/D where t is the wall thickness and D is overall diameter. (This is also true if the shell fails by crushing.) That is to say, if the shape is efficient, which one may suppose it is, the external pressure that causes failure is determined by the thickness of the wall relative to the overall size (Roark[10]). Consequently a linear increase in dimensions will keep the failure pressure constant. If a thin-walled buoyancy tank and the object it is supporting so as to be neutrally buoyant increase all dimensions linearly, neutral buoyancy is maintained. It is the case then, that if an animal such as *Nautilus* increases in size in a geometrically similar way, the depth to which it can descend before the shell collapses is unchanged.

A second case in which there seem to be no scaling problems in getting larger is that of burrowing bivalve molluscs with thin shell walls, such as *Tellina tenuis*. The shape of the shell does not change much during ontogeny. During burrowing, if the speed of locomotion is constant, the forces the shell is exposed to will be those of the resistance of the substrate to its passage and will be roughly proportional to the square of the linear dimensions. If we regard the shell as a pressure vessel then the factor t/D appears again, and as the "pressure" (force per unit area) is constant the stresses in shells of the same shape should be the same whatever the size of the animal. If on the other hand we consider the shell to be loaded as a beam by the sand exerting a pressure W, then the load at any point should still be independent of size. For consider a rectangular beam of length L, width b and thickness d: the maximum bending stress is proportional to Md/I, where M is the bending moment and I is the moment of inertia of the cross-section.[13] If the beam bends under a uniform pressure W, M is proportional to WbL^2 and I to bd^3. Hence the maximum stress is proportional to WL^2/d^2, which is independent of size for geometrically similar shells. Similarly, the deflection is proportional to L, and therefore the shape of the deflected shell is constant for all L. We have some information in my laboratory from a set of 150 *Tellina* shells. We found, first, that the ratio of shell mass to animal mass was effectively constant. We also found, by loading the shells in compression between flat platens, that the load the shells could bear increased slightly faster than the animals' *weight*. Since, if the above reasoning is correct, the forces due to locomotion are proportional to the surface area, the animals will if anything have a greater safety factor as they get larger even though they seem not to be showing any change in proportions. Other factors are no doubt important: for example, the larger animals probably move faster through the sand and are therefore exposed to greater stresses.

IMPACT LOADING

In addition to such cases as those discussed above, there is the general case of impact loading which may seem to produce no scaling problems. In the same way that the weight is the important feature of the static animal, so the kinetic energy is the important feature of the moving one. For geometrically similar animals, travelling at the same speed, the kinetic energy, $\frac{1}{2}mV^2$, will be proportional to mass. There may be other energy terms caused by the relative motion of the limbs etc., but we shall ignore these here. If an animal is brought to rest very rapidly, the problem is how the kinetic energy is dissipated. One can show that in general the ability of a skeleton to absorb kinetic energy without fracture is proportional to the volume of the skeleton. For a simple case, take a bar of linearly elastic material of length L, of cross section area A, tensile strength σ, and modulus of elasticity E. The maximum load it can bear in tension is $\sigma . A$, the values of L and E being irrelevant. On the other hand, the amount of energy it can absorb as strain energy is $\sigma^2 A . L/2E$.

Is this important? Is it likely that we can find parts of the skeleton or even whole skeletons which do not show the usual kind of scaling effects because they are designed, for the worst case, to resist impact? Probably not, for two reasons. First, and less important, if the extreme function of a skeletal element is to resist impact without breaking, then it should be pliant, because energy absorption is inversely proportional to stiffness. So energy absorbers, which should be pliant, are usually separate from and outside skeletal elements, which must be stiff to perform their function. One example is the skin. In rabbits' feet the thin layer of skin overlying the metatarsal increases the energy absorbed in breaking the bone by 40 % (Currey[2]). This energy-absorbing property of the soft tissues makes it very difficult to break skeletal elements which are buried in them by direct injury. Probably most injuries to skeletons caused by falls or mistakes during locomotion in animals with endoskeletons are the results of indirect injury—the energy-absorbing properties of the calf muscles are no use if the tibia is being rotated by a ski via a stiff boot, the metatarsals and the ankle bones.

Secondly, calculations such as those made above about energy absorption by the bar loaded in tension rest on insecure foundations, because they assume that the distribution of stresses in a body loaded in impact is the same as that in a body loaded statically. But if the body is at all massive there tends to be an extremely rapid build-up and decay of the stress at the point of impact, inertial effects become important, and the stress is distributed in a very "non-static" way through the body. Particularly, it tends to be high in the region of impact, and if the velocity of impact is high the stress becomes effectively independent of the volume of the body. It becomes necessary, if stresses are not to be high locally, for the duration of impact to be lengthened and this is one of the functions of the cartilage cap over the end of long bones in synovial joints, and soft tissues round an endoskeleton.

It is possible to suggest various scaling advantages and disadvantages of large size in relation to impact loads, but the situation is so complex that it is difficult to know where the balance of advantages would lie. It is most unfortunate that the analysis of dynamic loading is so difficult, because the skeletons of active animals like land mammals must be designed against dynamic loads, these being so much greater than static loads.

I should like to thank Dr David White and members of the Conference for their helpful criticism. This work was partially supported by the Science Research Council with a grant for the study of mechanical properties and fracture of invertebrate hard tissues.

REFERENCES

1. Buikstra, J. E. Healed fractures in *Macaca mulatta:* Age, Sex and Symmetry. *Folia primatol.* **23**, 140–148 (1975).
2. Currey, J. D. The effect of protection on the impact strength of rabbits' bones. *Acta Anat.* **71**, 87–93 (1968).
3. Currey, J. D. and Butler, G. The mechanical properties of bone tissue in children. *J. Bone Jt. Surg. A.* **57**, 810–814 (1975).
4. Denton, E. J. and Gilpin-Brown, J. B. On the buoyancy of the pearly nautilus. *J. mar. biol. Ass. U.K.* **46**, 723–759 (1966).
5. Gould, S. J. Muscular mechanics and the ontogeny of swimming in scallops. *Palaeontology,* **14**, 61–94 (1971).
6. Hepburn, H. R. and Joffe, I. Hardening of locust sclerites. *J. insect Physiol.* **20**, 631–635 (1974).
7. Hill, A. V. The dimensions of animals and their muscular dynamics. *Sci. Prog.* **38**, 209–230 (1950).
8. McMahon, T. Size and shape in biology. *Science,* **179**, 1201–1204 (1973).
9. Mc Mahon, T. Allometry and biomechanics: limb bones in adult ungulates. *Amer. Nat.* **109**, 547–563 (1975).

10. Roark, R. J. "Formulas for Stress and Strain." McGraw-Hill, New York (1965).
11. Schneider, H. J., King, A. Y., Bronso, J. L. and Miller, E. H. Stress injuries and developmental changes of lower extremities in ballet dancers. *Radiology*, **113**, 627–632 (1974).
12. Thayer, C. W. Adaptive features of swimming monomyarian bivalves (Mollusca). *Forma Et Functio*, **5**, 1–32 (1972).
13. Timoshenko, S. and Goodier, J. N. "Theory of Elasticity." McGraw-Hill, New York (1951).
14. Wainwright, S. A., Biggs, W. D., Currey, J. D. and Gosline, J. M. "Mechanical Design in Organisms." Edward Arnold, London (1976).

11. The Scaling and Mechanics of Arthropod Exoskeletons

HENRY D. PRANGE*

Department of Zoology, University of Florida, Gainesville, Florida 32611, U.S.A.

ABSTRACT

Measurements of animals ranging in body mass from 3 to more than 1000 mg indicate that the external linear dimensions of the body and limbs of the cursorial spider, *Lycosa lenta*, and the cockroach, *Periplaneta americana*, all scale very closely to the one-third power of body mass. In contrast to the relationship commonly found for the leg bones of vertebrates, the outside diameter of a leg segment of the exoskeleton does not increase, relative to its length, as body size increases.

The consequences of this scaling of the thickness of the leg and of the body mass to length suggest that mechanical similarity based upon a buckling model cannot be maintained in animals with exoskeletons. Calculations of the actual critical buckling load of the leg relative to the body weight support this conclusion.

The arthropods appear to maintain mechanical similarity in terms of bending moment or torque as long as it can be assumed that these quantities scale in proportion to body mass, not mass × length, which would give the same results as for buckling. Models based on this similarity also agree closely with the scaling of some endoskeletal elements of vertebrates such as the wing bones of birds. Possible explanations for these departures from expected mechanical similarities are discussed.

When consideration is given to the different types of stresses which may be important in the scaling of support structures and to the scaling of body mass to linear dimension it is possible to account for the observed scaling of skeletal mass to body mass.

McMahon[7] has suggested that the long bones of the vertebrates should be scaled to resist elastic buckling in a manner similar to a self-loaded column. In order to maintain this structural similarity, the length (l) and

*Present address: Physiology Section, I.U. School of Medicine, Bloomington, Indiana 47401.

G

the thickness (d) necessarily had to scale so that $d \propto l^{1.5}$. From this relationship he suggested that the body mass (M) should scale to the overall linear dimension so that $M \propto l^4$.

Currey[3] showed that the dimensions of several arthropods' legs were such that their legs too should fail by buckling. The scaling of the dimensions of the limbs of arthropods was discussed by Huxley[5] in his classic work on relative growth but his concern was largely with examples of sexual dimorphism rather than with those more related to structural support. I have examined the scaling of the dimensions of the limbs and body mass of arthropods to learn if the effects of the presumed mechanical criteria for scaling could also be found among animals with exoskeletons.

Among endoskeletal animals, development of structures may begin well before they are employed. For this reason ontogenetic scaling may not be a reliable means to ascertain the structural requirements of vertebrates. Among arthropods scale effects derived from structural criteria may be more readily observed in a single species. During their growth many arthropods moult through several instars; between each the rigid skeleton is cast off or absorbed. One might assume that, in the economy of evolution, the

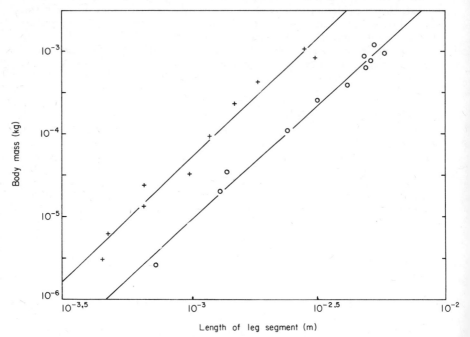

FIG. 1. Scaling of body mass to length of leg segment. For the wolf spider, *Lycosa lenta* (O), mass (kg) = 2485·3 (length, m)$^{2.820}$. For the cockroach, *Periplaneta americana* (+), mass (kg) = 32590 (length, m)$^{2.943}$. For both regression equations the correlation coefficient, r, is 0·99.

exoskeleton of each instar would fit specifically the design criteria necessitated by the stresses imposed by the body mass of that size only.

With this in mind, I made various measurements on instars of two species of cursorial arthropods, the wolf spider, *Lycosa lenta* (body mass range: 2·6 to 1218·3 mg), and cockroach, *Periplaneta americana* (body mass range: 3·1 to 1075·0 mg). Within each species I compared the length and width of the body (cephalothorax in the case of the spider) and the length and width of analogous leg segments of a walking leg. The scaling of all these measurements (Table I)

TABLE I. Scaling relationships of body and leg dimensions (m) where $y = ax^b$. The leg segment used in the case of the spider was the metatarsus of the third leg and in the case of the cockroach was the tarsus of the second leg.

Animal	y	a	x	b
Lycosa lenta	cephalothorax width	0·696	cephalothorax length	1·026
	leg segment length	0·466	cephalothorax length	1·039
	leg segment width	0·174	leg segment length	0·940
	wall thickness	0·0096	leg segment length	0·983
Periplaneta americana	body width	0·355	body length	0·959
	leg segment length	0·082	body length	0·944
	leg segment width	0·186	leg segment length	0·997
	wall thickness	0·0146	leg segment length	1·379

was essentially isometric (i.e. geometrical similarity was maintained). The relation of the length of the leg segment to body mass is shown in Fig. 1. In the case of the spider, $M \propto l^{2\cdot82}$; in the case of the cockroach, $M \propto l^{2\cdot94}$. Therefore, based upon the external dimensions, the body mass of these arthropods scales more nearly to l^3 (geometric scaling) than to l^4. This finding together with the isometry of the length and width of the leg segment suggests that buckling similarity is not maintained by these arthropods.

At this point I would like to leave arthropods briefly to mention another case of seemingly anomalous scaling I have found: that of avian long bones. For most vertebrates the long bones scale so that the width is proportional to the length raised to some power between 1·0 and 1·5. This relationship is maintained by the leg bones of birds; for example, in my measurements of avian femurs, $d \propto l^{1\cdot12}$. The wing bones of flying birds scale rather differently. Measurements from the same sample, which covered a range of body masses from 0·0036 to 10·773 kg, indicate that the avian humerus scales so that $d \propto l^{0\cdot81}$. Clearly these bones also do not scale in accord with the assumption of elastic buckling similarity.

Together with the data from the arthropods, the data on the bird wing bones led me to attempt to reconsider the effects of scaling of body mass on the dimensions of the support structures from the point of view that modes of failure other than buckling might be more important. Not all structures are subjected primarily to axial loading. There are many cases where the characteristic type of stress is bending or torsion.

These stresses should be related to body weight and dimensions. The scaling of the dimensions of support structures should then be a function of the type of stress imposed on them and the scaling of the body mass. It should be possible to examine the effects of scaling the dimensions involved on the equations which describe some of the more basic mechanical principles.

To allow generality, some assumptions about the systems concerned are necessary. First, that mechanical similarity or a constant safety factor is maintained for the type of stress the structure is scaled to withstand; second, that the elastic or shear modulus, whichever applies, remains constant within a series of similar structures from similar animals; and third, that one may use a characteristic linear dimension (l) to describe the scaling of both the body mass (M) and the thickness (d) of a skeletal element so that ($M \propto l^x$) and ($d \propto l^y$).

Thus, as some initial linear dimension l_0 is varied by a factor n, so that the resultant linear dimension $l_1 = nl_0$, the mass changes so that $M_1 = n^x M_0$; and the thickness changes so that $d_1 = n^y d_0$. In this notation the subscript $_0$ will indicate the initial set of dimensions and the subscript $_1$ indicates the set of dimensions which results after the initial set is scaled in accord with the assumptions given. Fitting these variables into the equations for the various kinds of stress allows one to solve for the relationship between x and y which must obtain to satisfy each particular mechanical similarity.

For the case of elastic buckling, the critical buckling load P_{cr}, is described by the equation

$$P_{cr} = \pi^2 EA/(l/K)^2$$

where E is the elastic modulus, A is the cross-sectional area, and K is a value derived from the area second moment, I, so that $K = (I/A)^{1/2}$. For mathematical simplicity I have solved these relationships first for the case of a solid rod rather than a tube. Some of the consequences of this simplification will be discussed later.

In this case, $A = \pi d^2/4$, $I = (\pi/4)(d^4/64)$ so that $K = (d^2/16)^{1/2}$; when these values are substituted into the buckling equation, we have

$$P_{cr_0} = \frac{\pi^3 E d_0^4}{16 l_0^2} \quad \text{and} \quad P_{cr_1} = \frac{\pi^3 E (n^y d_0)^4}{16 n^2 l_0^2}.$$

Since, to maintain mechanical similarity, we suppose that critical load is proportional to body mass, we have

$$P_{cr_1} = n^x P_{cr_0},$$

so that

$$\frac{n^x \pi^3 E d_0^4}{16 l_0^2} = \frac{\pi^3 E n^{4y} d_0^4}{16 n^2 l_0^2}.$$

Cancelling out similar terms, we find that

$$n^x = n^{4y-2}.$$

Solving for y gives

$$y = \frac{x + 2}{4}.$$

In cases where bending stresses may be more important than axial loading it might be assumed that the criterion of similarity would be a deflection, under the load of the body weight, which is proportional to length. At the end of a beam the deflection may be calculated from the equation

$$z = \frac{F l^3}{3EI}$$

where F is the load on the beam and z is the deflection. Assuming that load scales as does body mass and that z is proportional to l, a derivation similar to that used for the buckling case may be employed. Hence

$$F_1 = n^x F_0 \quad \text{and} \quad z_1 = n z_0.$$

Since

$$I_0 = (\pi/4)(d^4/64),$$

then

$$I_1 = (\pi/4)(n^{4y} d^4/64).$$

This relationship may be more simply written as

$$I_1 = n^{4y} I_0.$$

The equation for scaling x and y becomes

$$\frac{n F_0 l_0^3}{3EI_0} = \frac{n^x F_0 n^3 l_0^3}{3E n^{4y} I_0}.$$

Cancellation of similar terms and rearrangement gives

$$n^{4y} = n^{x+2} \quad \text{or} \quad y = \frac{x + 2}{4}.$$

A similar analysis for loading in the form of torsion may be performed. The assumption of similarity in this case is that the angle of torsion will remain constant. Torque may be calculated from the equation

$$T = \frac{G\phi\pi d^4}{2l},$$

where T is torque, G is the shear modulus and ϕ is the angle of torsion. Torque is the product of the load (F) and the length (l) of the lever arm (see discussion of Pennycuick's analysis[9] of the forces on a bird wing discussed below) supporting the load. F and l are assumed to scale as in the previous case so that,

$$T_0 = F_0 l_0 = \frac{G\phi\pi d_0^4}{2l_0}$$

and

$$T_1 = n^{x+1} F_0 l_0 = \frac{G\phi\pi n^{4y} d_0^4}{2n l_0}.$$

The similarity becomes

$$\frac{n^{x+1} G\phi\pi d_0^4}{2l_0} = \frac{G\phi\pi d_0^4 n^{4y}}{2n l_0}.$$

Cancellation and rearrangement again give

$$n^{x+2} = n^{4y} \text{ or, } y = \frac{x+2}{4}.$$

Thus, to maintain similarity under the conditions described the values of x and y should follow the same relationship irrespective of the type of loading.

In Fig. 2 the measured relationships of x and y are plotted along with the line for the equation $y = (x + 2)/4$ which was derived for the three types of loading. The line $y = (x + 1)/4$ is also plotted (see below).

The leg bones of birds (femur and tibiotarsus) scale so that the mass of the animal is proportional to $l^{2 \cdot 80}$. From buckling similarity the predicted value for the exponent of l which describes the scaling of d should be 1·20. For these bones the actual values are 1·12 and 1·13. Avian leg bones therefore scale roughly as would be expected for buckling (or bending) similarity which is reasonable since they are normally loaded either axially or by bending.

The lengths of the wing bones I measured scale to body mass so that $M \propto l^{2 \cdot 05}$. Greenewalt[4] reports similar scaling. The predicted relationship for similarity would be $d \propto l^{1 \cdot 01}$; the actual values I measured for the exponent were 0·81 and 0·80. Pennycuick[9] analyzed the stresses on the wing bones of a pigeon and found that, during flight, the humerus was subjected primarily to torsion and the ulna primarily to bending. The scaling of these elements, like that of the arthropod leg segments, is apparently anomalous in comparison with the initial assumptions.

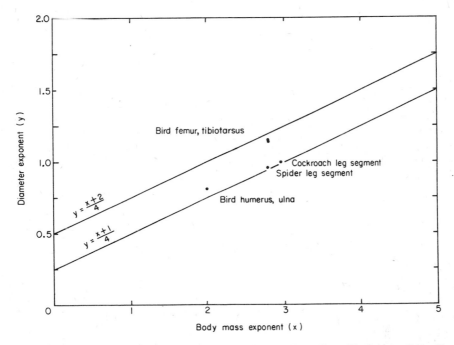

FIG. 2. Relationship of exponents of linear dimension for the scaling of body mass ($M \propto l^x$) and diameter of the skeletal element ($d \propto l^y$). For maintenance of buckling, bending or torsional similarity, $y = (x + 2)/4$.

If one chooses as a criterion of similarity that bending moment should scale with body *mass*, rather than mass × length, then the following derivation may be obtained. Bending moment (B_m) may be calculated from the equation

$$B_m = EI/R$$

where R is the radius of curvature at the point where the bending moment is calculated. With the assumption that R is proportional to l, and that the pertinent assumptions of the arguments above still obtain,

$$B_{m_0} = \frac{E\pi d_0^4}{64R},$$

$$B_{m_1} = \frac{E\pi(n^y d_0)^4}{64nR}$$

and

$$B_{m_1} = n^x B_{m_0}.$$

So

$$n^{4y-1} = n^x \qquad \text{and} \qquad y = \frac{x+1}{4}.$$

The formula given above for torque may be used for this derivation as well. In this case $T \propto F$ rather than $T \propto Fl$. As with bending moment,

$$T_0 = \frac{G\phi\pi d_0^4}{2l_0},$$

$$T_1 = \frac{G\phi\pi(n^y d_0)^4}{2nl_0}$$

and

$$T_1 = n^x T_0$$

So,

$$n^{4y+1} = n^x, \qquad \text{and} \qquad y = \frac{x+1}{4}.$$

This line has also been plotted in Fig. 2. When it is compared with the data it appears that the leg segments of arthropods (and the wing bones of birds) may scale as if to maintain a similarity of bending moment or of torque to body mass. It is not intuitively obvious why this scaling might be selected and it deserves further analysis.

Although the scaling of the external dimensions of the leg segments of either arthropod does not conform to that expected for buckling similarity, the tubular nature of the structure provides other possibilities for mechanical scaling. The diameter of the leg segment is isometric with its length and therefore does not contribute to scaling for buckling similarity; however, some increase in resistance to buckling can be gained from relative increases in the wall thickness. As can be seen from Table I, the wall thickness of the spider leg is also isometrically scaled to the length. There remains another possibility in this case for increased strength in the leg segment. Under the dynamic conditions of locomotion, the hydrostatic pressure in the hemolymph may rise as high as 60 kN m^{-2} (450 mmHg: see Parry and Brown,[8] Anderson and Prestwich[2]). This pressure might add increased rigidity if it causes the leg to act transiently as a hydrostatic skeleton during the peak stresses of activity. I tested this possibility experimentally and found that, when any effect could be detected, it provided an increase in strength of no more than 5% over the zero pressure condition. It appears that the mechanical reinforcement from the high hydrostatic pressures of the spider's hemolymph may only be in the resistance to kinking of the segment during severe bending, as suggested by Currey.[3]

The outside diameter of the leg segment of the cockroach is isometric with the length but the wall thickness shows a relative increase in scaling. In this case the assumption that the case for the solid rod is generally applicable to the scaling of a tubular structure is not correct. I have, therefore, derived the scaling requirements for wall thickness (t) necessary to maintain the mechanical similarities used in the first analysis. In this argument r_2 is the outside radius of the tube and r_1 is the inside radius. In this case

$$A = \pi(r_2^2 - r_1^2),$$

and

$$I = (\pi/4)(r_2^4 - r_1^4).$$

Since

$$r_1 = r_2 - t,$$

this value may be substituted into the equation for buckling to give

$$P_{cr} = \frac{\pi^3 E[r_2^4 - (r_1 - t)^4]}{4l^2}.$$

Assuming that

$$P_{cr_1} = n^x P_{cr_0}, \quad r_{2_1} = nr_{2_0} \quad \text{and} \quad t_1 = n^y t_0,$$

then,

$$\frac{E\pi^3[(nr_{2_0})^4 - (nr_{2_0} - n^y t_0)^4]}{(nl_0)^2} = \frac{n^x E\pi^3[r_{2_0}^4 - (r_{2_0} - t_0)^4]}{l_0^2}.$$

Cancelling and rearranging terms gives

$$n^x = \frac{n^4 r_{2_0}^4}{n^2[r_{2_0}^4} \frac{(nr_{2_0} - n^y t_0)^4}{- (r_{2_0} - t_0)^4]}.$$

A similar result is obtained for bending or torsion, when the moment is proportional to mass × length.

This is a difficult equation to solve generally for y in terms of x. However, since y and x can only have a limited range of values which are compatible with real animals, I have solved the equation empirically within that range. For comparison the solution for similarity of bending moment and torque, taken to be proportional to *mass*, have also been derived and both are given graphically in Fig. 3. The solutions vary somewhat as a function of the ratio r_2/t employed. In figure 3 the range shown for the ratio is from 5 (dashed line) to 15 (solid line). This includes the range found in the animals studied.

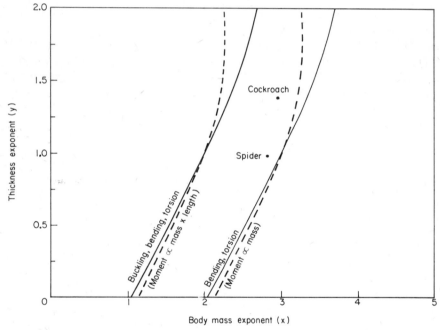

FIG. 3. Relationship of exponents of linear dimension for the scaling of body mass ($M \propto l^x$) and the thickness of the wall of a tubular support element ($t \propto l^y$) where the outside diameter is directly proportional to the linear dimension ($d \propto l^{1 \cdot 0}$). See text for derivations and equations. Broken lines: $r_2/t = 5$; solid lines: $r_2/t = 15$.

Some important conclusions may be drawn from comparison of the data with these derivations. In the case of the cockroach, although there is an increase in the relative wall thickness, it is not sufficient to allow it to scale for buckling similarity. It appears that, so long as the outside radius scales isometrically with the length of the leg segment, it is impossible for the animals within the range of r_2/t ratios studied to achieve buckling similarity with any reasonable value for body mass scaling.

At very large values of r_2/t it may be possible to approach buckling similarity. This condition would leave the animal with legs which were either much larger in diameter than those of the animals studied or much thinner in wall thickness. The former case would give rather bulky limbs; the latter case would give limbs which would be very weak. Values for a thickness exponent (y) greater than 2 give a wall which fills the lumen of the tube before any significant increase in linear dimension can occur.

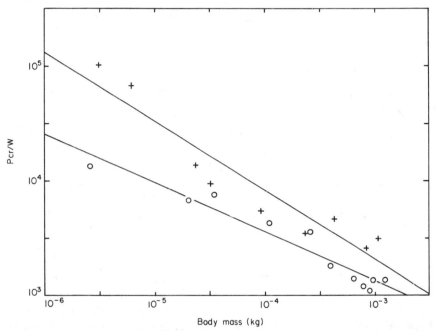

FIG. 4. Scaling of the ratio of critical buckling load (P_{cr}) of the leg segment to body weight (W) relative to body mass. For *Lycosa* (O), $P_{cr}/W = 69\cdot29$ (mass, kg)$^{-0\cdot429}$, $r = 0\cdot97$. For *Periplaneta* (+), $P_{cr}/W = 32\cdot63$ (mass, kg)$^{-0\cdot602}$, $r = 0\cdot95$.

From these arguments one may suggest that the only way that these animals can achieve sufficient resistance to buckling is to scale so that the initial absolute dimensions give extremely large critical buckling loads relative to the body weight and so that the decrease in relative buckling resistance which is inherent in the scaling of their leg dimensions still allows sufficient strength at the maximum body mass. I have used the value for E of 10^{10} Nm^{-2}, approximately that measured for locust cuticle by Jensen and

Weis-Fogh,[6] to calculate the relative critical buckling loads for the spider
and cockroach leg segments. These data are shown in Fig. 4. The maximum
values are very large and the minimum values, found, as suggested, for the
largest animals, appear to be sufficient to withstand any realistic loads which
one might assume to be derived from forces related to the body mass.

Perhaps the high minimum buckling strengths can be explained by the
observation that, to resist a given load in the bending mode, a column must
be more robust than it would have to be to withstand that load applied
axially. Hence, if the absolute requirements for bending resistance determine
the initial dimensions of the animal, the absolute requirements for buckling
strength may be surpassed throughout the range of body size even though
buckling similarity is not maintained. However, for this to be relevant,
bending moment would have to scale as body *mass*, not mass × length.

TABLE II. Scaling of skeletal mass to body mass. Where M_s is skeletal mass and M is
body mass, $M_s = aM^b$. The data on mammals are derived from 42 values collected
from the literature by John F. Anderson and myself; the data on birds were obtained
from our own measurements on 311 specimens; snake data are from Prange and
Christman,[10] and spider data are from Anderson.[1]

Animal	a	b	mass unit
Mammals	0·061	1·090	kg
Birds	0·065	1·071	kg
Snakes	0·020	1·174	g
Spiders	0·033	1·116	mg

Given the observation that the scaling of skeletal diameters and the body
mass can be related through their respective scaling to linear dimension, it
should be possible to predict the scaling of the skeletal mass (M_s) to the body
mass (M). The elements of the skeleton should scale, individually or in
combination, to maintain similarity to buckling, bending and torsional
stresses. In any case the diameter will scale in proportion to some power of
l between 1·0 and 1·25. If the skeleton is presumed to have a relatively constant
density and to scale its volume in proportion to ld^2, then the mass of the
skeleton should scale in proportion to l raised to some power between 3·0
and 3·5. Since the body mass is commonly found to be proportional to l^3, M_s
should be proportional to body mass raised to a power between 1·0 and 1·17.
If body mass were proportional to l^4, then M_s would scale to a power of M
between 0·88 and 1·00. Examples of the actual scaling of skeletal mass to
body mass are given in Table II. These data fit exactly into the range predicted when the body mass scales to l^3.

If one begins with the observation that the body mass scales generally to the third power of linear dimension, then it follows that d must be proportional to $l^{1.0-1.25}$. This seems to be the case in most animals: the scaling of the body mass determines the scaling requirements of the support structures. If, on the other hand, one assumes that organisms scale as does a self-loaded column, where the support structure also makes up the load, then d will scale in proportion to $l^{1.25-1.5}$ and the body mass determined by the scaling of the support structure, will scale to l^4. There is evidence from which to believe that this latter case may obtain for trees[7] but from my analysis it does not appear to do so for animals.

ACKNOWLEDGMENT

I am pleased to acknowledge the considerable assistance of my colleague, John F. Anderson, who collected the specimens and made many of the measurements in this study. Professors John D. Currey and T. J. Pedley have generously given me the benefit of their consultation from which I have learned much.

REFERENCES

1. Anderson, J. F. "Metabolic Rates in Spiders". Ph.D. Dissertation, University of Florida (1968).
2. Anderson, J. F. and Prestwich, K. N. The fluid pressure pumps of spiders. *Z. Morph. Tiere.* **81**, 257–277 (1975).
3. Currey, J. D. Failure of exoskeletons and endoskeletons. *J. Morph.* **123**, 1–16 (1967).
4. Greenewalt, C. H. Dimensional relationships for flying animals. *Smithsonian Misc. Coll.* **144** (2), 1–46 (1962).
5. Huxley, J. "Problems of Relative Growth", 2nd ed. Dover Publications, New York (1972).
6. Jensen, M. and Weis-Fogh, T. Biology and physics of locust flight. V. Strength and elasticity of locust cuticle. *Phil. Trans. Roy. Soc.* **B 245**, 137–169 (1962).
7. McMahon, T. Size and shape in biology. *Science,* **179**, 1201–1204.
8. Parry, D. A. and Brown, R. H. J. The hydraulic mechanism of the spider leg. *J. Exp. Biol.* **36**, 423–433 (1959).
9. Pennycuick, C. J. The strength of the pigeon's wing bones in relation to their function. *J. Exp. Biol.* **46**, 219–233 (1967).
10. Prange, H. D. and Christman, S. P. The allometrics of rattlesnake skeletons. *Copeia.* **3**, 542–545 (1976).

Comments on Dr Prange's paper

JAMES LIGHTHILL

Dr Prange has demonstrated a close approach to geometrical similarity among the skeletons from both the species studied. I want to emphasize that this is incompatible with any possibility that static stresses (stresses produced by gravity) are limiting.

Dr Prange proved this for buckling, bending and torsional stresses. Furthermore, in geometrically similar skeletons, the bending moments on limbs generated by the gravitational force mg on an animal of mass m must vary as mgl, leading to bending *stresses* varying as mgl^{-2}. Compressive stresses must, equally, take this form. In fact, the only quantity with the dimensions of stress that can be produced from the gravitational force mg and the animal's characteristic dimension l is mgl^{-2}. This means that in geometrically similar systems *every* static stress must be proportional to mgl^{-2}, and so, with density constant, to l. Such an increase of all static stresses in proportion to linear dimension throughout a series of skeletons of widely different dimensions would be incompatible with the view that static stress is limiting.

The only other stresses that might limit the skeletal design are dynamic stresses. These depend on the animal's mass m (instead of on its weight mg) and on some acceleration, f. The dynamic stress varies as mfl^{-2}. This could be limiting in a series of geometrically similar skeletons if they were designed against maximum accelerations f proportional to l^{-1}.

Note that this conclusion is precisely that derived from dimensional analysis in Wilkie's paper (pp. 34 and 35). The stress he mentions there is a dynamic stress (since he takes the animal's mass rather than its weight as fundamental). Thus, he establishes that geometrically similar systems in which the maximum dynamic stress takes some limiting value fixed by the properties of the material must have accelerations scaling as l^{-1}. He shows also that any time t then scales as l, while any velocity v is scale-independent.

(Note that if v is scale-independent, then accelerations proportional to v^2/l must scale as l^{-1}).

It follows that isometric skeletal scaling, as observed by Dr Prange in these arthropods, would be consistent with dynamic stresses being limiting if and only if characteristic velocities v were scale-independent. Wilkie emphasizes running velocities in this context. Another possibility, however, is that the skeletons of all the different instars are designed against damage by a hard external object striking the animal at velocities up to some particular constant value, v. Such a value might have emerged as a compromise between the selective advantages of (i) immunity against a large proportion of naturally occurring external-impact threats and (ii) retention for essential nonskeletal functions of some large proportion of the animal's total mass.

12. Scale Effects in Jumping Animals

H. C. BENNET-CLARK[*]

Department of Zoology, The University of Edinburgh, Scotland.

ABSTRACT

The height to which an animal can jump is proportional to the energy per unit mass that it can produce; superficially, it appears that all animals should jump to the same height. In smaller animals, the distance and the time of acceleration are less and so they must produce more power to reach the same height; below a body length of about 1 metre it is more economic to store the muscular energy before the jump and to release it more rapidly and at higher power during the jump impulse.

With decreasing size, the accelerations become greater and so greater forces are required from the muscles; this necessitates changes in the muscle geometry, in the shape and in the strength of the skeleton. As a result, the inertial losses tend to be relatively greater in smaller jumping animals.

As the time available for acceleration is less in smaller animals, the energy store must be able to deliver the energy more rapidly; the choice of storage material depends on its dynamic losses rather than on its storage capacity and the weight penalty of energy storage is usually negligible.

The effect of air resistance is greater in smaller animals and becomes the dominant influence in the smallest jumping animals; air resistance absorbs about 10% of the energy in a jump 1 m high by an animal 0·1 m long but absorbs 50% of the energy in an 0·1 m jump by an animal 1 mm long. The jump of the smallest animals permits them to move a greater proportion of their body length than that of larger animals but is associated with greater specialisation in the jumping mechanism.

INTRODUCTION: THE PROBLEM OF SIZE AND RANGE

Terrestrial animals of all sizes jump. The range may be many metres for a large mammal to a fraction of a metre for a small insect. In terms of the insect's size, the jump may be both more dramatic and more effective because the range can be over 100 times the animal's length and is hard to follow by eye.

*Present address: Department of Zoology, South Parks Road, University of Oxford, England.

To jump to a given height, h, an animal must accelerate itself upwards to a certain velocity, v, according to

$$h = \frac{v^2}{2g} \tag{1}$$

where g is the acceleration due to gravity and air resistance is neglected. Thus, to jump upwards 1 metre, an animal must move upwards at $4 \cdot 4$ m s^{-1}, whether it be a locust or a leopard. If it can attain twice the speed, it will reach 4 times the height.

A locust 50 mm long moving at such a speed will move through its own length in about $0 \cdot 01$ s and so will be hard to see and hard to catch, while a flea 2 mm long moves its own length in $0 \cdot 0005$ s and being tiny is doubly difficult to catch.

Though the jump of a small insect may appear remarkable, the range is less than that of most mammals, even though the relative amounts of muscle and the performance of the muscle may be very similar.

By definition, a jump results from a single muscular contraction and it is the work that can be done during this contraction that sets the upper limit on the range of all animals regardless of size. With smaller animals, the length of the limbs and hence the time available for acceleration is less. As a result, muscle power cannot be used directly, but must be stored.

Other factors conspire against the small animal. Because of the large accelerations, limitations may be set by the strength of the skeleton. Because of the surface area to volume (and weight) relationship, air resistance may absorb a significant proportion of the energy and set a practical limit on the performance.

THE ENERGETICS OF JUMPING

The height (and range) of a jump is determined by the initial velocity (Eqn (1)) which is in turn determined by the kinetic energy imparted to the animal by its muscles. This energy, E, is given by

$$E = \frac{mv^2}{2}, \tag{2}$$

where m is the animal's mass. Combining equations (1) and (2), one finds that the height of a jump, h, is given by

$$h = \frac{E}{mg}, \tag{3}$$

and so by the energy per unit mass produced by the muscles. If all animals have the same proportion of muscle capable of producing the same specific energy, they should all jump to the same height.

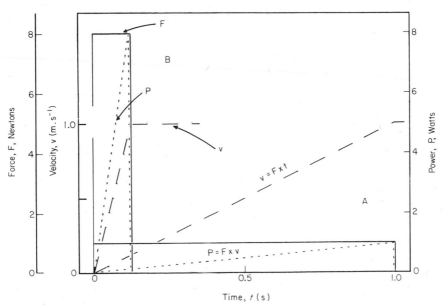

FIG. 1. Diagram to show the relation between force, velocity, power and time when a 1 kg mass is accelerated from rest to a velocity of 1 m s^{-1}. In case A, a constant force of 1 N is applied for 1 s and in case B, a constant force of 8 N is applied for 1/8 s. In both cases velocity and power rise linearly but, in case B, the power is 8 times that in case A.

To minimise the mass of the skeleton, it is advantageous for the force, F, to be constant throughout the jump impulse. As a result, velocity should rise linearly with time, since

$$v = \frac{Ft}{m}. \tag{4}$$

The power produced must also rise linearly (Fig. 1) to a maximum when the animal leaves the ground. This means that the power supply for the jump ideally should produce a constant force over a wide range of velocities; this, too, may not be compatible with direct muscular contraction in small animals.

As animals become smaller, so their limbs become shorter and the distance available for acceleration become less. As a result, the acceleration, a, and force, F, become greater according to

$$a = \frac{F}{m} = \frac{v^2}{2s}. \tag{5}$$

where s is the distance over which acceleration occurs. For this reason, it pays a jumping animal to have long legs as this reduces the forces.

Because of the shorter distance over which acceleration can occur, the time available for the jump impulse is less in a small animal (Fig. 1) and so not only is the force greater (Eqn (5)) but the peak power that must be produced, P, given by

$$P = mav = Fv \tag{6}$$

increases with decreasing size. At a certain size, it is no longer possible for this peak power to be provided by direct muscular contraction.

Because of the limitations set by the length of the limbs, if an animal wishes to jump higher, it must not only do more work (Eqn (3)) but must produce more force and so the peak power required increases as the 3/2 power of the height of the jump (from Eqns (5) and (6)). To jump twice as high, the animal must produce twice the energy and 2·8 times the peak power.

The height of jump, acceleration and specific power output for various jumping animals is given in Table I. Although the height of the jump is less in the smaller animals, the acceleration and power output tend to be greater.

POWER AND ENERGY PRODUCTION BY MUSCLE

Contracting muscle produces force and, by shortening, does external work on a load. Contraction velocity is maximal when the muscle is unloaded and falls to zero at the isometric force. In between, the force-velocity curve is a rectangular hyperbola (Hill[12]) and so neither force nor power vary linearly with velocity. Hill[13] has shown that for typical fast muscles, power output is maximal at about 1/3 isometric force and 1/3 peak velocity of shortening. Thus, if an animal uses its muscles at maximum power, it performs about 1/3 the maximal possible work of a single contraction and is limited to slower than maximal speeds of contraction. This may be of no consequence in a large animal where there is ample time for contraction (Table I) but smaller animals have to contract their muscles more rapidly in order to achieve the necessary acceleration in a shorter time. Furthermore, muscles with the highest contraction velocities have extensive excitation-contraction coupling systems which reduce the effective force produced. Hence the power output of muscle is limited to a value well below the theoretical maximum, and is usually below about 1 kW kg^{-1} (see Table II).

TABLE 1. Jumping performance of various animals.

Animal	Height of jump	Distance of acceleration	Time of acceleration	Mean acceleration	Peak power output	Source of raw data
Mammals						
leopard antelope	2·5 m	1·5 m	0·43 s	$16·0 \text{ m s}^{-2}$	115 W kg^{-1}	Hill[13]
lesser galago	2·25 m	0·16 m	47 ms	140 m s^{-2}	915 W kg^{-1}	Hall Craggs[10]
Insects						
locust adult	0·45 m	0·04 m	26 ms	110 m s^{-2}	330 W kg^{-1}	Bennet-Clark[4]
locust 1st instar	0·17 m	7 mm	8 ms	240 m s^{-2}	430 W kg^{-1}	Bennet-Clark and McGavin, unpublished
rat flea	0·10 m	0·5 mm	0·7 ms	2000 m s^{-2}	$2·75 \text{ kW kg}^{-1}$	Bennet-Clark and Lucey[5]

The energy produced by a muscular contraction is given by the area under the force length curve. The range of contraction is dependent on the relative lengths and overlap of the actin and myosin filaments (Gordon *et al.*[9]) while the peak force depends on the length of the myosin filaments (Huxley and Niedergerke[14]). While the myosin filament length seems to be about 1·6 μm in all vertebrate striated muscles, longer myosin filaments and sarcomeres are found in many arthropods and these muscles produce higher forces and higher energies of contraction (Table II, Fig. 2). The penalty, is that, as there appears to be a maximal rate of formation of cross-links between the actin and myosin filaments, the velocity of shortening for a long sarcomere muscle is lower though its force is higher. Thus, although

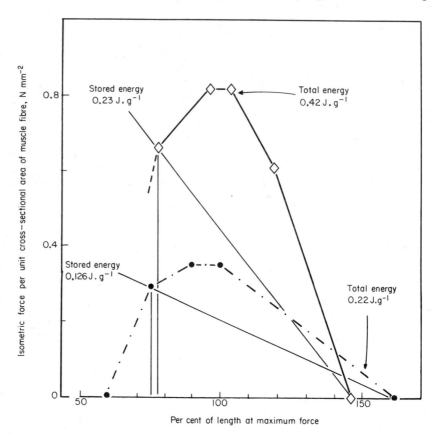

Fig. 2. Diagram of the isometric force–length relationships for two types of striated muscle. The energy produced by a slow contraction is given by the areas under the force-length curves and the maximum storable energy is given by the area of the triangles formed by the linear load lines. ●—·—● frog muscle, data from Gordon et al.[9] ◇—◇ Crayfish muscle, data from Zachar and Zacharova.[21]

TABLE II. Mechanical properties of various muscles.

Parameter and units	Animal and muscle	Temperature	Maximum value	Source
Isometric force, $N\,mm^{-2}$	Rat soleus	38°C	0·32	Wells[20]
	Locust extensor tibiae	30°C	0·8 calculated (1·2 at optimal length)	Bennet-Clark[4]
Maximum velocity, lengths s^{-1}	Rat anterior tibialis	38°C	5·7	Wells[20]
	Locust extensor tibiae	30°C	1·75	Bennet-Clark[4]
Power output in single contraction, $W\,g^{-1}$	Locust extensor tibiae	30°C	0·25 (average) 0·45 (peak)	Bennet-Clark[4]
	Rat anterior tibialis	38°C	0·57 (peak)	Calc. from Wells[20]
	Pigeon pectoralis	40°C	0·86 (est. peak)	Pennycuick and Parker[16]
Energy in single contraction, $J\,g^{-1}$	Frog leg	—	0·22	Calc. from Gordon et al.[9]
	Crayfish leg	—	0·42	Calc. from Zachar and Zacharova[21]
Stored energy, $J\,g^{-1}$	Frog leg	—	0·125	Fig. 2 Bennet-Clark[4]
	Locust extensor tibiae	30°C	0·08–0·15	Bennet-Clark and McGavin unpublished.

the energy of contraction of crayfish or locust muscle may be higher than for rat or frog muscle, the maximum power output is similar.

Since the height of a jump depends on the energy that can be produced, it becomes practicable to use a high energy muscle and to store the energy. This has other energetic advantages. The simplest energy stores are springs which more or less obey Hooke's law, so the strain is proportional to the stress or applied force per unit cross-sectional area of muscle fibre, which can be represented by a linear load line on the force-length curve for the muscle (Fig. 2). In this case, the energy that is stored is given by

$$\text{stored energy} = \frac{\text{peak stress} \times \text{peak strain}}{2} \qquad (7)$$

which is equal to the area of the triangle of which the load line is the hypotenuse. By suitable choice of a load line, over half the energy of contraction can be stored (Fig. 2) which means that the muscle produces $1\frac{1}{2}$ times as much as energy as when it contracts at maximum power (one third of the energy of contraction). It will be shown below that the weight of the energy stores is only a few per cent of the weight of the muscle whose energy is stored, so this mode of operation is energetically attractive.

Table II summarises some of the mechanical properties of muscle and provides a basis from which some of the problems of scaling can be examined.

ENERGY, POWER AND FORCE IN JUMPING

Although there is good circumstantial evidence that small animals store the energy for their jumps (eg. Bennet-Clark and Lucey,[5] Evans,[8] Bennet-Clark[4]) little attention has been given to the scaling of power limitations.

From Eqns (1), (5) and (6), the height of the jump, h, is given by

$$h = \left(\frac{2sp}{m}\right)^{2/3} \cdot \frac{1}{2g} \qquad (8)$$

where s is the distance of acceleration, P is the peak power output of the muscle, m is the animal's mass and g is the acceleration due to gravity. It may reasonably be assumed that s is proportional to the animal's length, l, and that P is proportional to the mass of muscle which is expected to be between 5 and 20% of the animal's total mass. Thus h is proportional to $l^{2/3}$, with a constant of proportionality which depends on the specific power output of the muscle. This is shown graphically in Fig. 3. If we assume a high specific power output of 1 kW kg^{-1} for muscle, a practical upper limit for direct muscle action is given by the 100 W kg^{-1} line; the leopard operates in this condition but the smaller jumping animals use more power than can be supplied by their muscles directly. This suggests that they must store the muscle energy and release it rapidly through a power amplifier.

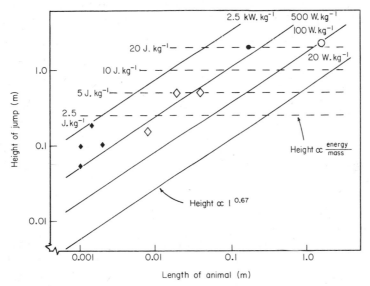

FIG. 3. Graph of height of jump against animal length for various mammals and insects. The graph also shows the required specific energy and specific power to produce the jump assuming that the animals accelerate through a distance equal to their body length and that there is no air resistance. The points show observed jumps for fleas ◆ (original data and from Bennet-Clark and Lucey[5]); for locusts and grasshoppers ◇ (original data and from Bennet-Clark[4]); for the lesser Galago ● (data from Hall-Craggs[10]); (for leopard or antelope ○ (data from Hill[13]).

The energy required for jumping a given height is directly proportional to the mass (Eqn (3)) and is independent of size. This is also shown on Fig. 3 and it can be seen that the jump of both the leopard and the lesser galago are compatible with the energy that could be produced by muscles weighing some 16% of the total body weight. The smaller jumps of the smaller animals suggests that a smaller proportion of the body weight is muscle; the extensor muscles of locust hind legs are only about 5% of the total weight (Bennet-Clark[4]).

Thus it appears that the smallest jumping animals, far from jumping well, actually jump badly and fail to exploit their muscles adequately. Part of this may be attributed to the fact that as the size decreases, the accelerations and forces rise (Table I); whereas a leopard need only produce forces of 1·6 times its body weight, with correspondingly low demands on the skeleton, a flea jumping 1/10 as high produces forces of 200 times its body weight. Not only must this force be withstood by the skeleton but the internal organs must be firmly anchored to withstand the high accelerations.

The requirement for increased force with decreasing size affects the muscle geometry. Whether the energy is produced directly or is stored, the force is transmitted through the activated jumping muscle (see Bennet-Clark and Lucey,[5] Evans[8]). Thus, as an animal becomes smaller, so the force required from unit mass of its muscles becomes greater.

TABLE III. Effect of size on force required from muscle in jumping animals. It is assumed that the jump height is 1 m, that the animal accelerates over this distance and that 1/10 of the body weight is muscle capable of producing 1 N mm^{-2}.

Body length	Body weight	Acceleration	Force produced by muscle	Effective diameter: length ratio of muscle
1 m	50 kg	$1\,g$	100 N kg^{-1}	30:1
1 mm	0·5 mg	$1000\,g$	10^5 N kg^{-1}	1:1

The effect of this is shown in Table III. The force required for a given jump is inversely proportional to size and affects the practical geometry of the animal's muscles.

In the cat family, the muscles are long and thin and operate at a high mechanical advantage so their effective length to diameter ratio is high. In locusts, the jumping muscles are pinnate and the effective length: diameter ratio is 1:3·1 but, since the mechanical advantage of the leg is about 30:1, the final ratio is around 20:1 (calculated from Bennet-Clark[4]). In fleas, which do not jump nearly as high as is suggested in Table III, the length to diameter ratio of the muscles is again about 1:1 and, as they operate at a mechanical advantage of below 10:1, the final ratio is between 5 and 10:1 (calculated from Bennet-Clark and Lucey[5]) and the animal is accelerated at about $100\,g$ (see Table III). The extreme condition is found in click beetles where the final length to diameter ratio is about 2:1 (calculated from Evans[8]) and the acceleration is $400\,g$.

The high force requirement has effects on skeletal design which may restrict the performance of small animals. Because of the large area of muscle insertion, large specialised areas of strong skeleton are required, which increases the weight of the skeleton. Because of the increasing forces on either side of the extending limb, this must be made relatively stronger and heavier with decreasing size. This is seen in the click beetle *Athous* which jumps to a height of 0·3 m by bending its body in the middle over a period of about 0·6 ms and a distance of 0·75 mm (Evans[7]). The two parts of the body have an energy of rotation but since this does not contribute to the jump impulse, it is wasted.

In this case, the body can be regarded as a very short strong extending limb; about 40% of the stored muscle energy is lost in rotating the body (Evans [8]). In a locust, where the time of acceleration for a similar jump is 26 ms and the distance over which the impulse is given is 40 mm, a long tubular limb extends with a total energy loss of about 1 per cent of the total (Bennet-Clark[4]). In the first case, the losses approach a level where they become unacceptable and in the second case they are negligible. Such losses clearly reduce the economic height of jump in a small animal.

THE EXPLOITATION OF STORED ENERGY

Energy storage offers the possibility of the rapid release of the energy of contraction of a slowly contracting muscle with an increase in power that can be produced. On release, the force available from the store falls as energy is returned (Eqn 7). When such a store is coupled to a limb in which the mechanical advantage falls as it extends, the force produced at the end of the limb may increase and the power rise during extension (Fig. 4). This approaches the ideal suggested on p. 187 (Fig. 1).

Because the force in the energy store is highest before the release of energy, the safety factors in the store and associated skeletal elements can be lower than where energy is produced directly and the muscle force may rise abruptly if the animal slips or strikes an obstacle. In a locust which stores energy for jumping in the apodeme of the extensor tibiae muscle and semi-lunar processes from which the hind tibia is suspended, the safety factor in these regions is about 1·2 (Bennet-Clark[4]). In a pigeon, which uses direct muscle power to fly, the safety factor of the flight muscle tendons is 5·1 and of the wing bones 3·7 (Pennycuick and Parker[16]).

My recent estimates of the strength of the locust tibia, based on the thickness of the cuticle, its known tensile strength (Jensen and Weis-Fogh[15]) and the bending moment to which it is subjected in jumping, suggest that the safety factor is between 1·5 and 2 along the whole of its length. Such a low safety factor is unexpected but has the advantage of minimising the weight of the skeleton and the inertial losses in the system. Locusts and grasshoppers with only one hind leg are often found—they jump half as far—so skeletal failure is less disastrous than it would be in a bird.

One of the functions of energy storage is to produce a more rapid movement than could be produced directly. The concomitant of this is that the

antagonist muscles will be extended rapidly during the jump; I have calculated that the flexor tibiae muscle of the locust extends at 30 lengths s^{-1} during
the jump and in a kick can be as much as 100 lengths s^{-1} (Bennet-Clark[4]).
This is far faster than the contraction velocity of the muscle and it seems
likely that if the activated muscle was forcibly extended at this rate, it would
be damaged or, at best, absorb a part of the jump energy. The flexor muscle
in the locust is provided with a catch (Heitler[11]) which does not release to
allow extension of the leg in jumping until the muscle is nearly fully relaxed
so extension is antagonised only by the relaxed flexor muscle. Other types
of catches have been reported for click beetles (Evans[7]) and fleas (Bennet-
Clark and Lucey,[5] Rothschild *et al.*[17]) and it is likely that they exist in all
jumping insects.

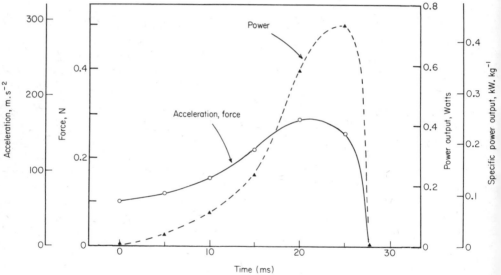

FIG. 4. Graph of force and power against time for the jump impulse of an adult locust weighing
1·7 g. The energy is stored before the jump at a lower power. The power and force both rise
during the impulse; this may be compared with Fig. 1.

Catches have not been described in jumping mammals and, where energy
storage occurs, the limb extension appears to be prevented by the direct
action of the flexor muscles and initiated by relaxation of the flexors. Since
the time available for this relaxation is probably more than the relaxation
time of the muscles (see Table I), a catch is probably unnecessary and thus
is only found in smaller animals where the period of leg extension is shorter
than that of the relaxation of the muscles.

PROPERTIES OF ENERGY STORES

An ideal energy store for muscular energy would be matched to the force-length characteristics of the muscle, would return all the energy and would impose no weight penalty. The geometry of the energy store should be such that it does not require elaborate specialisations of the more rigid skeletal elements to which it is attached.

TABLE IV. Mechanical properties of energy stores

Material	Tendon collagen	Locust apodeme	Resilin	Steels
Occurrence	Vertebrates	Insect	Insect	
Ultimate tensile strength	$100 \, \text{N mm}^{-2}$	$600 \, \text{N mm}^{-2}$	$3 \, \text{N mm}^{-2}$	450–2700 N mm^{-2}
Young's Modulus	$1000 \, \text{N mm}^{-2}$	1.9×10^4 N mm^{-2}	$2 \, \text{N mm}^{-2}$	2.1×10^5 N mm^{-2}
Elastic elongation	10% plus	3.2%	140% plus	0.45–1.3%
Specific energy storage	$5 \, \text{J g}^{-1}$ plus	$7.5 \, \text{J g}^{-1}$	$2.1 \, \text{J g}^{-1}$ plus	0.125– $1.4 \, \text{J g}^{-1}$
Source	Alexander[3]	Bennet-Clark[4]	Jensen and Weis-Fogh[15]	Various

The simplest and most economical way to use a material as an energy store is in straight tension. This can be seen in the leg tendons of many mammals, notably in the kangaroo which has a cylindrical Achilles tendon, 0·35 m long and 15 mm in diameter, used in the bouncing locomotion to conserve the kinetic energy of the up-and-down motion (Dawson and Taylor[6]). The collagen tendons store a total of about 800 J (see Table IV) which is sufficient to store the energy of a 2 m high jump or to accelerate the 40 kg animal to a speed of $6.3 \, \text{m s}^{-1}$. In locusts, the extensor apodeme or tendon of the hind tibia is a long thin strap weighing 0·5 mg per leg which stores a large part of the jump energy (Bennet-Clark[4]). Since this strong cuticle is stiffer than collagen (Table IV), its use necessitates a high-force pinnate muscle and a high mechanical advantage in the femoro–tibial articulation. However, in both cases, the weight of the store is small, 0·4% of the weight of the kangaroo and only 0·3% of the weight of the locust.

As the weight penalty of energy storage is so small, it is sometimes convenient to adopt a store whose geometry is less efficient but which allows simplifications of skeletal or muscular design. The semilunar processes of the locust hind leg, from which the tibia is suspended, act in bending as energy stores for the jump. In bending, shear forces become appreciable and since the shear modulus of a material is less than its tensile or Young's modulus (see Alexander[1]), less energy can be stored; 5 mg of the highly sclerotised cuticle of the semilunar processes store a similar amount of energy to the 1 mg of linearly oriented cuticle of the extensor apodeme (Bennet-Clark[4]). However this small penalty is offset by the fact that a relatively long movement is obtained from a short length of a stiff material. Similar considerations apply to the use of helical and spiral steel springs (see Table IV) to tailor the force–length relationships of the store to the task.

If the material of the energy store has a low Young's modulus, so that it is relatively soft, it may also be convenient to use it in short thick pads in shear. This is seen in fleas, where the muscle forces tend to be high and where, because of its rapid release of energy, the protein resilin (Weis–Fogh[19]) is used as the energy store (Bennet-Clark and Lucey[5]). The advantage of this may be that the rigid skeletal elements on either side of the store can be relatively simple compression members and that, since the shear modulus is less than the tensile modulus, smaller demands are placed on the junction between the rigid and compliant regions; the engine mountings of motor cars use rubber in shear in blocks which have metal to rubber bonds and to which similar design considerations apply.

Thus practical energy stores depart from the ideal. The choice of the material depends also on the losses that can be accepted. In a mammal, the time available for the impulse is long (Table I) and a strong fibrous hydrated protein, which probably has a high loss factor at high rates of recovery, is acceptable. In insects, the material may be highly sclerotised, stiff, brittle cuticle, as used by locusts, where up to 35% of the energy may be lost in a kick lasting 3 ms but only 5% is lost in the normal jump impulse which lasts 26 ms (Bennet-Clark[4]). In fleas and click beetles, where the jump impulse lasts less than 1 ms, the energy is stored in resilin (Bennet-Clark and Lucey,[5] Sannasi[18]). Resilin has been tested at up to 200 Hz and, above about 50 Hz, the energy absorbed does not exceed 5% (Jensen and Weis-Fogh[15]) so in this respect, resilin is an ideal storage material for small insects.

The material and the geometry of the energy store appear to be controlled by biological as well as by engineering considerations. Resilin does not occur in vertebrates, but elastin may have a similar function since it has a low Young's modulus and its use appears to be confined to short thick ligaments and blood vessels. Again, resilin is a protein and so will be attacked by the moulting fluid at ecdysis while sclerotised cuticle will be less affected, leaving the insect incapacitated for a shorter period at each moult; resilin, with all its advantages, only appears to be used in small adult jumping insects.

EFFECTS OF AIR RESISTANCE

A small animal of the same proportions has a larger surface area per unit mass than a large animal. This means that the aerodynamic drag, which may be negligible in a large animal, becomes a dominant factor in a tiny one (for reviews, see Alexander[2, 3]).

When account is taken of the air resistance, the equation of motion of the animal becomes

$$m\frac{\mathrm{d}^2 x}{\mathrm{d}t^2} + \tfrac{1}{2}\rho C_D A \left(\frac{\mathrm{d}x}{\mathrm{d}t}\right)^2 + mg = 0 \tag{9}$$

where x is the distance of the centre of mass above the ground, A is the

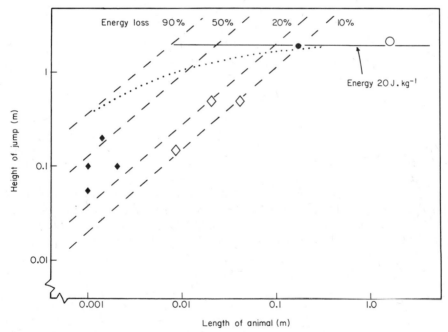

FIG. 5. Graph of the height of jump against animal length to show the effect of air resistance. The dashes show the calculated height for a given energy loss to air resistance, the continuous line shows the height of jump from a specific energy of 20 J kg⁻¹ before allowance is made for air resistance, while the dotted line shows the height when account is taken of air resistance. The points show the same data as in Fig. 3.

frontal area of the animal, m is its mass, ρ is the air density, C_D is the drag coefficient and g is the acceleration due to gravity. The height of the jump, h, is then given by

$$h = \frac{m}{\rho A C_D} \cdot \log_e \left(\frac{\rho C_D A v^2}{2mg} + 1 \right) \tag{10}$$

where v is the initial velocity of the centre of mass. The difference between the height calculated from equations (1) and (10) gives the energy lost to air resistance and is plotted in Fig. 5. For simplicity, it is assumed that the coefficient of drag of the larger animals is unity; this is not unreasonable as most are roughly cylindrical and apparently unstreamlined. In the case of fleas, the Reynolds number is below 100 so that the drag coefficient tends to be above unity and velocity dependent, and the losses may be greater than the graph suggests.

The effect of air resistance is clearly negligible in the case of the leopard; in locusts and the lesser galago, the energy loss is small and probably acceptable; but in fleas air resistance appears to operate as a limiting factor to the height of the jump. It is worth noting that, when account is taken of air resistance, the specific energy for a 0·3 m high jump by a flea is 20 J kg^{-1} and that the same specific energy propels a leopard to a height of 2·5 m.

CONCLUSIONS

Various scale effects reduce the height to which small animals can jump and other effects necessitate changes in the geometry of limbs and muscles and in the materials used as energy stores. Even so, the more important of these effects do not act in direct proportion to the length or mass of the animal so that, while the largest animals only jump a few times their own length, the smallest may jump some hundreds of times their length and thus far surpass their normal running speed. Large animals are not highly specialised for jumping as opposed to running but the smaller jumping insects are so highly specialised that walking performance is impaired; specialised they may be, but fleas and grasshoppers are extremely successful.

ACKNOWLEDGMENT

I am grateful to G. Alder for deriving the equations dealing with the effects of air resistance and for his advice on their application to this problem.

REFERENCES

1. Alexander, R. McN. "Animal Mechanics." Sidgwick and Jackson, London (1968).
2. Alexander, R. McN. "Size and Shape." Edward Arnold, London (1971).
3. Alexander, R. McN. The mechanics of jumping by a dog (*Canis familiaris*) *J. Zool., Lond.* **173**, 549–573 (1974).
4. Bennet-Clark, H. C. The energetics of the jump of the locust, *Schistocerca gregaria. J. exp. Biol.* **63**, 53–83 (1975).
5. Bennet-Clark, H. C. and Lucey, E. C. A. The jump of the flea: a study of the energetics and a model of the mechanism. *J. exp. Biol.* **47**, 59–76 (1967).
6. Dawson, T. J. and Taylor, C. R. Energetic cost of locomotion in kangaroos. *Nature, Lond.* **246**, 313–314 (1973).
7. Evans, M. E. G. The jump of the click beetle (Coleoptera, Elateridae)—a preliminary study. *J. Zool., Lond.* **167**, 319–336 (1972).
8. Evans, M. E. G. The jump of the click beetle (Coleoptera: Elateridae)—energetics and mechanics. *J. Zool., Lond.* **169**, 181–194 (1973).
9. Gordon, A. M., Huxley, A. F. and Julian, F. J. The variation in isometric tension with sarcomere length in vertebrate muscle fibres. *J. Physiol., Lond.* **184**, 170–192 (1966).
10. Hall-Craggs, E. C. B. An analysis of the jump of lesser galago (*Galago senegalensis*). *J. Zool., Lond.* **147**, 20–29 (1965).
11. Heitler, W. J. The locust jump: specialisations of the metathoracic femoral-tibial joint. *J. comp. Physiol.* **89**, 93–104 (1974).
12. Hill, A. V. The heat of shortening and the dynamic constants of muscle. *Proc. R. Soc. Lond.* **B 126**, 136–195 (1938).
13. Hill, A. V. The dimensions of animals and their muscular dynamics. *Sci. Prog., Lond.* **38**, 209–230 (1950).
14. Huxley, A. F. and Niedergerke, R. Structural changes in muscle during contraction. Interference microscopy of living muscle fibres. *Nature, Lond.* **173**, 971–973 (1954).
15. Jensen, M. and Weis-Fogh, T. Biology and physics of locust flight V: strength and elasticity of locust cuticle. *Phil. Trans. R. Soc.,* **B 245**, 137–169 (1962).
16. Pennycuick, C. J. and Parker, G. A. Structural limitations on the power output of the pigeon's flight muscles. *J. exp. Biol.* **45**, 489–498 (1966).
17. Rothschild, M., Schlein, Y., Parker, K., and Sternberg, S. Jump of the oriental rat flea *Xenopsylla cheopis* (Roths). *Nature, Lond.* **239**, 45–48 (1972).
18. Sannasi, A. Resilin in the cuticle of a click beetle. *J. Georgia ent. Soc.* **4**, 31–32 (1969).
19. Weis-Fogh, T., A rubber-like protein in insect cuticle. *J. exp. Biol.* **33**, 668–684 (1960).
20. Wells, J. B. Comparison of mechanical properties between slow and fast mammalian muscles. *J. Physiol., Lond.* **178**, 252–269 (1965).
21. Zachar, J. and Zacharova, D. The length-tension diagram of single muscle fibres of the crayfish. *Experientia,* **22**, 451–452 (1966).

PART III

AQUATIC LOCOMOTION

13. Introduction to the Scaling of Aquatic Animal Locomotion

T. Y. WU

California Institute of Technology, Pasadena, California, U.S.A.

ABSTRACT

This introductory lecture attempts to discuss the scaling problems related to the locomotion of aquatic animals in two major categories characterized by low and high Reynolds numbers of swimming motion. In the low-Reynolds-number flow regime it first gives a brief survey of flagellar and ciliary locomotion in order to ascertain the different key parameters that play a role. The hydromechanical and physiological performance of a group of micro-organisms, of different sizes but otherwise similar in their organization from the scaling point of view, is examined.

In the high-Reynolds-number category, discussions of the scaling problems include the carangiform and lunate-tail locomotion of different groups of fishes and cetaceans, with consideration of hydromechanical efficiency and physiological function. The discussion is based on the data of both comparative zoology and dynamical similarity.

INTRODUCTION

It is most gratifying to see this Cambridge Symposium take on the main theme of *scaling and similitude* in animal locomotion. Its importance for the future development of biological science has become increasingly recognized. A timely exploration of this theme, which is the objective of this meeting, will certainly add strength to and point out new, rewarding directions for current research activities such as those presented at the International Symposium on Swimming and Flying in Nature held in Pasadena in July 1974.

It is a great privilege to have been called upon by Professor Lighthill and Professor Weis-Fogh to give an introductory talk on the influence of size on the different types of aquatic locomotion found in various groups of animals.

In accepting the call I fully realize that the total range of subject-matter to be covered is very extensive and it would be hard to emulate Professor Lighthill's elucidation of the scaling aspects of some 10 groups of aquatic animals at the 1973 Duke University Conference. I feel, nevertheless, that I have benefited from this exciting challenge, for already I have found that important and difficult problems in fluid mechanics have been illuminated by studies in metabolism and chemical energy conversion, and vice versa. For example, we can actually make a better estimate of the drag of a fish from biological measurements than from direct flow measurement, because the latter is formidably hard. This underlines the importance of close collaboration between scientists from different fields engaged in interdisciplinary research on topics such as animal locomotion. In this regard I would like to acknowledge with appreciation the co-operation of the two groups at Cambridge, under Sir James Lighthill and Professor Torkel Weis-Fogh respectively, as well as that of many other colleagues from all over the world.

I would like to discuss the scaling problems of aquatic animal locomotion in two major categories, characterized respectively by low and high Reynolds numbers of the swimming motion, which are broadly equivalent to aquatic animals of microscopic and macroscopic size. In the low-Reynolds-number regime I shall give a brief survey of flagellar and ciliary propulsion in order to ascertain the key parameters underlying the hydromechanical and physiological performance of the two groups of micro-organisms. Here the limited range of animal sizes, swimming speeds, and beat frequencies observed in specific groups, coupled with the very limited data available on their energy conversion processes, show that a lot more information is required before a good understanding can be reached.

In the high-Reynolds-number category, I shall concentrate on the scaling problems of carangiform and lunate-tail locomotion of different groups of fishes and cetaceans, considering their hydromechanical efficiency and physiological functions on the basis of data from both comparative zoology and dynamical similarity. This is an area where experimental results are relatively abundant, and there has recently been considerable research activity, so that a more refined investigation of the scaling problem is possible. Current studies combining both physiology and dynamics promise further progress which may have far-reaching effects, as I hope to point out later.

LOCOMOTION OF MICRO-ORGANISMS

The two major kinds of microscopic organisms to be discussed here are the flagellates and the ciliates. In general they have a maximum linear dimension smaller than 1 mm, with a majority of them being much smaller. Flagellates propel themselves through an aqueous medium by performing undulatory movements of a long thin flagellum (or several flagella) in the form of a planar, helical, or a general three-dimensional wave. Ciliary propulsion utilizes the co-ordinated effort of a large number of flagella-like organelles, called cilia, which are distributed over the cell membrane, and wave in a manner known as 'metachronism'. Both flagellates and ciliates swim in aqueous media, attaining a characteristic velocity, V, in such a way that the Reynolds number

$$Re = Vl/\nu \tag{1}$$

is very low. Here $\nu = 0.01 \, \text{cm}^2 \, \text{s}^{-1}$ is the kinematic viscosity of water, and l is the length of the organism. Typical values of the Reynolds number are around 10^{-3} for spermatozoa, 10^{-6} for most bacteria, and in the range $0.01 < Re < 1$ for various ciliates (the Reynolds number based on ciliary length l_c is considerably smaller). At these low Reynolds numbers, all the inertial effects, both of the liquid and of the organism, become insignificant. Consequently, hydromechanical calculations of flagellar and ciliary propulsion are generally based on neglecting the inertial effects, leaving the viscous stress and pressure as the only quantities that need be considered in the momentum balance for given swimming movements.

From the laws of motion, the net momentum and net angular momentum of the fluid surrounding a self-propelling micro-organism, swimming at constant velocity, must be zero. However, the backward drag force and the equal and opposite thrust acting on a swimming micro-organism cannot be separated, as they can for a large aquatic animal (as we shall see), and must be dealt with together. For a long, thin organelle such as a flagellum or a cilium, the forces tangential and perpendicular to its centerline, denoted by F_s and F_n respectively, which it exerts on the fluid while moving with tangential and normal velocities V_s and V_n, depend mainly on its length l and not on its transverse dimensions. Calculations of these forces can be considerably simplified by means of the "resistive-force theory", which was first developed for flagellar applications by Hancock[33] and Gray and Hancock,[31] was extended to helical movements by Chwang and Wu,[25] and was recently refined further by Lighthill.[48, 49] It provides a simple relationship between the force and the velocity in the form

$$F_s = C_s V_s l, \qquad F_n = C_n V_n l, \tag{2}$$

where C_s and C_n are the resistive force coefficients which depend on the transverse dimensions and wavelength of the body but not on its velocity. Both C_s and C_n are proportional to the (dynamic) viscosity coefficient of the fluid, μ (equal to the kinematic viscosity, ν, times the density $1 \, \text{g cm}^{-3}$).

The ratio of C_s to C_n, or

$$\gamma = C_s/C_n \tag{3}$$

plays an important role in swimming propulsion since its value (always <1) determines what type of undulatory motion will give the optimum forward velocity or optimum efficiency, as first elucidated by Lighthill (Refs 43, p. 442; 48, p. 122). I shall try to recapitulate his argument by considering the ideal case of *zero-thrust swimming* of a flagellum (such as that of a sperm detached from the cell body, or of a spirochete). I shall deal solely with the translational motion, leaving the possible angular motion and all further details to the literature. (The same argument can be applied, qualitatively at least, to larger elongated animals.)

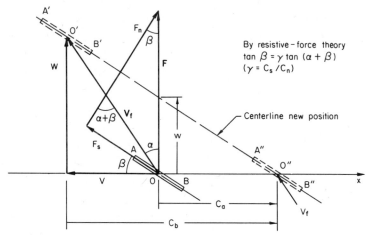

FIG. 1. A resistive-force theory diagram. A segment AB of an elongated body in undulatory motion moves with velocity $\mathbf{V}_f$ and exerts a force $\mathbf{F}$ on the fluid in a zero-thrust propulsion. Its lateral velocity is W relative to the animal's mean position and w relative to the water. The wave velocity is c_a relative to the water and c_b relative to the body.

Figure 1 shows a segment AB of an elongated animal supposed to be performing a plane or helical wave, and achieving a swimming velocity V in the negative x-direction. In the ideal case of zero thrust (or zero x-component force), the force $\mathbf{F}$ exerted by the segment on the fluid can only be in a transverse direction as shown. Nevertheless there is a forward tangential component F_s along the centerline of the body when the latter is inclined to the path of swimming at an angle β. (Figure 1 may be envisaged as in the plane of a planar wave, or, in the case of a helical wave, in the plane tangential to the helical movement of the segment in question.) The segment, under the combined traction of F_s and the normal component of force, F_n, will move to a new position $A'B'$ with velocity $\mathbf{V}_f$ which has the forward x-component V and a transverse component W. Clearly, we have

$$F_s = F \sin \beta, \qquad F_n = F \cos \beta, \tag{4}$$

and

$$V_s = V_f \sin (\alpha + \beta), \qquad V_n = V_f \cos (\alpha + \beta), \tag{5}$$

where α is the angle between $\mathbf{F}$ and $\mathbf{V}_f$. Substituting (4) and (5) in (2) yields

$$\tan \beta = \gamma \tan (\alpha + \beta). \tag{6}$$

The angle α vanishes when $\beta = 0$ or $\pi/2$, and achieves a maximum when

$$\beta = \beta_m = \arctan \gamma^{1/2}, \tag{7}$$

giving

$$V_{\max}/W = \tan \alpha_{\max} = \tfrac{1}{2}(1 - \gamma)/\gamma^{1/2}. \tag{8}$$

In the meantime the constant phase of the body wave has been propagated backwards to reach position $A''B''$ (which would now be occupied by a new segment of the body) with wave (phase) velocity c_a relative to the water, or with wave velocity

$$c_b = c_a + V \tag{9}$$

in the body frame of reference. Since $W/c_b = \tan \beta$ (see Fig. 1),

$$V_{\max}/c_b = \tan \alpha_m \tan \beta_m = \tfrac{1}{2}(1 - \gamma). \tag{10}$$

The lateral velocity of the segment relative to the water, w, is equal to $W(c_a/c_b)$, which when $\beta = \beta_m$ assumes a value such that

$$\frac{w}{W} = \frac{c_a}{c_b} = \tfrac{1}{2}(1 + \gamma). \tag{11}$$

The above results agree with Lighthill's (Ref. 48, p. 123). An important consequence is that for fixed W (or c_b), $V_{\max}$ decreases with increasing γ, and vanishes at $\gamma = 1$, which means self-propulsion in the present case (of zero thrust) would be impossible when $C_s = C_n$.

A more general situation arises when a flagellum or a ciliary system has a passive cell body to propel so that each segment must on average have its force $\mathbf{F}$ directed both sidewise and somewhat forward. The efficiency may then be defined as the ratio of useful work done by the forward component of $\mathbf{F}$ in propelling the animal forward at speed V, to the power required which is $\mathbf{F} \cdot \mathbf{V}_f$. The values of $V_{\max}$ and V at the maximum efficiency can also be calculated. Lighthill[48] showed that the value of V at maximum efficiency $(\eta = \eta_{max})$ is

$$V/c_b = 1 - \gamma^{1/2} \quad \text{and} \quad \eta_{max} = (1 - \gamma^{1/2})^2. \tag{12}$$

This value of V is not much less than V_{max} (Eqn 10), even when $\gamma = 0.2$.

Detailed analysis has produced the following expressions for the force coefficients in the low Reynolds number case

$$C_s = \frac{2\pi\mu}{\ln(2q/b) - 1/2}, \quad C_n = \frac{4\pi\mu}{\ln(2q/b) + 1/2}, \tag{13}$$

where b is the cross-sectional radius of the thin segment, and q is a quantity depending on body shape and certain wave parameters. Lighthill[49] contended that for flagellar waves with wavelength λ it is accurate to take

$$q = 0.09\lambda, \tag{14}$$

and to delete the term $-1/2$ in the denominator of the expression for C_s. (The values of q for rigid slender bodies of various shapes are given by Batchelor,[8] Tillett,[61] Cox[26]). The value of γ, according to these formulas, lies in a range between 0·6 and 0·7 and is not close to 0·5 as earlier theories have commonly assumed. The corresponding efficiency is therefore about 0·05 for flagellar propulsion, which is rather low, and $V/c_b \simeq 0.2$. Lighthill[46] noted that for swimming snakes γ is likely to be about 0·1, leading to $\eta \simeq 0.5$ and $V/c_b \simeq 0.7$.

Another important feature of flagellar propulsion is that the maximum inclination of a plane flagellar wave, or the pitch angle of a helical flagellar wave, should, for maximum speed, be

$$\beta_{max} \simeq 40° \tag{15}$$

(from Eqn 7) for γ around 0·6. This figure is close to the value for minimum power output, and comparable values are indeed observed.

With this introduction we now proceed to discuss the problems of flagellar and ciliary propulsion separately.

Flagellar Propulsion

The mechanical power needed for a flagellar organism of length l_f to swim at velocity V may be readily estimated by applying the resistive theory, giving

$$P_f = F_s V_s + F_n V_n = C_n(\gamma V_s^2 + V_n^2)l_f \simeq 50\mu V^2 l_f, \tag{16}$$

where the constant factor 50 is borne out by detailed calculation to be typical of flagellar movement. It is based on the observation that there is not much difference in diameter (about 0·2 μm) between the flagella of different species (excluding the bacterial flagella), and λ/b is typically between 100 and 200, though the length l_f may vary considerably from one species to another. Additional power is further required to overcome the drag of a passive cell body (of a three-dimensional form with length l_b say), if any, which is

$$P_b = \text{const. } \mu V^2 l_b. \tag{17}$$

The value of P_b is generally smaller than that of P_f.

To proceed with a scaling analysis we need to know the chemical energy input per flagellar beat (say E_c); this can in principle be obtained from the measured rate of dephosphorylation of ATP (the chemical compound adenosine triphosphate) by the flagellum. Such information, however, is extremely scarce. Under the circumstances it has been suggested (see Alexander,[1] p. 40) that for protozoic flagella having the well-known arrangement of 9 + 2 pairs of micro-fibrils running along them, the maximum chemical energy input of a flagellum is likely to be proportional to its length l_f (or,

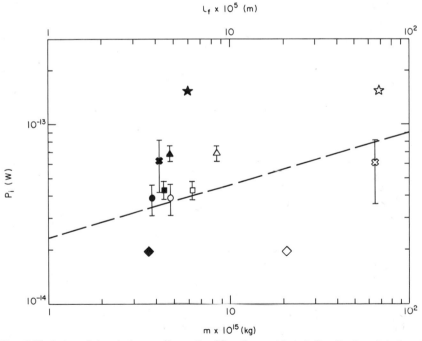

FIG. 2. Variation of chemical power input P_i of flagellates with their flagellar length l_f (in solid symbols) and mass m (in open symbols). $\times$ —*Colobocentrotus atrafus*; $\bigcirc$—*Strongylocentrotus purpuratus*; $\square$—*Lytechinus pictus*; $\triangle$ *Ciona intestinalis*; $\diamondsuit$—*Chaetopterus variopedatus*; $\star$—Bull sperm. Experimental data from Brokaw and Gibbons.[24] The slope of the dotted regression line is 0·28.

equivalently, to the mass of the flagellum since its cross-sectional diameter is uniform and constant). From this assumption and Eqn (16) is would follow that the swimming velocity V is scale-independent.

The experimental data which have made this preliminary study possible are taken from Brokaw and Gibbons[24] who investigated the process of converting the chemical energy of ATP into the mechanical work of sliding between the doublet tubules of the axoneme, and hence of generating particular flagellar waves, in six different species of spermatozoa. Figure 2 shows the dependence of chemical power input of the flagellates, P_i, on the flagellar length l_f (which is 3 to 15 times the cell-body length) and the total mass m of the organism. Although a regression line can be drawn by the method of least squares for the $\log P_i - \log m$ relation, with slope 0·28, the data are considered to be too scattered for a confident conclusion to be drawn. The short range of flagellar length l_f covered by these species makes determination of the $P_i - l_f$ relation difficult, but, from these results at least, P_i is clearly far from being proportional to l_f as predicted.

The difficulties seem to stem from the apparent lack of correlation between l_f and m, as can be seen from Fig. 2. This is in fact not surprising since the thrust required from each segment of a long flagellum to balance the drag force on the cell body is very small. Thus changes in mass, and hence drag force, will involve only very slight modification of the propulsive mechanism, which will still be approximately described by the zero-thrust theory outlined above. This implies that the flagellar length l_f (or the total cell length l) may not be the pertinent parameter to use in scaling; perhaps future experimental results will help settle this point.

We present in Fig. 3 the variation of swimming velocity V with m, for which a least square fit yields $V = \text{const.}\, m^{-0.04}$, ie. V is virtually independent of body mass. This is in sharp contrast with the case of large animal locomotion, as we shall see in the discussion of the scaling of fish swimming.

The nondimensional quantity

$$\mathscr{E} = P/mgV, \tag{18}$$

where P stands for either the chemical power input or the mechanical power required, and g is the acceleration of gravity, provides a useful measure of the relative merit of different propulsive systems. If an animal expends power $P = dE/dt$ to keep moving along a straight level path at speed $V = dx/dt$, where x is distance measured along the path, the ratio $P/mgV = (mg)^{-1} dE/dx$ therefore gives the energy expenditure to transport unit weight unit distance. The parameter $\mathscr{E}$ may thus be called the "specific energy cost" of transport (cf. the "net cost of transport" as defined by Alexander). It has been used by von Karman and Gabrielli[66] to evaluate the comparative merits of 14 classes of transportation vehicle, and by Schmidt-Nielsen,[57] Tucker,[63, 65]

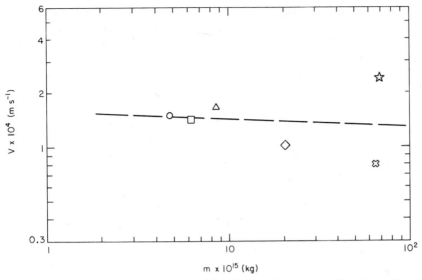

FIG. 3. Variation of swimming velocity of flagellates with their mass. The slope of the dotted regression line is −0·04. For other legends see caption to Fig. 2.

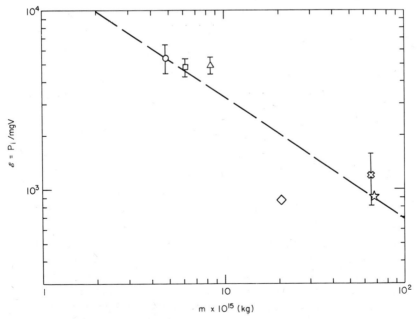

FIG. 4. The specific energy cost of flagellates versus their mass. The slope of the dotted regression lines is −0·68. For other legends see caption to Fig. 2.

and others for studies of comparative physiology. Figure 4 shows the specific energy cost of the six flagellates in question, which may be represented, by a least-square fit, as $\mathscr{E} = $ const. $m^{-0.68}$. This $\mathscr{E}-m$ relationship is quite similar to those of other aerial, aquatic and terrestrial animal groups, but the slope -0.68 of the regression line is steeper than most of them already known (see Tucker [65]).

Summarizing this preliminary study of flagellar propulsion, we have found

$$P_i \simeq \text{const. } m^{0.28}, \qquad V = \text{const. } m^{-0.04} \tag{19a}$$

and

$$\mathscr{E} = P_i/mg = \text{const. } m^{-0.68}. \tag{19b}$$

In view of the difficulties involved in such metabolism experiments, we should appreciate the value of these experimental results, although only six species have been studied, and not too much confidence can be placed in the quoted slopes of the regression lines. The results do throw enough light to make the following theoretical argument attractive.

Suppose we accept the assumptions that (i) the propulsive velocity V of a flagellum under the zero-thrust condition depends only on the wave form and beat frequency and is not appreciably affected by the presence of a cell body, and (ii) the power P_f required for maintaining a *steady* flagellar motion (hence $P_f = $ const.) forms a basal reference above which the power P_b required for propelling the cell body (Eqn 17) will vary when the cell body size l_b is varied. Then the *changes* in the power input, P_i, would scale according to

$$P_i \propto l_b \propto m^{1/3}, \quad \text{and hence} \quad \mathscr{E} = P_i/mgV \propto m^{-2/3}. \tag{20}$$

These relationships are in close agreement with (19a) and (19b). It will be of interest to acquire more experimental data for further studies along this line.

Ciliary Propulsion

Blake and Sleigh[13] and Holwill[34a, b] have given excellent surveys upon the mechanics and physiology of ciliary propulsion, and their introductory lectures on the subject at the 1974 Caltech Symposium were of especial value. They pointed out that ciliary length, pattern of movement and rate of beat are important factors in determining the effectiveness of propulsion of fluids by cilia, whether for ciliated protozoa or metazoa, or for the inner organs of higher animals. Ciliates are generally known as very impressive swimmers when judged by their specific speeds (body lengths travelled/s). However, the four types of metachronal pattern in which ciliary movements are organised (symplectic, antiplectic, dexioplectic, and laeoplectic) are so different from each other that it would be inappropriate to study ciliary efficiency collectively, as these authors and others have pointed out. This is also true for studies of scaling problems in ciliary propulsion.

Theoretical analysis of ciliary propulsion began when Taylor[60] introduced the "envelope model", based on the idea that the concerted action of large numbers of cilia on the fluid can be represented by the waving motion of an impermeable inextensible sheet enveloping the cilia tips. This model has been extended to cover an extensible sheet, with an estimate of the inertia effect by Tuck,[62] and has been applied by Blake[10, 11] to evaluate the motion of *Opalina*. It was pointed out by Lighthill[46] that it is only to symplectic motion that such a model should really be applied. In order to treat other types of metachronism, especially those such as antiplectic motion, in which the cilia tips may be far apart during the motion, Blake[12] developed a sublayer model in which each cilium was modelled by an elongated body attached to a flat cell surface.

Another simplified model, called the "traction-layer" model, has been investigated, at the suggestion of Wu,[77] by Keller.[39] It is based on the concept that the discrete forces of a system of cilia can be represented by an "equivalent" continuous distribution of an unsteady body force within the volume of the ciliary layer, provided that the spacings between adjacent cilia are small compared with the ciliary length, l_c. Conversion from the discrete to the continuum force is based on the resistive theory which gives the relationship between the original force exerted by a cilium, and its velocity relative to the flow produced by the ciliary system as a whole.

Using this procedure the model has been applied by Keller to an infinite plane ciliary layer with the body force represented by a Fourier series

$$f(x, y, t) = \sum_{n=0}^{N} f_n(y) \exp\left[in(kx - \omega t)\right] \qquad (-\infty < x < \infty, 0 < y < l_c) \qquad (21)$$

where $N \leqslant \infty$, $\lambda = 2\pi/k$ is the metachronal wavelength, ω is the radian frequency, and f vanishes at the cell surface ($y = 0$) as well as outside the cilia layer ($y \geqslant l_c$). The functions $f_n(y)$ can be given a polynomial representation or be determined by numerical methods. The Stokes equation with this force term included can be solved to give the flow velocity for particular movements of the cilia. A related optimisation problem is to find the external mean force distribution $f_0(y)$ that will give the minimum power required, P_0, to maintain a rate of working on the fluid which is less than P_0 by a fixed amount; this is equivalent to finding the minimum power required for fixed mean square of f_0, averaged across the cilia layer. The result for a four-term polynomial representation of $f_0(y)$ is shown in Fig. 5. Thus we find the mean flow velocity at the outer boundary of the cilia layer as

$$u_0(l_c) = 0{\cdot}52A(\omega l_c)(kl_c) = 0{\cdot}52Ac(2\pi l_c/\lambda)^2, \tag{22}$$

where $c = \omega/k$ is the metachronal wave velocity and A is proportional to the amplitude of the dimensionless force distribution. The slope of $u_0(y)$ at $y = 0$ shows that there is a viscous skin friction at the cell surface which is counterbalanced by the fluid force acting on the cilia over the same area of the surface.

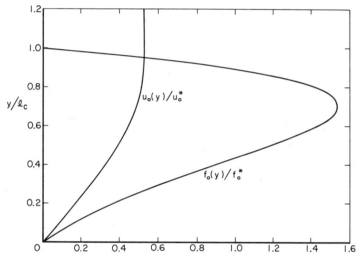

FIG. 5. The optimum continuous force distribution giving minimum power required for fixed mean square force, and the corresponding velocity profile. $f_0^* = A\mu\omega/k^2$ and $u_0^* = A\omega k l_c^2$ where l_c is the ciliary length, ω is the radian frequency and k the wave number of the ciliary wave, and A is a constant making the y-integral of $(f_0/f_0^*)^2$ unity.

The first harmonic velocity profile, $u_1(y)$, has been evaluated by Keller[39] in a few cases for which the wave parameters are available. The results for *Paramecium multimicronucleatum* are given in Fig. 6, from which we note that the contribution from the second harmonic is already very small. We further note from Fig. 6 that like the mean velocity $u_0(y)$, $u_1(y)$ is also proportional to $(kl_c)^2$, or $(l_c/\lambda)^2$.

Experimental investigation of the scaling relation (22) has not yet been fully carried out. Furthermore, a complete examination of the scaling of ciliary propulsion cannot be made without physiological data on the chemical energy input, which are also not yet available. In addition, the hydromechanical analysis must be carried out for a finite shape, in the manner indicated by Blake[10] and Brennen.[17]

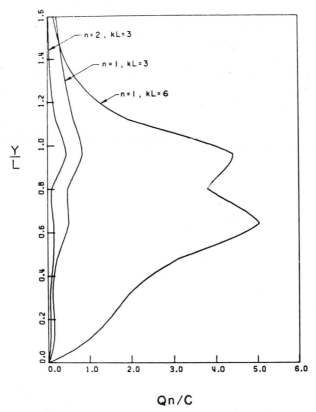

$\dfrac{Y}{L}$

Qn/C

FIG. 6. Amplitude profile for the first and second harmonic x-components of flow velocity (Q_n, non-dimensionalised with respect to the metachronal wave speed $c = \omega/k$) for *Paramecium* with $ka = 0{\cdot}80$ and $kb = 1{\cdot}60$. (a and b are the row and column spacings between cilia, and L here denotes ciliary length).

LOCOMOTION OF LARGE AQUATIC ANIMALS

The swimming motions of large aquatic animals are characterized from the hydrodynamicist's point of view by large values of the ratio of typical inertial forces to typical viscous forces in the surrounding water, i.e. of the Reynolds number, $\mathrm{Re} = Vl/v \gg 1$. Here V is the mean swimming speed, l the length of the animal, and v the coefficient of kinematic viscosity of the fluid. In general, the Reynolds number lies in the range $10^4 < \mathrm{Re} < 10^8$ for various fishes and cetaceans swimming normally. At such high Reynolds numbers the effects of viscosity are mostly confined to a thin boundary layer adjacent to the body surface. This is especially true for the streamlined body shapes of such animals in undulatory swimming motion, when flow separation is not observed to occur, and only a very thin wake is formed behind the body. In such cases the boundary layer will continually grow with distance along the fish, but its maximum thickness (at the tail end) is generally not more than a few per cent of the body thickness. Viscous effects can therefore be neglected in analysing the flow outside this boundary layer. The pressures acting on the body are those predicted by inviscid theory, because the boundary layer is too thin to support a pressure gradient across it. These pressures generate a thrust force which must overcome the drag force in order to maintain a constant forward velocity. The drag force in this case arises mainly from the viscous shear stresses exerted by the fluid on the body surface; the contributions of "form drag" (resulting from the distal thickening of the boundary layer and possibly from flow separation), of "induced drag" (due to the shedding of vortex sheets), and of "wave resistance" (energy lost through making waves when swimming at the surface) are generally insignificant for these animals.

Thus the hydromechanical problem of fish locomotion can be conveniently divided into two parts: the drag of a streamlined body, analysed by calculating the viscous shear stresses in the boundary layer, and the generation of an equal and opposite thrust, determined from an analysis of the potential flow outside the boundary layer. If the mean thrust is denoted by T, and the mean drag force by D, we have, in steady swimming,

$$T = D. \tag{23}$$

Power Expenditure and Energy Balance

From the physical standpoint the minimum power necessary to transport the body at a given velocity V is determined by the resistance of the medium, the efficiency of the mode of locomotion, the energy consumption of the particular type of power plant that delivers the mechanical power, and many other factors. The principle of energy conservation may be expressed as

$$DV = \eta P, \tag{24}$$

which states that the animal's rate of working against the resistance D is provided by the power input P, used with efficiency η. For aquatic animals the variations of potential energy are generally unimportant, and in any case can be considered separately as for aerial and terrestrial animals.

The power input for locomotion can also be determined by measuring the animal's metabolic rate and by applying standard energy conversion factors. Thus, when P in Eqn (24) is taken to be the mechanical power, η is the hydrodynamic efficiency η_h of the propulsion. When P is taken instead to be the metabolic rate (the rate at which biochemical energy is released), then η is equal to $\eta_h\eta_c$, where η_c is the efficiency with which biochemical energy is converted to muscle power.

Metabolic Rate

The extensive literature on the metabolism and swimming speeds of fish makes it plain that the physical details of an experiment must be clearly specified, in order that its results may be interpreted correctly (see, for example) Winberg,[75] Drabkin[28] and Brett.[20, 21] Of major importance in the study of the metabolic rate of a fish are its level of activity and the temperature of the water (since water holds less dissolved oxygen at elevated temperatures). Such factors as preconditioning (a period of fasting and exercise prior to the test, so that none of the fish's energy is spent in digesting or absorbing food during the test), the state of maturity and sex of the specimen, and flow conditions during the experiment, are also important. We can give only a brief account here.

(i) Level of Activity

The level of activity in swimming fish is hard to characterize in general. However, for the salmon, a migratory fish, Brett[20, 21, 22] (see Fig. 7) found three levels of performance: sustained (speeds that can be maintained almost indefinitely), prolonged (speeds maintained for 1 or 2 hours with a steady effort, but leading to fatigue) and burst (speeds achieved at maximum effort lasting for only about 30 seconds). The physiological basis of each of these activity levels is different, and itself depends on body size and environmental temperature.

The relationship between swimming speed and metabolic rate can also be evaluated in terms of the rate of working by the muscle, as discussed in other lectures at this Symposium. Lighthill[47] reasoned that the efficiency with which muscles produce power increases in proportion to the frequency ω of contraction. Hence in both carangiform and lunate-tail swimming η_c should be proportional to the swimming velocity V. Thus, when the mechanical power required for maintaining the speed V is proportional to V^3, the fish consumes energy at a metabolic rate

$$P = P_b + \alpha V^2, \tag{25}$$

where P_b is its basal metabolic rate and α is a constant. It therefore follows that P/V, proportional to the specific energy cost $\mathscr{E}$ (Eqn 18), is given by

$$P/V = P_b/V + \alpha V, \tag{26}$$

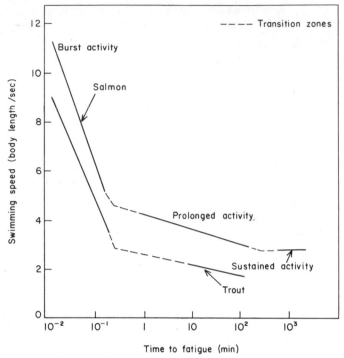

FIG. 7. Swimming endurance of sockeye salmon and rainbow trout. Transition zones between different levels of activity are shown with broken lines. (After Brett[19, 23]).

which has a minimum at the speed $V_m = (P_b/\alpha)^{1/2}$. A speed for which the specific energy cost is a minimum has been found by Brett[21] for salmon (see Fig. 8), by Tucker[64] for small parrots, and was also noted by Weihs[74] in his theoretical study. Figure 8 shows the basic curve of $\mathscr{E}$ against V for sockeye salmon; their ocean migrant speed is very close to the minimum energy cost, and their speeds in Fraser River migration ($\sim 1\cdot 2\ \mathrm{m\ s^{-1}}$) are somewhat lower than their lowest burst speed.

(ii) Fish Size and Metabolic Rate

The relationship between body mass and the metabolic rate of fish has received considerable attention, particularly for the resting state (for expository reviews see Kleiber,[40] Schmidt-Nielsen[58]). The equation which describes this relationship is of the general form

$$P = am^b, \tag{27}$$

(see Eqn 19) or in the logarithmic form

$$\log P = \log a + b \log m, \qquad (28)$$

where m is the body mass, and a and b are coefficients independent of body size. Gray[29] first tackled this problem by a different approach; he directly considered the total mass of muscle, m', so that $P = a'm'$, and derived the

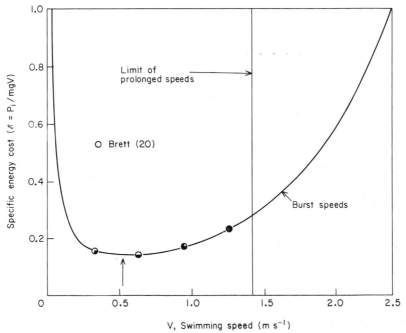

FIG. 8. Specific energy cost $\mathscr{E}$ as related to swimming speed, for adult sockeye salmon. Ocean migration occurs slightly above the speed at minimum $\mathscr{E}$ (shown with an arrow), which is about $\frac{1}{2}V_{crit}$(⊖), while river migration takes place nearly at V_{crit}(●). For the definition of V_{crit} see text and the caption to Fig. 9. (Adapted from Brett[21, 23]).

power factor a' relating to unit mass of fish muscle from that of a rowing athlete. Application of this method by Bainbridge[2, 3, 4, 5] to several species of fish led to the equation $m' = 0.005l^{2.9}$, where l is the length of the fish in cm when m' is measured in g.

Varous authors have considered body mass m directly. Winberg[75] examined 266 cases for freshwater fish and obtained an average value of $b - 0.81$ (at 20°C). Heusner et al.[34] investigated three species and derived an overall mean of $b = 0.73$ (at 25°C). There are other reports giving a rather wide range of values, varying all the way from near the "surface law" ($b = 0.67$)

to direct proportionality ($b = 1$). The difficulty in seeking a generally applicable value seems to be associated with that met in measuring the drag of a fish.

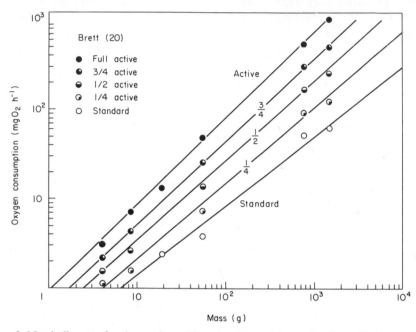

FIG. 9. Metabolic rate of sockeye salmon (*Oncorhunchus nerka*) as a function of body mass at various levels of swimming activity (15°C). The standard level was determined by extrapolation to zero swimming speed of the oxygen consumption vs. speed relation. The line at full activity corresponds to the maximum speed which could be sustained for 60 min, V_{crit}. Intermediate fraction levels correspond to swimming speeds equal to $\frac{3}{4}$, $\frac{1}{2}$, and $\frac{1}{4}$ of V_{crit} respectively.

Brett[18, 20, 21] has shed considerable and much-needed light on the problem by distinguishing between the different levels of activity. He provided the following results for the sockeye salmon (*Oncorhynchus nerka*), with oxygen consumption measured in mg O_2 h^{-1}, and body mass in g (see also Fig. 9):

$$
\begin{aligned}
\log a &= -0.632 & b &= 0.775 \pm 0.145 \quad \text{(standard)} \\
&= -0.523 & &= 0.846 \pm 0.145 \quad (\tfrac{1}{4}\,\text{max}) \\
&= -0.357 & &= 0.890 \pm 0.145 \quad (\tfrac{1}{2}\,\text{max}) \qquad (29) \\
&= -0.223 & &= 0.926 \pm 0.145 \quad (\tfrac{3}{4}\,\text{max}) \\
&= -0.050 & &= 0.970 \pm 0.053 \quad \text{(max)}
\end{aligned}
$$

Here the fractions are referred to the maximum speed which could be sustained for 60 min in fresh water at 15°C (V_{crit}), and his standard metabolic rate is obtained by extrapolating the O_2-consumption versus velocity curve to zero swimming velocity. This approach appears to make the study on the scaling of swimming velocity more systematic, as we now proceed to discuss.

Scaling of Swimming Velocity

In order to use the principles of dynamical similarity, we first express the drag D on a fish swimming at speed V in terms of its drag coefficient C_D, as follows:

$$D = \tfrac{1}{2}\rho V^2 S C_D \tag{30}$$

where ρ is the water density, S the fish surface area (proportional to l^2) and C_D depends on the Reynolds number Re and the body shape. Substitution of this expression and the metabolic rate relationship (27) into the energy equation (24) then yields

$$V^3 \propto l^{3b-2}/C_D. \tag{31}$$

Two cases can be examined by assuming that for carangiform and lunate-tail swimming, in both of which the fish is well-streamlined, C_D is the same, apart from a shape factor, as that of a flat plate of the same area at the same Reynolds number. That is, C_D is proportional to $\text{Re}^{-1/2}$ or $\text{Re}^{-1/5}$ according as the boundary layer is laminar or turbulent. This leads to the scaling law:

$$V = \text{const. } l^\beta, \tag{32}$$

with

$$\beta = \tfrac{3}{5}(2b - 1) \quad \text{(laminar)}$$

$$\beta = \tfrac{3}{14}(5b - 3) \quad \text{(turbulent)}$$

$$\beta = b - \tfrac{2}{3} \quad (C_D = \text{const.})$$

The last case (of quite large but constant C_D) corresponds to the situation

TABLE I

C_D \ β \ b	0·78 (standard)	0·85 ($\tfrac{1}{4}$-max)	0·89 ($\tfrac{1}{2}$-max)	0·93 ($\tfrac{3}{4}$-max)	0·97 (max)	1·00
laminar	0·33	0·42	0·47	0·51	0·56	0·60
turbulent	0·19	0·26	0·31	0·35	0·40	0·43
$C_D = \text{const.}$	0·11	0·18	0·22	0·26	0·30	0·33

for separated flow past a blunt body with a broad wake formation, or to the case when flow separation occurs in the cross-flow past an undulating fish body. If Brett's values for b are adopted we obtain Table I for β. The last column, for $b = 1$, is included to show the values of the index β when the metabolic rate is proportional to body weight, and the last row, for $C_D = $ const., is for the hypothetical case in which the flow is fully separated.

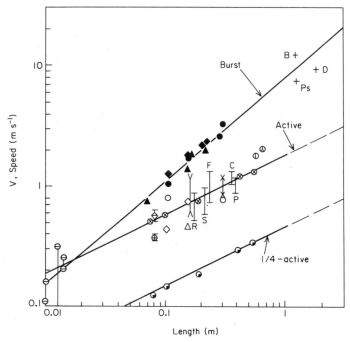

FIG. 10. Variation of swimming speed with body length at different levels of activity. *Burst speed data*: ◆ dace, ▲ goldfish, ● trout, B+ barracuda, P_s + porpoise, D + dolphin (Bainbridge[5]). *Full activity data*: open symbols (Bainbridge[5]); ⊗ sockeye salmon (15°C) (Brett[20]); ◇ bass, ⊕ coho salmon (Dahlberg *et al.*[27]); ⊢⊣ C cod, R redfish, F winter flounder, S sculpin, P pout (Beamish[9]); ⟩—⟨ goldfish (Smit *et al.*[59]); ⊖ larval anchovy (Hunter[36]); ×—× herring (Jones[38]); ① salmon, ⊗ trout (Paulick and DeLacy[55]). $\frac{1}{4}$-*activity data*: ◐ sockeye salmon at 15°C (Brett[20]).

A collection of existing experimental data is given in Fig. 10. In the log V vs log l plot for (i) Brett's 60-min. maximum activity results for salmon at 15°C, (ii) Beamish's[9] results for 6 species of fish over a range of temperatures (8–14°C) and endurance spans, and (iii) a few other sources (see Fig. 10 caption), the slope of the mean regression line is 0·5 (i.e. $\beta = $ 0·5). This result led Brett[20, 21] to propose "that the swimming speed is proportional to $l^{0.5}$, indicating a decrease in the relative ability to maintain a sustained speed as size increases". It is of interest to point out that the speed at the $\frac{1}{4}$-max activity level ($b = 0.85$) of the sockeye salmon is also proportional to $l^{0.5}$ (see Fig. 10). There are, however, other reports (Brett,[22] Magnuson,[50] Hunter,[35] Hunter and Zweifel[37]) indicating that a better overall average value for β is 0.6.

Considerable attention has been given to the burst speeds which, for various species of streamlined fish, tend to be nearly independent of both size and temperature. A representative plot is included in Fig. 10, showing Bainbridge's[5] results for 4 species of fish and two cetaceans; the slope of this line gives a value of β equal to 0.88. Similar values have been obtained by others, such as $\beta = 0.94$ for herring (at $\sim 12°C$) by Blaxter and Dickson.[16] In fact, a speed of 10 lengths per second has been regarded by several authors as a "common rule" for the maximum burst speed of many streamlined fish.

At this point a comparison between experiment and the foregoing similitude calculation may have far-reaching implications. First, on the basis of the observed sustained speeds (with the index $\beta = 0.5$) and the corresponding metabolic-rate index $b = 0.97$, it may be argued by implication (see the fourth column of Table I) that the boundary layer adjacent to the *swimming* fish would be at most only partially turbulent up to Reynolds number of 10^6 (the upper range covered by the data). Furthermore, there is no evidence to suggest any flow separation. If this argument can be established, the study of the metabolism of swimming fish will yield most valuable results for hydrodynamicists, and may enable them to overcome the difficulties in making measurements in the boundary layer over an undulating body and hence estimating the drag force. Further, we note that the rule of burst-speed = $10\,l\,s^{-1}$ would correspond to a value of b equal to 1.33 (for laminar boundary layers), and even higher values when the flow is turbulent. Whether this implies a substantial depletion of stored energy or whether it has some other significance remains to be explored.

Specific Energy Cost

The results obtained above for the scaling of metabolic rate and swimming velocity may now be used to evaluate the scaling of the specific energy cost, $\mathscr{E} = P/mgV$. From (18), (27) and (32) we immediately deduce that

$$\mathscr{E} = P/mgV = \text{const.}\, m^{-\gamma} \tag{33}$$

where

$$\gamma = 1 - b + \beta/3. \tag{34}$$

Using Table I we obtain for γ the following result (Table II):

TABLE II.

C_D $\diagdown$ γ $\diagdown$ b	0·78 (standard)	0·85 ($\frac{1}{4}$-max)	0·89 ($\frac{1}{2}$-max)	0·93 ($\frac{3}{4}$-max)	0·97 (max)	1·00
laminar	0·34	0·30	0·27	0·24	0·22	0·20
turbulent	0·29	0·24	0·21	0·19	0·16	0·14
C_D = const.	0·26	0·21	0·18	0·16	0·13	0·11

Figure 11 presents the experimental data, derived from those given in Figs 9 and 10, showing how the specific energy cost varies with body size for fish swimming at different levels of activity. The solid lines are regression lines obtained by a least-square-error fit, giving $\gamma = 0.34$ at $\frac{1}{4}$-max, 0·29 at $\frac{1}{2}$-max, 0·25 at $\frac{3}{4}$-max, and 0·21 at full activity. These values of γ and the observed mean of $\beta = 0.5$ are plotted against b (which may be regarded as a measure of the level of activity) in Fig. 12. Also shown are the similarity predictions of β and γ given by (32) (or Table I) and (34) (or Table II) for the three distinct reference states characterized by a laminar boundary layer, a turbulent boundary layer and separated cross flow. This comparison

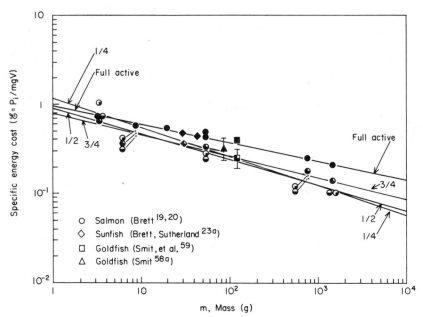

Fig. 11. Relation between specific energy cost $\mathscr{E}$ and body size for fish swimming at different levels of activity. Solid lines are regression lines obtained from data of Brett[20] for salmon at 15°C. For other legend and symbols see Fig. 9.

strongly suggests that the boundary layers of these swimming fish would be largely laminar at these high Reynolds numbers ($8 \times 10^3 <$ Re $< 1.4 \times 10^6$ being the range covered by the test cases), with at most only a small (presumably distal) part turbulent, and that the specimens tested were swimming at somewhere about $\frac{3}{4}$ of their maximum sustainable activity.

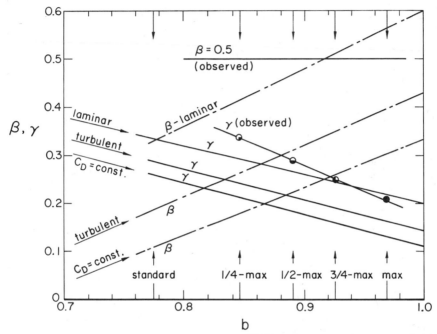

FIG. 12. Comparison of the observed values of β (Brett[20, 21]) and γ (see Fig. 11) with the similartiy predictions (Eqs (32), (34)) based on the three distinct reference states characterized by a laminar boundary layer, a turbulent boundary layer and C_D = const. (for the case of separated cross flows).

The specific energy cost of transport has also been discussed in the recent studies of Schmidt-Nielsen[57] and Tucker[65] for the locomotion of various animals. For comparison we find, from Schmidt-Nielsen's Fig. 1(a) and Tucker's Fig. 2, that the observed values of γ for fish are

$$\gamma = 0.30 \qquad \text{(Schmidt-Nielsen)}, \qquad (35)$$

$$\gamma = 0.27 \qquad \text{(Tucker)}. \qquad (36)$$

These values of γ are in accord with the present finding when the level of activity lies between about $\frac{1}{2}$ and $\frac{3}{4}$ of the maximum performance.

With particular reference to fish size we note from Fig. 11 that the regression lines of $\mathscr{E}$ at different levels of activity cross each other at a certain mass. This cross-over suggests that some (specific) energy would be saved if smaller fish were to swim at speeds corresponding to greater activity and larger fish were to cruise at lower relative speeds.

Scaling of Tail Beat Frequency and Body Wavelength

Bainbridge[2, 4, 7] observed that the maximum amplitude of the tail beat of fishes in carangiform motion is approximately 0·2 of the fish length l, with no noticeable dependence on other parameters. The "reduced frequency" $\omega l/V$, where ω is the radian frequency, is scale independent, and takes values around 10. Observations by other authors are essentially in agreement with Bainbridge's. According to Hunter and Zweifel,[37] however, the reduced frequency of jack mackerel was found to decrease slightly with increasing size (with a mean value of $\omega l/V$ equal to 11 for $l = 4·5$ cm, decreasing to 7·44 for $l = 27$ cm), while the amplitude remains constant about the value of 0·23 l.

A theoretical estimate of the scaling of ω can be obtained by applying slender-body theory (Lighthill,[42] Wu,[76] Newman and Wu[53] and Newman[52]) which provides the following approximate relation for the mean thrust

$$T \propto \rho b_0^2 \omega^2 a_0^2 (1 - V/c), \tag{37}$$

where a_0 represents the amplitude of oscillation of the fish tail, b_0 its characteristic depth, and $c = \omega/k = f\lambda$ denotes the wave velocity of the wave (of frequency f and wavelength λ) which is propagated distally along the body. The dimensionless coefficient multiplying (37) is either constant or very nearly scale-independent. If we assume that c/V (or, equivalently, the hydromechanical efficiency) is scale-independent, equating T and the drag D (see Eqn 23) yields

$$\omega l/V = \text{const.} \, C_D^{1/2} \tag{38}$$

in which we have taken a_0 and b_0 to be proportional to l and S to l^2. Thus, for laminar flow, with V proportional to $l^{1/2}$ (see Eqn 32 and its sequel), we obtain

$$\omega l/V = \text{const.} \, l^{-3/8}. \tag{39}$$

Though this estimate is somewhat crude, it is in qualitative accord with the observed data of Hunter and Zweifel. (The scaling of $\omega l/V$ would be weaker than $l^{-3/8}$ for smaller $\beta < 0·5$). The corresponding scaling of ω is

$$\omega = \text{const.} \, l^{-7/8} = \text{const.} \, l^{-0·88}. \tag{40}$$

Further, if c/V is again assumed to be scale-independent (which implies that the hydromechanical efficiency is likewise scale-independent), it then follows that the wavelength λ will scale as

$$\lambda = c/f = (V/fl)(c/V)l = \text{const. } l^{11/8}. \qquad (41)$$

It should be of interest to examine these scaling relations when the necessary data are available.

In passing we note that the wing beat frequency of many species of birds tends to be proportional to $m^{-0.6}$ or to $l^{-0.78}$ (Greenewalt,[32] Alexander (Ref. 1, p. 7)) which is quite close to the expression (40) for fish.

Viscous Resistance

As we noted earlier, reasonable accuracy in the measurement of metabolic-rate, coupled with direct observations of pertinent parameters such as the fish mass, length, swimming velocity, and the frequency and wavelength of its tail beat, has enabled us not only to carry out the similarity study but also yields valuable information about the drag coefficient. This information appears to contain indirect evidence concerning both the functional dependence of C_D on the Reynolds number Re and, even, its order of magnitude, which in turn may throw light on whether there is transition from laminar to turbulent flow in the boundary layer and on the possible occurrence of flow separation in unsteady flows. To acquire such information by direct measurement for freely swimming fish has proved too difficult a task, and it is worth commenting on this.

The great difficulties involved in measuring the drag on a swimming fish directly has forced researchers to assume that it is the same as the drag on an equivalent straight rigid body. Therefore many attempts have been made to measure the viscous drag of either a mechanical model, or a paralyzed or pickled fish, in steady flow in wind or water-tunnels, towing tanks, or water tanks for drop-down tests; another approach has been to observe the retardation of a fish in glide (for brief reviews see Gray[30a], Newman and Wu[53]). Generally speaking, the measured values of the drag coefficient C_D show wide scatter over a range of a few times to tens of times the drag coefficient for turbulent skin friction on a flat plate of the same area at the same Reynolds number. However, there are reports (Lang and Daybell[41], Webb[73]) indicating that the measured drag is close to that of an equivalent mechanical model. Drastic improvement in experimental accuracy is urgently needed, but difficult to achieve.

I would like to remark that the basic assumption on the equivalence of swimming drag to model drag may not be generally valid. This is because the pressure field around a swimming fish must differ considerably from that about a rigid model, and this difference will fundamentally alter the boundary layer calculation. At an appropriate frequency and phase, the undulatory motion may delay both the transition to turbulence and the flow separation.

CONCLUSION

In the field of aquatic animal locomotion the number of groups of animals which are fundamentally distinct, from either the hydromechanical or the physiological point of view, has been placed by Lighthill[43] to be about 10. The Reynolds numbers achieved in their locomotion span the range from 10^{-6} for eukaryotic flagella to about 10^8 for the giant blue whale. The different modes of propulsion include the flagellar and ciliary wave motions of protozoa, the undulatory movements in anguilliform, carangiform and lunate-tail swimming of many fish, the flying mode adopted by rays, turtles and penguins, the jet propulsion of squids, and so forth. Equally vast in variety are their distinct physiological characteristics and behavior. Nevertheless, by studying the different scaling problems some quite definite relations between the key parameters have emerged, and these seem to exhibit certain underlying principles. The field is extremely complex, and I have concentrated on only a few points where the problems are relatively easy, and have pointed out some hopeful possibilities for the future. I am quite sure that the conspicuous gaps which I have left in my discussion will be closed by other speakers with their enlightening observations. May I reiterate my conviction that collaborative studies in this multidisciplinary subject can be surprisingly rewarding.

ACKNOWLEDGMENT

I would like to express my deep gratitude to Professor Charles Brokaw, Dr Allan Chwang, Dr Howard Winet, Dr Anthony Cheung, and Mr George Yates not only for the extremely valuable discussions I have had with them, but also for their splendid support in providing me with important knowledge and material throughout the course of preparation of this work. I am also most thankful to Helen Burrus for her experienced assistance in many ways.

This work was jointly sponsored by the Office of Naval Research and the National Science Foundation. Their continuous support is gratefully acknowledged.

REFERENCES

1. Alexander, R. McNeill. "Size and Shape". Studies in Biology No. 29. Edward Arnold Ltd., London (1971).
2. Bainbridge, R. The speed of swimming of fish as related to size and to the frequency and amplitude of the tail beat. *J. exp. Biol.* **35**, 109–133 (1958a).
3. Bainbridge, R. The locomotion of fish. *New Scient.* **4**, 476–478 (1958b).
4. Bainbridge, R. Speed and stamina in three fish. *J. exp. Biol.* **37**, 129–153 (1960).
5. Bainbridge, R. Problems of fish locomotion. *In* "Vertebrate Locomotion", (Harrison, J. E., ed.), *Symp. Zool. Soc. Lond.* **5**, 13–32 (1961).
6. Bainbridge, R. Training, speed and stamina in trout. *J. exp. Biol.* **39**, 537–555 (1962).
7. Bainbridge, R. Caudal fin and body movement in the propulsion of some fish. *J. exp. Biol.* **40**, 23–56 (1963).
8. Batchelor, G. K. Slender-body theory for particles of arbitrary cross-section in Stokes flow. *J. Fluid Mech.* **44**, 419–440 (1970).
9. Beamish, F. W. H. Swimming endurance of some northwest Atlantic fishes. *J. Fish. Res. Board Can.* **23**, 341–347 (1966).
10. Blake, J. R. A spherical envelope approach to ciliary propulsion. *J. Fluid Mech.* **46**, 199–208 (1971a).
11. Blake, J. R. Infinite model for ciliary propulsion. *J. Fluid Mech.* **49**, 209–222 (1971b).
12. Blake, J. R. A model for the micro-structure in ciliated organisms. *J. Fluid Mech.* **55**, 1–23 (1972).
13. Blake, J. R. and Sleigh, M. A. Mechanics of ciliary locomotion. *Biol. Rev.* **49**, 85–125 (1974).
14. Blake, J. R. and Sleigh, M. A. Hydromechanical aspects of ciliary propulsion. *In* "Swimming and Flying in Nature". Plenum Publ., New York (1975).
15. Blaxter, J. H. S. Swimming speeds of fish. *FAO Fish. Rep.* **62**, **2**, 69–100 (1967).
16. Blaxter, J. H. S. and Dickson, W. Observations on the swimming speeds of fish. *J. Cons. Int. Explor. Mer.* **24**, 474–479 (1959).
17. Brennen, C. An oscillating-boundary-layer theory for ciliary propulsion. *J. Fluid Mech.* **65**, 799–824 (1974).
18. Brett, J. R. The energy required for swimming by young sockeye salmon with a comparison of the drag force on a dead fish. *Trans. R. Soc. Can.* **1**, Sec. IV, 441–457 (1963).
19. Brett, J. R. The respiratory metabolism and swimming performance of young sockeye salmon. *J. Fish Res. Board Can.* **21**, 1183–1226 (1964).
20. Brett, J. R. The relation of size to the rate of oxygen consumption and sustained swimming speeds of sockeye salmon (*Oncorhynchus nerka*). *J. Fish. Res. Board Can.* **22**, 1491–1501 (1965a).
21. Brett, J. R. The swimming energetics of salmon. *Sci. Am.* **213**, 80–85 (1965b).
22. Brett, J. R. Swimming performance of sockeye salmon (*Oncorhynchus nerka*) in relation to fatigue time and temperature. *J. Fish. Res. Board Can.* **24**, 1731–1741 (1967a).
23. Brett, J. R. "Salmon". *In* "Encyclopaedia of Science and Technology" McGraw-Hill. 348–349 (1967b).
23a Brett, J. R. and Sutherland, D. B. Respiratory metabolism of Pumpkinseed (*Lepomis gibbosus*) in relation to swimming speed. *J. Fish. Res. Board Can.* **22**, 405–409 (1965).

24. Brokaw, C. J. and Gibbons, I. R. Mechanisms of movement in simple flagella and cilia. *In* "Swimming and flying in Nature", Plenum Publ., New York (1975).

25. Chwang, A. T. and Wu, T. Y. A note on the helical movement of micro-organisms. *Proc. R. Soc. Lond.* **B178**, 327–346 (1971).

26. Cox, R. G. The motion of long slender bodies in a viscous fluid. Part 1. General theory. *J. Fluid Mech.* **44**, 791–810 (1970).

27. Dahlberg, M. L., Shumway, D. L. and Doudonoff, P. Influence of dissolved oxygen and carbon dioxide in the swimming performance of large-mouthed bass and coho salmon. *J. Fish. Res. Board Can.* **25**, 49–70 (1968).

28. Drabkin, D. L. Imperfection: Biochemical phobias and metabolic ambivalence. *Perspectives Biol. Med.* **2**, 473–517 (1959).

29. Gray, J. Studies in animal locomotion. *J. exp. Biol.* **13**, 192–199 (1936).

30. Gray, J. How fish swim. *Sci. Am.* **197**, 48–54 (1957).

30a Gray, J. "Animal Locomotion". Weidenfeld and Nicolson, London (1968).

31. Gray, J. and Hancock, G. J. The propulsion of sea-urchin spermatozoa. *J. exp. Biol.* **32**, 802–814 (1955).

32. Greenewalt, C. H. Dimensional relationships for flying animals. *Smithson Misc. Collns.* **144**(2), 1–46 (1962).

33. Hancock, G. J. The self-propulsion of microscopic organisms through liquids. *Proc. R. Soc. Lond.* **A127**, 96–121 (1953).

34. Heusner, A., Kayser, C., Marx, C., Stussi, T. and Harmelin, M. L. Relation entre le poids et la consommation d'oxygene. II. Etude intraspecifique chez le poisson. *C.R. Soc. Biol.* **157**(3), 654 (1963).

34a Holwill, M. E. J. Hydrodynamic aspects of ciliary and flagellar movement. Chapt. 8 in "Cilia and Flagella", (M. A. Sleigh, ed.), 143–175. Academic Press, New York (1974).

34b Holwill, M. E. J. The role of body oscillation in the propulsion of micro-organisms. *In* "Swimming and Flying in Nature". Plenun Publ., New York (1975).

35. Hunter, J. R. Sustained speed of jack mackerel. *Fish. Bull.* **69**, 267–271 (1971).

36. Hunter, J. R. Swimming and feeding behaviour of larval anchovy *Engraulis mordax. Fish. Bull.* **70**, 821–838 (1972).

37. Hunter, J. R. and Zweifel, J. R. Swimming speed, tail beat frequency, tail beat amplitude and size in jack mackerel, *Trachurus symmetricus* and other fishes. *Fish. Bull.* **69**, 253–266 (1971).

38. Jones, D. J. Theoretical analysis of factors which may limit the maximum oxygen uptake of fish. *J. theor. Biol.* **32**, 341–349 (1971).

39. Keller, S. R. "Fluid Mechanical Investigations of Ciliary Propulsion". Ph.D. Thesis, Calif. Inst. of Technology, Pasadena, Calif. (1975).

40. Kleiber, M. Prefatory chapter: An old professor of animal husbandry ruminates. *Ann. Rev. Physiol.* **29**, 1–20 (1967).

41. Lang, T. G. and Daybell, D. A. Porpoise performance tests in a seawater tank. NAVWEPS Rep. 8060, NOTS TP 3063. Naval Ordnance Test Station, China Lake, Calif. (1963).

42. Lighthill, M. J. Note on the swimming of slender fish. *J. Fluid Mech.* **9**, 305–317 (1960).

43. Lighthill, M. J. Hydromechanics of aquatic animal propulsion. *Ann. Rev. Fluid Mech.* **1**, 413–445 (1969).

44. Lighthill, M. J. Aquatic animal propulsion of high hydromechanical efficiency.

   J. Fluid Mech. **44**, 265–301 (1970).
45. Lighthill, M. J. Large-amplitude elongated-body theory of fish locomotion. *Proc. R. Soc. Lond.* **B179**, 125–138 (1971).
46. Lighthill, M. J. "Aquatic animal locomotion". *In* Proc. 13th Intern. Congr. IUTAM (August, Moscow). Springer Verlag, Berlin (1972).
47. Lighthill, M. J. Scaling problems in aquatic locomotion. Symposium at the Duke University, 1973 (unpublished).
48. Lighthill, M. J. "Mathematical Biofluiddynamics", SIAM, Philadelphia (1975).
49. Lighthill, M. J. Flagellar hydrodynamics. The John von Neumann Lecture, *SIAM Reviews,* 1975.
50. Magnuson, J. J. Hydrostatic equilibrium of *Euthynnus affinis,* a pelagic teleost without a gas bladder. *Copeia,* 56–85 (1970).
51. Magnuson, J. J. Comparative study of adaptations for continuous swimming and hydrostatic equilibrium of scombroid and xiphoid fishes. *Fish. Bull.* **71**, 337–356 (1973).
52. Newman, J. N. The force on a slender fish-like body. *J. Fluid Mech.* **58**, 689–702 (1973).
53. Newman, J. N. and Wu, T. Y. A generalized slender-body theory for fish-like forms. *J. Fluid Mech.* **57**, 673–693 (1973).
54. Newman, J. N. and Wu, T. Y. Hydromechanical aspects of fish swimming. *In* "Swimming and Flying in Nature". Plenum Publ., New York (1975).
55. Paulick, G. J. and DeLacy, A. C. Swimming ability of upstream migrant silver salmon, sockeye salmon and steelhead at several water velocities. Univ. Wash. Coll. Fish. Tech. Rep. 44 (1957).
56. Schmidt-Nielsen, K. Energy metabolism, body size, and problems of scaling. *Fed. Proc.* **29** (4), 1524 (1970).
57. Schmidt-Nielsen, K. Locomotion: Energy cost of swimming, flying, and running. *Science,* **177**, 222–228 (1972a).
58. Schmidt-Nielsen, K. "How Animals Work". Cambridge University Press (1972b).
58a Smit, H. Some experiments on the oxygen consumption of goldfish (*Carassius auratus* L) in relation to swimming speed. *Can. J. Zool.* **43**, 623–633 (1965).
59. Smit, H., Amelink-Houtstaal, J. M., Vijverberg, J. and von Vaupel-Klein, J. C. Oxygen consumption and efficiency of swimming goldfish. *Comp. Biochem. Physiol.* **39A**, 1–28 (1971).
60. Taylor, G. I. Analysis of swimming of microscopic organisms. *Proc. R. Soc. Lond.,* **A209**, 447–461 (1951).
61. Tillet, J. P. K. Axial and transverse Stokes flow past slender axisymmetric bodies. *J. Fluid Mech.* **44**, 401–417 (1970).
62. Tuck, E. O. A note on a swimming problem. *J. Fluid Mech.* **31**, 305–308 (1968).
63. Tucker, V. A. Energetic cost of locomotion in animals. *Comp. Biochem. Physiol.* **34**, 841–846 (1970).
64. Tucker, V. A. Aerial and terrestrial locomotion: A comparison of energetics. In Bolis, L., Schmidt-Nielsen, K. and Maddrell, S. H. P. (eds.), "Comparative Physiology", North-Holland (1973).
65. Tucker, V. A. The energetic cost of moving about. *Am. Scientist,* **63**, No. 4, 413–419 (1975).
66. von Karman, Th. and Gabrielli, G. What price speed? Specific power required for propulsion of vehicles. *Mech. Engng.* **72**, 775–781, 1950 Thurston Lecture (Collected works of Theodore von Karman, IV, 399–414) (1950).

I

67. Webb, P. W. "Some Aspects of the Energetics of Swimming of Fish with Special Reference to the Cruising Performance of Rainbow Trout". Ph.D. Thesis, Univ. Bristol, Bristol, England (1970).
68. Webb, P. W. The swimming energetics of trout. I. Thrust and power output at cruising speeds. *J. exp. Biol.* **55**, 489–520 (1971a).
69. Webb, P. W. The swimming energetics of trout. II. Oxygen consumption and swimming efficiency. *J. exp. Biol.* **55**, 521–540 (1971b).
70. Webb, P. W. Effects of partial caudal-fin amputation on the kinematics and metabolic rate of underyearling sockeye salmon (*Oncorhynchus nerka*) at steady swimming speed. *J. exp. Biol.* **59**, 565–581 (1973a).
71. Webb, P. W. Kinematics of pectoral fin propulsion in *Cymatogaster aggregata*. *J. exp. Biol.* **59**, 697–710 (1973b).
72. Webb, P. W. Pisces (zoology) bioenergetics. McGraw-Hill, Encycl. Sci. Technol. Yearb. 1973, 333–336 (1974).
73. Webb, P. W. Hydrodynamics and energetics of fish propulsion. Bulletin 190, Fish. Res. Board. Can., Ottawa, Canada (1975).
74. Weihs, D. Mechanically efficient swimming techniques for fish with negative buoyancy. *J. Marine Res.* **31**, 194–209 (1973).
75. Winberg, G. G. Rate of metabolism and food requirements of fishes. Bylorussian State University. Minsk 251 p. (Transl. Fish. Res. Bd. Canada, No. 194) (1956).
76. Wu, T. Y. Hydromechanics of swimming propulsion. Part 3. Swimming and optimum movements of slender fish with side fins. *J. Fluid Mech.* **46**, 545–568 (1971).
77. Wu, T. Y. Fluid mechanics of ciliary propulsion. Proc. 10th Anniv. Meeting Soc. Eng. Sci., North Carolina State U. Raleigh, North Carolina (1973).

14. Low Reynolds Number Undulatory Propulsion in Organisms of Different Sizes

M. E. J. HOLWILL

Physics Department, Queen Elizabeth College, London, England.

ABSTRACT

Studies of undulatory propulsion at low Reynolds numbers have involved observations of motility in bacteria, eukaryotic cells, spermatozoa and small worms. Throughout the considerable range of sizes represented by these organisms it is found that, with few exceptions, the ratio of wave amplitude to wavelength is essentially that which corresponds to a maximum hydrodynamic efficiency, proportional to the ratio of the square of the propulsive velocity to the power expended against viscous forces. For organisms which propagate three dimensional waves, the ratio of the diameter of the cell body to that of its propelling filament also has an optimal value in terms of propulsive efficiency. Some cells seem to have evolved mechanisms to optimise this ratio while for others the ratio appears to be disadvantageous. The number of waves observed on the flagella of micro-organisms varies considerably among cell types, being one or two for many protozoa and six or more for bacteria and certain sperm. For bacteria, at the lower end of the size-scale, the waveform is probably determined by structural considerations, since the flagellum seems to be a more or less rigid helix, but the sperm tails, which can be a millimetre or more long, may support several waves for hydrodynamic reasons. Preliminary comparative studies of the fluid flow fields induced by flagella and small worms provide information about the effects of solid boundaries (such as a glass coverslip) on motility, and suggest that a reappraisal of certain experimental results may be necessary.

INTRODUCTION

Studies of undulatory propulsion at low Reynolds numbers have involved investigations of motility in bacteria, eukaryotic cells, spermatozoa and

233

small worms. In the first three systems, motility is induced by the activity of one or more whip-like appendages known as flagella which are attached to a cell body and produce the motive force. The flagellar wave movements of eukaryotic cells and spermatozoa may be two- or three-dimensional and may or may not be symmetric; these cells propagate waves of active bending for which the motive force is believed to be derived from a sliding filament mechanism (Satir,[22] Warner and Satir[24]). There is evidence (Berg[2]) to suggest that bacterial flagella are rigid helices which, during movement, are rotated by means of a "motor" at the base. This view is not universally accepted (see e.g., Calladine[4]), and alternative mechanisms consider waves of bending propagated by sequential deformations and movements of the sub-units which comprise the flagellum. For small worms, which are essentially isolated filaments, the wave propagation is achieved by muscular effort and the undulations are planar (Gray and Lissman[9]).

TABLE I. Parameters associated with undulatory propellers.

	Bacterial flagella	Eukaryotic flagella	Sperm tails	Small worms
Diameter (μm)	0·02	0·2	0·2–1·0	20–40
Length (μm)	1–5	10–100	10–1000	1000
Reynolds Number (based on diameter)	10^{-6}	10^{-3}	10^{-3}–10^{-2}	10^{-1}

The systems to be discussed thus have quite different mechanisms by which bending is achieved and also show considerable variation in size. Individual bacterial flagella can be as small as 10 nm in diameter and a few μm in length, although there is a tendency for bacteria to bear many flagella and for these to aggregate during motility, thereby having the appearance, but possibly not the effect, of a filament many times the radius of an individual organelle. The majority of eukaryotic flagella and a number of sperm tails have a diameter of about 0·2 μm with lengths varying from tens to hundreds of μm. Some sperm, in particular those of mammals and insects, have significantly greater diameters than this (see e.g., Afzelius,[1] Phillips[21]) while worms have diameters in excess of 10^{-1} mm. Thus, there is a range of sizes in terms of filament diameter from 10 nm to 10^{-1} mm, a ratio of 1:10^{4}, while the variation in length is considerably less, giving a ratio of at most 1:100 between the extremes. This information is summarized in Table I, together with estimates for the Reynolds numbers based on filament diameter. The Reynolds number, which expresses the order of magnitude of forces derived from mass fluid movements (inertial forces) to those arising from viscous action, is very low so that in the systems to be discussed the propulsive forces are predominantly viscous.

A wave is characterized by specifying its shape (e.g. sinusoidal) together with the frequency, wavelength and wave amplitude; these are the parameters which, with the filament dimensions, are used in hydrodynamic calculations of the thrust and power developed by an undulating filament. In making comparisons between systems of differing size, it is convenient to express equations in non-dimensional terms wherever possible and the

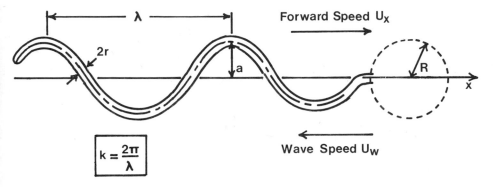

Parameters used to describe undulatory propulsion

FIG. 1. Diagrammatic representation of a system propelled by an undulating filament showing symbols used in the text.

hydrodynamic analyses of low Reynolds number undulating propulsion provide expressions containing the ratios of amplitude (a) to wavelength (λ) and of propulsive speed (u_x) of organism or fluid to wave speed (u_w) (see Fig. 1). A number of theoretical investigations which consider the major types of wave observed on micro-organisms have been made, and it is possible from them to estimate the propulsive velocity and power expenditure for two- and three-dimensional waves of uniform and non-uniform character (Taylor,[23] Gray and Hancock,[8] Holwill and Miles,[15] Coakley and Holwill,[7] Chwang and Wu[6]). In developing the appropriate equations, much use has been made of the concept of force coefficients which were first successfully applied to flagellar movements by Gray and Hancock[8] (see also the above paper by Wu). In using the coefficients it is assumed that each element of an undulating system generates the same force as it would if it were part of an infinitely long cylinder moving with the same velocity. In their analyses, Gray and Hancock[8] considered those coefficients (C_T, C_N) characterized respectively by motion along and normal to the cylinder axis, and obtained from Hancock's[10] earlier work the relationships

$$C_N = 2C_T \tag{1}$$

and

$$C_T = \frac{2\pi\mu}{\ln(2\lambda/r) - \frac{1}{2}} \tag{2}$$

where μ is the fluid viscosity and r is the filament radius (Fig. 1). It has been shown experimentally that the ratio C_N/C_T lies closer to 1·8 than to 2 (Brokaw[3]) and recently Lighthill[17] has developed a theoretical analysis of flexible filaments which predicts the smaller ratio. The logarithm term of Eqn (2) is an approximation and improvements have been suggested by several authors, but all have a form similar to that of Eqn (2) in that they depend on a ratio of length to radius. Since the ratio appears in a logarithm, the force coefficients are relatively insensitive to its value which, together with the large experimental errors incurred in observing motility of micro-organisms, probably accounts for its successful use in predicting propulsive speeds (Fig. 2). In testing the validity of refinements to Eqns (1) and (2) the study of a variety of organisms with different slenderness ratios will be valuable.

Over the range of sizes available there appears to be no correlation between propulsive speed expressed in wavelengths per second and the wavelength itself, although satisfactory comparisons are not possible because of the range of head sizes propelled. Even for organisms with rather small heads, no correlation is obvious, although the ratio of propulsive speed to wave speed is roughly constant at 0·2. For organisms with larger head sizes, the ratio is reduced, as expected.

HYDRODYNAMIC EFFICIENCY

If we consider the undulatory system from a purely hydrodynamic stand-point, we can calculate a hydrodynamic efficiency. This is equal to the rate at which an external force would have to do work against the drag, if it were pulling the animal rigidly through the water at speed U_x, per unit of power, P, expended by the animal in propelling itself at the same speed (see Wu's paper). Since the drag is proportional to the speed (at low values of the Reynolds number), the efficiency is proportional to U_x^2/P. For both planar and three-dimensional waves this quantity passes through a maximum when plotted as a function of $2\pi a/\lambda$ (Fig. 3; Holwill and Burge,[14] Chwang and Wu[6]). Experimentally it is found that, over a wide range of organism sizes, the value of $2\pi a/\lambda$ is about 1 and thus lies close to the optimal value predicted

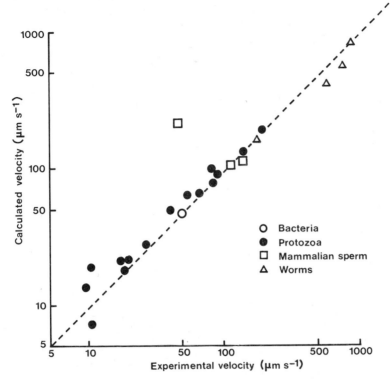

FIG. 2. Comparison between experimentally observed velocities and theoretically predicted velocities for a number of organisms. ○ Bacteria; ● Eukaryotes; □ Mammalian sperm; △ Small worms.

TABLE II. Values of $2\pi a/\lambda$ for undulatory propellers.

Propeller		$2\pi a/\lambda$	Ref.
Bacterial flagellum		0·8–1·6	Liefson[17]
Eukaryotic flagellum—smooth		1·1–1·9	Holwill[11]
	hispid	1–3·9	Holwill[11]
Sperm tails		1 05–1·6	Holwill[11]
Small worms		0·9–1·5	Gray and Lissman[9]

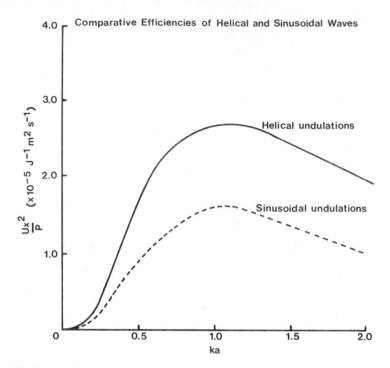

FIG. 3. Variation of propulsive efficiency (expressed as ratio of the square of the propulsive velocity, U_x, to power expended, P) with $2\pi a/\lambda = ka$.

in Fig. 3 (Table II). This is presumably an example of evolutionary optimisation but it is interesting that many eukaryotic cells when exposed to media of differing viscosities alter their waveforms but maintain the optimal ratio of amplitude to wavelength. These organisms have the ability to adapt efficiently to a situation rarely encountered *in vivo*, and it is tempting to speculate that a feedback system exists in these cells to relay information about the environment to the mechanism which bends the organelle.

For a system propelled by helical waves, Chwang and Wu[6] have shown that an optimal ratio exists between the radius of the head and that of the propelling filament. The optimum value depends on the amplitude/wavelength ratio and for the flagellated systems under discussion lies between

15 and 40. The radius of the flagellar bundle of certain bacteria is such that the ratio lies in this range, so it appears that a sufficient number of individual flagella aggregate to achieve optimal hydrodynamic efficiency. Again, it is possible that hydrodynamic considerations during evolution determined the optimal number of flagella for certain bacteria, but some cells with single organelles have a head/tail ratio of over 100 (see Weibull[25]) which is well outside the optimal range, so this possibility should be considered carefully with other evidence. The details of flagellar motion in bacteria are still controversial and rather few experimental observations of "normal" motility have been made, so that further studies are needed before it is possible to make a general statement about propulsive efficiency in these cells.

Not many experimental observations have been made on eukaryotic cells propelled by helically undulating flagella and in view of the fact that practically all flagella with the 9 + 2 microtubular structure have a diameter of 0·2 μm, and the effective cell diameter of organisms propelled by helical waves is of order 20 μm, it might be expected that these organisms operate at less than optimal efficiency. However, the flagellum of *Euglena viridis* is known to bear fine hairs and it is believed that these are wrapped about the flagellum *in vivo*. As noted by Holwill[13] this effectively increases the diameter of the propelling filament to give a head/tail diameter ratio in the optimal region. Other organisms are known to bear extra flagellar structures and it may be that their purpose is to increase the effective filament diameter to give optimal hydrodynamic efficiency. It would be interesting to study the motile behaviour of some of these cells to discover whether they operate at maximal efficiency.

From their studies of power expenditure, Chwang and Wu[6] conclude that it is more advantageous (in terms of propulsive velocity relative to power input) to use helical waves than planar waves for propulsion when the ratio of body radius to propeller radius is greater than 10, while plane waves are more efficient when this ratio is less than 5. The analysis is not sufficiently accurate to determine the best type of propeller in the range $5 < R/r < 10$. Experimentally it is found that bacteria and some eukaryotic cells with helical flagella have $R/r > 10$, and for many spermatozoa and the small worms which are propelled by planar undulations the ratio is less than 5. However, some eukaryotes and sperm with $R/r > 10$ are propelled by plane waves, thus indicating that other factors, possibly associated with the internal cell structure, determine the wave type appropriate to a particular organism.

The flagella of many eukaryotic cells and the small worms adopt a bend shape which contains about one wavelength during movement (Holwill,[11] Gray and Lissman[9]) while bacterial flagella and insect sperm often bear many waves (Weibull,[25] Phillips[21]). The wave shapes are, as noted above, optimal from the point of view of hydrodynamic efficiency as defined earlier. The power required to sustain an undulation is roughly proportional to the square of the amplitude (Carlson,[5] Holwill and Burge[14]) and to the viscosity, if the beat frequency remains unaltered. If the rate at which energy is supplied from within the system is maintained constant, the variations in wave pattern among organisms and the modifications induced by changes in viscosity may reflect efforts by the organisms to optimise their power outputs.

FLOW FIELDS

In my laboratory, John Lunec has designed an experiment based on scaling to solve a problem in fluid flow around smooth flagella. Experiments showed that the fluid near a smooth flagellum performed a circulatory movement close to the flagellar envelope (Lunec[19]), a result not predicted theoretically. By using small worms in viscous fluids to give dynamical similarity, he was able to position them with a micro-manipulator and found that the circulatory pattern occurred when the worm was close to a plane boundary. Although he is still quantifying the system, the presence of a microscope slide and coverslip clearly have significant effects upon the fluid flow, and may well affect such parameters as the propulsive force and power expenditure. Results from this investigation may require the re-appraisal of experiments performed on micro-organisms close to surfaces.

CONCLUDING REMARKS

A study of scaling undulatory propulsion at low Reynolds numbers raises more questions than it answers. While it is interesting that the widely differing systems (both in size and type) adopt configurations of maximum efficiency in hydrodynamic terms, the mechanisms by which they have done so are not clear. There is considerable evidence to suggest that the internal machinery which bends the system can respond, in certain cases, to alterations in the environment and models have been produced to indicate how external changes can affect the internal behaviour of eukaryotic flagella (Lubliner and Blum,[18] Miles and Holwill[20]). These models are specific to flagella and contain many variable parameters, including, for example, the flexural rigidity of the system and its internal viscosity, and a discussion in terms of scaling would be fruitless in the light of the information available. Some authors (see Holwill, Ref. 12 for references) have used large scale models of micro-organisms to study hydrodynamic aspects of their movement, but the information so obtained is limited and provides no insight into the mechanisms which are used to bend organelles or to control their movements.

From a survey of the literature it is apparent that a considerable amount of experimental information is available describing the movement of eukaryotic cells and spermatozoa, but few studies have been made of bacterial motility and of the movement of small worms. The available material suggests that throughout the size range the majority of organisms adopt patterns which yield optimum efficiency in hydrodynamic terms, but it would be of interest to obtain more data at the extremes of the size range so that further comparative studies could be made.

REFERENCES

1. Afzelius, B. A. (Ed.) "The Functional Anatomy of the Spermatozoon". Pergamon Press, Oxford (1975).
2. Berg, H. C. Bacterial behaviour. *Nature*, **254**, 389–392 (1975).
3. Brokaw, C. J. Bending moments in free-swimming flagella. *J. exp. Biol.* **53**, 445–464 (1970).
4. Calladine, C. R. *Nature* **249**, 385 (1974).
5. Carlson, F. D. The motile power of a swimming spermatozoon. Proc. 1st Nat. Biophys. Conf. 443–449 (1959).
6. Chwang, A. T. and Wu, T. Y. A note on the helical movement of micro-organisms. *Proc. Roy. Soc. Lond.* **B178**, 327–346 (1971).
7. Coakley, C. J. and Holwill, M. E. J. Propulsion of micro-organisms by three-dimensional flagellar waves. *J. theor. Biol.* **35**, 525–542 (1972).
8. Gray, G. and Hancock, G. J. The propulsion of sea-urchin spermatozoa. *J. exp. Biol.* **32**, 802–814 (1955).
9. Gray, J. and Lissman, H. W. The locomotion of nematodes. *J. exp. Biol.* **41**, 135–154 (1964).
10. Hancock, G. J. The self-propulsion of microscopic organisms through liquids. *Proc. Roy. Soc. Lond.* **A217**, 96–121 (1953).
11. Holwill, M. E. J. Physical aspects of flagellar movement. *Physiol. Rev.* **46**, 696–785 (1966).
12. Holwill, M. E. J. Scale effect and model micro-organisms. *Nature*, **226**, 1046–1047 (1970).
13. Holwill, M. E. J. Hydrodynamic aspects of ciliary and flagellar movement. In "Cilia and Flagella". Sleigh, M. A. (Ed.). Academic Press, London (1974).
14. Holwill, M. E. J. and Burge, R. E. A hydrodynamic study of the motility of flagellated bacteria. *Arch. Biochem. Biophys.* **101**, 249–260 (1963),
15. Holwill, M. E. J. and Miles, C. A. Hydrodynamic analysis of non-uniform flagellar undulations. *J. theor. Biol.* **31**, 25–42 (1971).
16. Liefson, E. "Atlas of Bacterial Flagellation." Academic Press, New York (1960).
17. Lighthill, M. J. Flagellar Hydrodynamics. The John von Neumann Lecture, 1975. *SIAM Review* (1976).

18. Lubliner, J. and Blum, J. J. Model of flagellar waves. *J. theor. Biol.* **34**, 515–534 (1972).
19. Lunec, J. Fluid flow induced by smooth flagella. *In* "Swimming and Flying in Nature". Wu, T. Y., Brokaw, C. J. and Brenne, C. (Eds) Plenum, New York (1975).
20. Miles, C. A. and Holwill, M. E. J. A mechanochemical model of flagellar activity. *Biophys. J.* **11**, 851–859 (1971).
21. Phillips, D. M. Structural variants in invertebrate sperm flagella and their relationship to motility. *In* "Cilia and Flagella", Sleigh, M. A. (Ed.) Academic Press, London (1974).
22. Satir, P. The present status of the sliding microtubule model of ciliary motion. *In* "Cilia and Flagella", Sleigh, M. A. (Ed.) Academic Press, London (1974).
23. Taylor, G. I. The action of waving cylindrical tails in propelling microscopic organisms. *Proc. Roy. Soc. Lond.* **A211**, 225–239 (1952).
24. Warner, F. and Satir, P. The structural basis of ciliary bend formation. Radial spoke positional changes accompanying microtubule sliding. *J. Cell Biol.* **63**, 35–63 (1974).
25. Weibull, C. *In* "The Bacteria", Gunsalus, I. C. and Stanier, R. Y. (Eds). Academic Press, New York (1960).

15. Methods of Ciliary Propulsion and their Size Limitations

M. A. SLEIGH and J. R. BLAKE

Department of Biology
University of Southampton, England
and CSIRO Division of Mathematics and Statistics, Canberra, Australia.

ABSTRACT

Organisms propelled by cilia lie in a distinct size range with lengths usually between 20 μm and 2 mm; ctenophores propelled by comb plates extend to lengths of about 10 cm. This range overlaps at the lower end of the size scale with flagellated organisms and at the upper end with bodies propelled by muscular activity. Data from new studies on ciliated organisms throughout the size range are introduced and are discussed from the point of view of scaling. Most ciliated bodies are propelled at about 1 mm s^{-1}, so that the velocity of propulsion in body lengths s^{-1} decreases with increasing size. Velocities of propulsion depend upon ciliary length and beat frequency, which remain fairly constant through the whole size range.

Two Reynolds numbers are appropriate to ciliary propulsion, one based on the cilia and the other on the body; both are usually much less than one, indicating that ciliated bodies swim by using viscous mechanisms instead of the inertial effects associated with flying and swimming of larger bodies. The velocity of propulsion, bending moment, force and rate of working are scaled with respect to the beat frequency, cilia length and viscosity of the ambient medium. From these parameters an efficiency is defined in which the rate of working of an equivalent inert body moving at the swimming speed of the organism is compared with the rate of working of all the cilia. This efficiency decreases rapidly as body length increases indicating that the most suitable size for ciliated bodies is below about 350 μm. The inefficiency at longer lengths is due to the vast numbers of cilia, often in excess of 10^6. The minimum effective length of a cilium determines the lower limit of size of ciliated bodies at about 20 μm. The swimming of the "ciliated" *Opalina* conforms better with the series of flagellated organisms than with ciliated ones, and the ctenophores are exceptional in exploiting inertial features of compound cilia with a relatively high Reynolds number.

243

INTRODUCTION

Many small organisms propel themselves by flagella or cilia. These two types of organelle are both cellular projections whose motion in the surrounding fluid generates viscous forces that provide propulsive thrust. The distinction between the two types of organelle is that flagella cause propulsion by the propagation of waves along their length, while the pronounced unilateral motion of a cilium is the main source of its propulsive forces.

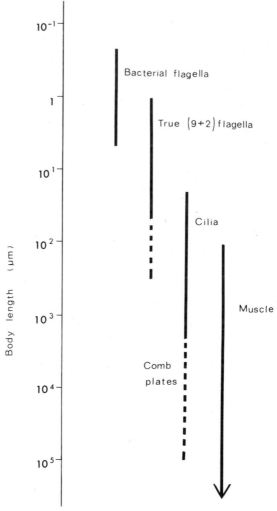

FIG. 1. Scale of linear dimensions of organisms that swim by use of bacterial flagella, true flagella, cilia and muscles.

The range of body sizes within which flagellar and ciliary locomotion is used is indicated in Fig. 1. Bacterial cells about 0·2 to 5·0 μm in diameter are propelled by special types of "bacterial flagella", only about 20 nm in diameter. True flagella and cilia are larger, around 250 nm in diameter, and possess similar geometry and almost identical internal organization wherever they occur. Small algae and protozoa 1 to 50 μm long may swim by one or two flagella, and some larger cells are propelled by many flagella. Cilia normally occur in large numbers; many protozoa, small worms, rotifers and larval stages of many animals in the size range between about 20 μm and 2 mm may swim using hundreds or thousands of cilia, and ctenophores of lengths up to 100 mm or more use large compound swimming cilia. Many small animals use cilia to creep or glide over surfaces, but the extent to which this should be regarded as swimming on a layer of fluid (usually mucus-enriched) is unresolved. Fluids are propelled over stationary surfaces by cilia of many animals, e.g. in the human lung and reproductive tracts, and around the gills and food collecting appendages of invertebrate animals of most groups.

In a survey of the animal kingdom it is seen that organisms of some groups are propelled by flagella, in other groups cilia are used, and members of yet other groups use muscles for propulsion. Seldom does one find more than one means of swimming practised among the members of a single taxonomic group, although there are exceptions, notably the use of different modes of swimming at different stages in the life cycle of an animal, which may have a ciliated larva and a muscular adult form. Small individuals and the juvenile stages of many groups swim by muscles, including organisms, in the range of a few hundreds of microns, that use paddle-like limbs, and others, in about the millimetre range, that are propelled by body undulations or by water-jets from a contractile bell. Clearly the range of sizes of organisms propelled by cilia overlaps extensively with that of flagellates at the lower end and with that of muscular swimmers at the upper end. We know of no estimates of relative efficiency of ciliary swimming in comparison with flagellar or muscular swimming in the regions of overlap. Information on flagellar propulsion will be found in the paper by Holwill in this volume, and in several papers in the proceedings of the symposium on "Swimming and Flying in Nature" (Brennen, Brokaw and Wu[5]); data on the swimming of microscopic animals by muscular means is in general very scanty.

This paper is concerned with those organisms that swim using cilia. We shall discuss the range of sizes in which ciliary propulsion is efficient and the different ways in which cilia are used for swimming by different organisms. We thereby hope to illustrate those features of the length, arrangement and use of cilia that may be relevant to the size of the organism propelled and the efficiency of propulsion.

THE GENERAL CHARACTERISTICS OF CILIA AND THEIR ACTIVITY

Cilia are cylindrical projections from the cell surface and are constructed of a bundle of longitudinal protein fibrils (the axoneme) enclosed within a membrane that is continuous with the cell membrane. The axoneme has a characteristic structure of two separate, central, microtubular fibrils surrounded by a ring of 9 double microtubular fibrils. Changes in the shape of the cilium are believed to be brought about by active sliding of these outer doublets along one another as a result of the formation of cross bridges between adjacent doublets (Satir[13]). Sliding is restrained by the presence of radial connections between the doublets and the central fibrils (Warner and Satir[20]). The diameter of a cilium is normally about 0·25 μm, but the length is variable from about 5 μm to 1200 μm or more. The longer cilia are aggregated to form compound structures; simple cilia are seldom more than 15–20 μm long, whilst compound cilia containing up to a hundred or so ciliary axonemes have lengths in the range of about 25–100 μm, and the longest compound cilia ranging up to 1 mm or more in length have hundreds of thousands of component units. Such compound cilia may commonly have the shape of a tapering cylinder, or, in the largest examples, may be formed into plates of paddles.

The ciliary beat cycle typically consists of two phases, an effective stroke in which the cilium remains fairly straight and swings stiffly through an arc, bending only in the basal region, and a recovery stroke in which a region of bending is propagated along the cilium to the tip. The beat patterns of different cilia vary in detail, a diversity of examples having been described previously (Sleigh[16, 17]). The effective stroke of the *Paramecium* cilium is almost planar, but in the recovery stroke the cilium swings to one side and the distal part moves around its attachment to complete the cycle. The paddle-shaped compound cilium (comb-plate) of *Pleurobrachia* performs a planar and conventional beat cycle in spite of its large size and rectangular surface. The cilia of *Opalina* propagate a bending wave in a manner that makes them appear rather more like flagella.

Cilia normally occur in large numbers, and the cilia of an active ciliated surface are coordinated into metachronal waves which travel across the surface. The wave crests represent cilia that are all in much the same phase of their cycle of beat (usually the effective stroke), and the troughs represent cilia at another phase of the cycle (usually the recovery stroke). The detailed structure of the waves is very variable. Four basic patterns of metachronism were recognised by Knight-Jones[9] according to the relation of the direction of propagation of the wave to the direction of the effective stroke of the beat. The form of the metachronal waves has an influence upon the character of the fluid motion that can be generated. The symplectic metachronism of *Opalina* gives the ciliated surface the appearance of a continuous undulating envelope, from which free cilia do not project; the antiplectic metachronism of the ctenophores makes the comb rows appear something like paddle wheels; and the diaplectic waves may appear like the thread ridges on a continuously rotated screw. Frequently the cilia form a continuous cover over the surface of the cell or cells, but, as the use of the cilia becomes more specialized, there is a tendency for the cilia to occur in narrower bands or even single rows. The cilia of such narrow bands or rows normally propel fluids across the band, but in specialized cases the effective stroke of the ciliary beat may be directed along the band or row.

CILIATED ORGANISMS AND THEIR SWIMMING SPEEDS

A selection of organisms that swim by cilia, covering the range from below 50 μm to about 0·5 m, is illustrated in Fig. 2. Some data on the dimensions and movement of these organisms and their cilia are presented in Table I, and give order of magnitude indications of the effectiveness of cilia in propulsion. While the table covers the whole spectrum of organisms that use cilia for swimming, the large majority of ciliated protozoa, rotifers and marine larvae have lengths between about 50 μm and 500 μm. Relatively few of the larger species in these groups have lengths up to 2 mm, and the majority of swimming acoelomate worms (like *Convoluta* and *Stenostomum*) that are propelled by cilia are not much longer than this. The ctenophores form a distinct group, with lengths in the range of about 1–10 cm, that use giant comb-plate cilia for swimming; *Cestus* is a very large ctenophore that swims by muscular undulations, but uses its comb-plates for orientation movements.

The swimming velocity of different ciliated organisms is quoted in two ways in Table I. All organisms studied, except *Opalina* and *Pleurobrachia*, swim at much the same velocity in μm s^{-1}, irrespective of size. It follows then that in terms of body lengths per second the velocity should decrease with increasing size. This is shown to be true by the data presented in Fig. 3. Outstanding exceptions from the general trend indicated by the 45° line are *Opalina* and *Pleurobrachia*. We have seen that in *Opalina* both the pattern of beating and the metachronism are unusual, showing features that may explain the unexpectedly slow locomotion, and in *Pleurobrachia* the use of giant compound comb plate cilia gives a different character to the swimming (the plate effectively produces a jet as it beats against the wall). Features that limit the use of cilia for swimming at either end of the range of sizes will be discussed below. It is interesting that if we add examples of flagellates to this graph we obtain a second 45° line at the left, which will pass close to the

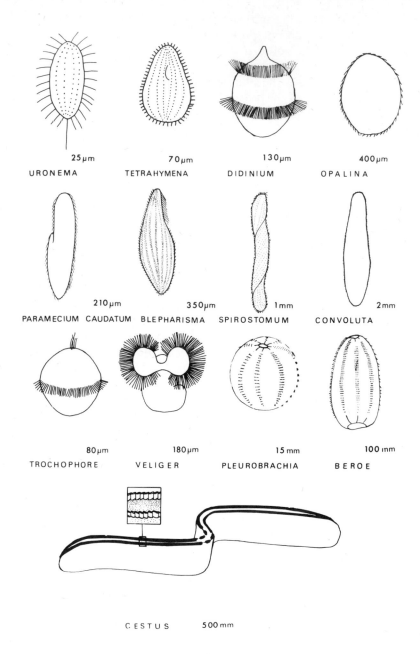

FIG. 2. Some representative organisms propelled by cilia, including those for which data is given in Table I.

TABLE I. Dimensions of features of the body and cilia for representative examples of the range of ciliated organisms.

Name	Group	Body length (L) μm	Swimming‡ velocity μm s⁻¹	Swimming‡ velocity body lengths s⁻¹	R_b	Number of body cilia (n)	Cilia length (l) μm	R_c	nl	Metachronism
Uronema	Protozoa	25	1150	46	10^{-2}	200*	5	10^{-3}	10^3	?
Tetrahymena	Protozoa	70	480	7	10^{-2}	500*	7	10^{-2}	$3 \cdot 5 \times 10^3$	?
Didinium	Protozoa	130	1250	10	10^{-1}	1750	12·5	10^{-2}	$2 \cdot 2 \times 10^4$	Dexioplectic
Paramecium	Protozoa	210	1000	5	10^{-1}	5000	12	10^{-2}	6×10^4	Dexioplectic
Veliger (Aplysia)	Mollusc larva	150	?	?	—	200	56†	10^{-1}	$1 \cdot 1 \times 10^4$	Laeoplectic
Blepharisma	Protozoa	350	600	2	10^{-1}	7000*	7·5	10^{-2}	5×10^4	?
Opalina	Protozoa	400	50	0·12	10^{-2}	10^5	15	10^{-2}	$1 \cdot 5 \times 10^6$	Symplectic
Spirostomum	Protozoa	1000	1000	1	1	10^5	12	10^{-2}	$1 \cdot 2 \times 10^6$	?
Convoluta	Flatworm	2000	600	0·3	1	7×10^6	7·5	10^{-2}	5×10^7	Laeoplectic
Pleurobrachia	Ctenophore	15000	75000	5	10^3	$200(10^8)$	500†	10	$10^5 (10^{10})$	Antiplectic

*Excludes compound cilia; †indicates compound cilia; ‡figures for swimming velocity are "maximal" values, particularly for *Pleurobrachia*.

point for *Opalina*. A third line would occur at the right if we include larger ctenophores like *Beroe*.

It will come as little surprise to see that larger organisms have more cilia, but the relative constancy of ciliary length in most examples indicates a tendency to retain short simple cilia as a means of propulsion even in the larger ciliated forms. Compound cilia tend to be used where the ciliary apparatus has a specialised function in the collection of food as well as or

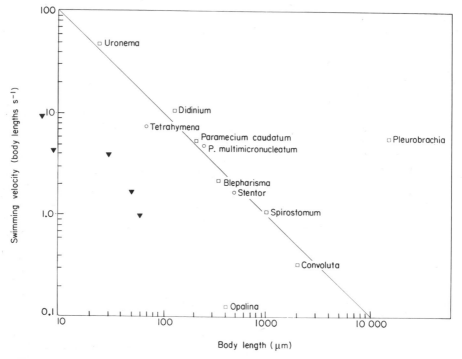

FIG. 3. The relationship between swimming velocity (in body lengths per second) and body length for some ciliated organisms. Square symbols indicate new data, circles indicate data from the literature and triangles indicate a range of flagellated organisms, included for comparison.

instead of swimming. The comb plates of *Pleurobrachia* are again an exception. One might regard the product of ciliary length and ciliary number as a measure of the total energetic machinery of the organism, and it is interesting that a graph of total ciliary length against swimming velocity compares favourably with the straight line with a gradient of $-\frac{1}{2}$ drawn in Fig. 4. Again *Opalina* is an exception, and *Pleurobrachia* comes far from the line if one considers total axoneme length, but lies near the line if one regards the comb plate as a single ciliary unit.

SIZE LIMITATIONS

As observed by Lighthill,[10] the undulatory mode is the principal means of aquatic propulsion; it is relatively insensitive to scale, whereas ciliary propulsion is limited to small organisms where the viscous forces are dominant. Fig. 1 indicates that ciliated organisms exist in a range of sizes from 20 μm to 2 mm and those with comb plates extend up to about 10 cm. We shall now consider the fluid dynamical and internal mechanical constraints that may determine this range.

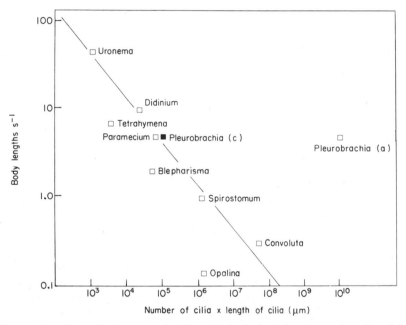

FIG. 4. The relationship between swimming velocity (in body lengths per second) and the total ciliary length (*nl*) for some ciliated organisms. The *Pleurobrachia* (a) value is based on the number of axonemes present, and the (c) value regards the whole comb plate as the ciliary unit.

Propulsion by cilia has interested biologists for several centuries, but it is only in the last few decades that any meaningful attempts have been made to understand the fluid dynamics of ciliary propulsion. Prandtl[12] expressed an interest in the problem, and the theoretical analyses presented by Taylor[18, 19] and Hancock[7] led to an application of hydrodynamic theory to flagellar motion by Gray and Hancock[6] which has provided a foundation for more recent research on propulsion by flagella and cilia. Much of this work has been reviewed and extended by Blake and Sleigh,[3, 4] where the reader will also find information not discussed here because it is less directly related to the scaling aspects of ciliary function.

The external dynamics of organisms propelled by cilia are generally dominated by the viscous forces. We can define two Reynolds numbers, one relevant to the cilia, R_c, the other concerned with the whole organism, R_b, as follows:

$$R_c = \frac{\rho \sigma l^2}{\mu}, \qquad R_b = \frac{\rho U L}{\mu}.$$

Here L and l are the lengths of the body and of the cilium respectively, σ is the radian frequency, U is the velocity and μ and ρ are respectively the dynamic viscosity and density of the fluid. Values of R_c and R_b for various organisms can be found in Table 1. Except for the larger comb plates R_c is almost always much less than one, indicating the dominance of viscous effects. The body Reynolds number R_b is not as important as R_c; the exterior flow field around the organism is nearly irrotational, since the cilia bring the fluid from the zero velocity no-slip condition up to the required irrotational "slip velocity" (see Blake[2]). Viscous effects damp out any time or spatial variations extremely quickly in the exterior flow field (in a fraction of a metachronal wave length). The fluid dynamical equations are linear, and it follows that drag and lift (or normal and tangential resistance coefficients) are linearly dependent on velocity, not on the square of velocity as occurs at the higher Reynolds numbers of flying and fish swimming. Increasing the viscosity of the medium does not change the essential mechanics of the ciliary beat, since the Reynolds number is already much less than one, but the beat pattern and metachronism can change considerably (Sleigh,[14, 15] Machemer[11]).

One might expect from the earlier discussion that the important dynamical aspects of ciliary propulsion should depend on the dimensions and frequency of beating of the individual cilia and not so much on the dimensions of the organism (see Fig. 3). Scaling analysis of the motion of a single cilium with respect to its length, radian frequency and viscosity gives:

$$U \propto \sigma l, \qquad F \propto \sigma l^2 \mu, \qquad M \propto \sigma l^3 \mu \qquad \text{and} \qquad P \propto \sigma^2 l^3 \mu,$$

where U is the velocity of propulsion, F the force, M is the bending moment in the cilium, and P is the rate of working of a single cilium. The bending moments and rates of working rise rapidly with length (l^3) in comparison to the linearity of the velocity, so on both structural and energetic grounds the cilia would need to be relatively short. The length of a cilium could also be limited by the need for nutrients like ATP to diffuse along it; however reasonable estimates show that this would not become a limiting factor until $l \approx 1$ mm. Typical values for the velocity are around 1 mm s^{-1}, the bending moment around 10^{-16}–10^{-13} Nm and the rate of working of the order of 10^{-16}–10^{-14} W for each cilium.

There appears to be a lower limit to the length of cilia of about 5 μm which is consistent with the observed minimum radius of curvature of 1·5–2·0 μm. The smallest ciliate studied, *Uronema*, has cilia near this minimum length. Interestingly enough the smallest cilia found in mammalian lungs are about 5 μm long. The length of cilia, l, is usually much less than that of the body, L (i.e. $l \ll L$), which is the converse of the situation in uniflagellates. For effective propulsion by cilia we need n, the number of cilia, to be much larger than the ratio of L/l (i.e. $n \gg L/l$); this, in our experience, is always found to be the case (see Table I) and we should not expect ciliated organisms to be smaller than about 20 μm. The size of the largest flagellates overlaps with that of the smallest ciliates, but the flagellates move at velocities which are about ten times smaller. Cells propelled by one or two flagella may be limited to lengths of no more than about 50 μm for structural reasons, because the flagella may not be capable of maintaining the bending moments required to propel larger bodies at reasonable speeds. Therefore, larger cell bodies have either large numbers of flagella (e.g. *Trichonympha*) or many cilia. The undulatory pattern of locomotion shown by flagella is a higher mode of beating than the pendular pattern shown by cilia, and it is interesting that the beat pattern of *Paramecium* cilia can be converted to a flagellar form by increasing the viscosity (see Sleigh[14, 15] Machemer[11]).

We can define an hydrodynamic efficiency η by comparing the rate of working for an inert cell body moving at the same velocity as the organism with the dissipation due to all the cilia on the cell body as follows:—

$$\eta = \frac{K\mu L U^2}{nP},$$

where K is a non dimensional constant depending on shape. By substitution of the scale factors into this formula we find

$$\eta \propto \frac{L}{nl}.$$

In Fig. 5 we have plotted this efficiency against the length of the organism. It is observed that the efficiency decreases approximately as L^{-1}. This is attributed to the relatively constant cilia concentrations on the body surface (0·05–0·4 cilia μm^{-2}); hence n is directly proportional to the cell surface area, which has dimensions L^2. *Pleurobrachia* does not conform with this conclusion because of the small number of comb plates. It is clear from this figure that the more efficient organisms exist in the size range of 20–350 μm, e.g. *Uronema*, *Tetrahymena*, *Didinium*, *Blepharisma* and *Paramecium*; larger

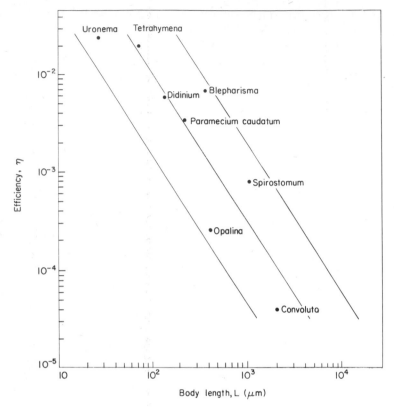

Fig. 5. The relationship of propulsive efficiency, as defined in the text, with body length.

organisms expend much more energy because of the vastly increased numbers of cilia, although they do reduce this by decreasing ciliary lengths and frequencies (e.g. *Spirostomum, Convoluta*).

From the scaling analysis we see that a longer or faster-moving cilium will encounter more fluid resistance and would be likely to bend backwards with a loss of propulsive efficiency; this tendency can be overcome if several ciliary units move together as a compound structure, in which not only is there enhanced stiffness, but an increased quantity of energetic machinery. Simple cilia with a propulsive function are seldom more than 15–20 μm long, whilst compound cilia with tens or hundreds of component cilia may reach lengths of 50–100 μm, and the giant comb plate cilia may be more than 1 mm long and contain 10^5 or more ciliary units. The formation of compound units extends the versatility of cilia as propulsive organelles, yet the pattern of beating of the simple *Paramecium* cilium is quite similar to that of the comb plate of *Pleurobrachia* which may be 50 times as long.

Baba[1] has shown for the abfrontal cilia of *Mytilis edulis* that the flexural rigidity increases with the square of the radius (not the fourth power, as one would expect if it were a solid body) indicating its dependence on the number of axonemes in the compound body. Typical values for the flexural rigidity are in the range 10^{-21}–10^{-22} Nm^2 (Hiramoto[8]). Clearly there is a structural or other internal limitation on cilia size.

It is desirable, if inertial effects are to be avoided, for the ciliary Reynolds number of organisms with cilia covering all or substantial areas of their body to be much less than one. If the ciliary mechanism is to operate in this favourable regime, the cilia length should be less than about 100 μm since most cilia beat at a frequency in the range 10–20 Hz. With this upper length limit the velocity of propulsion would be restricted to below about 1 cm s^{-1}. On the other hand, the largest ciliated organism are the ctenophores, which exploit the inertial advantages of large compound cilia in the form of comb plates.

The remarkable consistency in velocities of propulsion (with the exception of *Opalina*) is achieved because the cilia beat at different frequencies, are of differing lengths and are distributed and metachronally co-ordinated in varying ways. We will not consider this in any more detail, but a fuller discussion of the mechanics of ciliary beating can be found in Blake and Sleigh.[3, 4]

In conclusion, cilia are only effective for the swimming of small organisms less than 1 mm long; ciliary propulsion in the larger organisms is very slow in comparison to other methods of swimming, but ciliary movement over surfaces and the compounding of cilia into comb plates have been two evolutionary answers to the fluid dynamic limitations of cilia as locomotor organelles.

REFERENCES

1. Baba, S. Flexural rigidity and elastic constant of cilia. *J. exp. Biol.* **56**, 459–467 (1972).
2. Blake, J. A finite model for ciliated micro-organisms. *J. Biomech.* **6**, 133–140 (1973).
3. Blake, J. R. and Sleigh, M. A. Mechanics of ciliary locomotion. *Biol. Rev.* **49**, 85–125 (1974).
4. Blake, J. R. and Sleigh, M. A. Hydro-mechanical aspects of ciliary propulsion. *In* Brennen, C., Brokaw, C. J. and Wu, T. Y., Proceedings of Intern. Symp. on "Swimming and Flying in Nature", Pasadena, July 1974. Plenum Publ. Co., New York (1976).
5. Brennen, C., Brokaw, C. J. and Wu, T. Y. (eds.) Proceedings of International Symposium on "Swimming and Flying in Nature", Pasadena, July 1974. Plenum Publ. Co., New York (1976).

6. Gray, J. and Hancock, G. J. The propulsion of sea urchin spermatozoa. *J. exp. Biol.* **32**, 802–14 (1955).
7. Hancock, G. J. The self propulsion of microscopic organisms through liquids. *Proc. R. Soc. Lond.* **A217**, 96–121 (1953).
8. Hiramoto, Y. Mechanics of ciliary movement. *In* Sleigh, M. A., (ed.), "Cilia and Flagella", Academic Press, London (1974).
9. Knight-Jones, E. W. Relations between metachronism and the direction of ciliary beat in Metazoa. *Q. J. microsc. Sci.* **95**, 503–21 (1954).
10. Lighthill, M. J. Hydromechanics of aquatic animal propulsion. *Ann. Rev. Fluid. Mech.* **1**, 413–46 (1969).
11. Machemer, H. Ciliary activity and origin of metachrony in *Paramecium*: Effects of increased viscosity. *J. exp. Biol.* **57**, 239–59 (1972).
12. Prandtl, L. "The Essentials of Fluid Dynamics". Blackie, London (1952).
13. Satir, P. The present status of the sliding microtubule model of ciliary motion. *In* Sleigh, M. A., (ed.), "Cilia and Flagella", Academic Press, London (1974).
14. Sleigh, M. A. "The Biology of Cilia and Flagella". Pergamon, London (1962).
15. Sleigh, M. A. Some aspects of the comparative physiology of cilia. *Am. Rev. Resp. Dis.* **93**, 16–31 (1966).
16. Sleigh, M. A. Patterns of ciliary beating. *Symp. Soc. exp. Biol.* **22**, 131–50 (1968).
17. Sleigh, M. A. Patterns of movement of cilia and flagella. *In* Sleigh, M. A. (ed.), "Cilia and Flagella", Academic Press, London (1974).
18. Taylor, G. I. Analysis of swimming of microscopic organisms. *Proc. R. Soc. Lond.* **A209**, 447–61 (1951).
19. Taylor, G. I. Analysis of swimming of long and narrow animals. *Proc. R. Soc. Lond.* **A214**, 158–83 (1952).
20. Warner, F. D. and Satir, P. The structural basis of ciliary bend formation. Radial spoke positional changes accompanying microtubule sliding. *J. Cell Biol.* **63**, 35–63 (1974).

16. Unsolved Problems of Interest in the Locomotion of Crustacea.

JOHN H. LOCHHEAD

Department of Zoology, University of Vermont, Burlington, Vermont, U.S.A. *

ABSTRACT

Crustaceans are notable for their proficiency in a wide variety of types of locomotion. Yet we have almost no sophisticated knowledge of the mechanisms involved. The present paper discusses a few examples, particularly of crustaceans that swim, highlighting some of the problems in need of research.

Crustaceans that walk may be terrestrial or aquatic, or may push through sand or mud. They face problems relating to their size, the weight carried, the resistance against which they must push, and the number of legs used.

Talitrid amphipods can jump in air to heights over twenty times their body length. Their performance and the mechanism used invite investigation.

The swimming mechanisms of Crustacea are very varied. Discussed in the present paper are the backward leaps of lobsters, the propeller-like use of swimming paddles by portunid crabs, the fast spurts of the copepod *Diaptomus*, apparently achieved with the aid of the thoracic limbs, the rowing of *Daphnia*, the slow glides of *Diaptomus*, resulting mainly from fast rotary beats of the antennae, and the varied gliding movements of the fairy shrimp *Chirocephalus*, produced by rotary beats of the thoracic exopodites.

A Reynolds number approaching 10^6 may apply for very large lobsters. But for most crustaceans much lower numbers apply, ranging down to values below one. The effects of inertia and viscosity thus differ widely among the different species.

In crustaceans that swim, the propulsive limbs are usually fringed with swimming setae. A study of the performance characteristics of these setae would be of great interest.

The purpose of the present paper is to call attention to the need for research on the locomotory mechanisms of Crustacea. Within this group locomotion

* Present address: 49 Woodlawn Rd., London S.W.6, England.

is more varied and often more efficient than in any other animals. Yet studies of crustacean locomotion have been largely descriptive, with only occasional use of experimental methods.

The subject was reviewed by me in 1961 (Lochhead[15]). Since that time my interests have been in a quite different area, so I probably have missed some important publications. Three recent papers that deserve mention are those by Burrows and Hoyle,[2] Macmillan,[17] and Pond.[20]

Emphasis in the present paper will be on mechanisms used for swimming. However, a few words should be said about the problems involved in walking and jumping.

Some crustaceans walk on land, others under water, and there are some that push their way through mud or sand. The number of legs used varies with the species, and no doubt a wide variety of gaits will be found, responding to all those factors now familiar to us from the work of Manton on terrestrial arthropods. It might be expected that crustaceans under water would use lower gears than are employed in air, because they encounter greater fluid resistance. However, at least in the crayfish, it seems that a more important factor is the greatly reduced weight on the legs when under water (Pond[20]).

Species that move through mud or sand range in size from tiny copepods, that fit between the sand grains, up to quite large crabs. It would be of interest to discover at what size thixotropy in the wet sand first becomes an important factor.

Especially fascinating are the crabs, which have exploited a number of advantages in moving sideways.

Although not many crustaceans jump, talitrid amphipods are star performers, able to jump in air to heights over twenty times their body length (Smallwood[21]). There have been some interesting discussions of the factors that influence jumping ability in animals of different sizes (see, for example, Alexander[1] and Bennet-Clark (this Symposium)). At the moment we do not know where talitrids fit in, when compared with other animals.

Turning now to the swimming mechanisms used by Crustacea, these are extremely varied. There is space here to consider only a few examples, but I hope to provide a glimpse into some of the problems that await experimental investigation. A few outrageous calculations will be made, based on flimsy data. The intention, however, is merely to give a preliminary idea of some dimensions of interest to a potential investigator.

Largest among swimming Crustacea are some of the lobsters. These animals escape from danger by a single, quick, ventral flexure of the tail fan, resulting in a fast backward spurt through the water. Speeds are known to reach $1 \cdot 2 \, \text{m s}^{-1}$ (Lindberg,[13] for *Panulirus*), and possibly may reach $8 \, \text{m s}^{-1}$ (Herrick[9]). Occasional specimens of *Homarus americanus* grow to a length of $0 \cdot 6 \, \text{m}$. For such an animal swimming backwards at $1 \cdot 5 \, \text{m s}^{-1}$, the

Reynolds number would be 900,000. Thus it is just possible that a lobster might reach the critical Reynolds number at which its drag coefficient is a minimum, a possibility unlikely for any other crustacean.

Among large Crustacea, by far the best swimmers are the portunid crabs. These animals swim sideways by the powerful action of specialized paddles on the last pair of walking legs. The paddles are oval, flattened, and in some cases slightly spoon-shaped. They project above the body and beat back and forth synchronously, at right angles to the direction of locomotion (Fig. 1). Although unproven, it seems evident that the paddles act like variable pitch propellers, equally effective during the backstroke and the forestroke (Kühl[11]). The body is spindle-shaped, providing equal streamlining for going either to the left or to the right. Whether or not the body supplies any lift is not known.

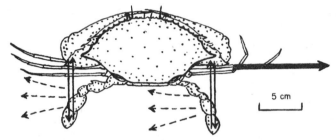

FIG. 1. A portunid crab. The heavy arrow indicates the direction of swimming; thin-line arrows, the direction of propulsive limb-beats; broken-line arrows, water currents.

The portunid *Polybius* has been recorded swimming at $1·33$ m s^{-1} (Kühl[11]). It is certain, however, that these animals can swim much faster, for they can overtake and catch mackerel in the open sea (Kükenthal and Krumbach[12]). Reynolds number, assuming a body width of 15 cm and a speed of $1·33$ m s^{-1}, would be 200 000. At this relatively high value any surface roughness would be expected to cause turbulence, so it is not surprising that the bodies of portunids are typically very smooth.

For the paddles, about $1·5$ cm wide in the direction of water flow, the Reynolds number would be 20 000.

In an animal 15 cm wide, each paddle perhaps travels about 15 cm in one full beat. At a frequency of $10·5$ beats per second (Kühl[11]), the average speed of the paddles works out at $1·6$ m s^{-1}. For an animal swimming at $1·33$ m s^{-1}, the ratio of swimming speed to paddle speed thus may be about 5:6. This approximate value does not seem unreasonable in comparison with a ratio of 3:4 for flight velocity versus wing-tip speed reported for *Drosophila* (Chadwick[5]).

Portunids can also use their paddles to swim backwards. However, the mechanism is then quite different from that which is used for swimming sideways. In particular, the paddles act as oars rather than as propellers, and the left and right legs beat alternately instead of synchronously.

Future work on the swimming of portunids should yield interesting results. Our present knowledge of this topic owes much to the work of Kühl,[11] but that work should now be confirmed and extended. How, for instance, do the paddles alter their amplitude, angle of attack, and frequency of beat, relative to the load carried and the size of the animal?

A very different example is that of the copepod *Diaptomus*, which can make short spurts at speeds reported to exceed $20\,\mathrm{cm\,s^{-1}}$ (Lowndes[16]). That is one hundred times the body length of the animal, per second, and certainly one would like to have confirmation of such a remarkable performance.

The mechanism used for these spurts is in dispute, but I shall present only the views of Storch,[22] because only he has provided much evidence to support his ideas.

Storch used cinematography and his findings were essentially as follows. During rests between jumps, *Diaptomus* floats with its antennules extended laterally, probably stiffened by turgor, while its five pairs of biramous thoracic limbs point forwards, bunched together close under the body (Figs 2 and 4). A few milliseconds before a jump the antennules become limp, due probably to the opening of proximal valves. The thoracic limbs then

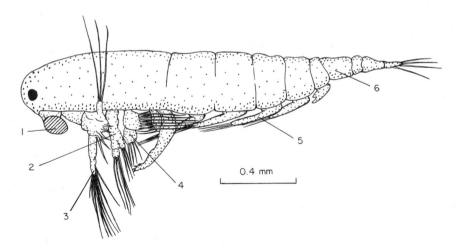

FIG. 2. *Diaptomus*, side view. *1*, cut stump of antennule; *2*, mandibular palp; *3*, endopodite of antenna; *4*, endopodite of first maxilla; *5*, 3rd thoracic limb; *6*, 1st segment of abdomen.

beat back strongly one after the other, each pair starting only after the backstroke of the pair behind has been completed. At the same time, the antennules hinge back passively and the abdomen acts as a rudder and stabilizer. After the five pairs of thoracic limbs have finished their power strokes, they all swing forward together and the resting position is resumed.

The duration of one entire beat of all the thoracic limbs was found by Storch to be $\frac{1}{60}$ s. He supposed that this time interval could be divided into $\frac{1}{360}$ s for each of the five backstrokes and $\frac{1}{360}$ s for the recovery stroke. In an animal 2 mm long, the tip of each thoracic leg perhaps swings a distance of about 0·8 mm. If that distance is covered in $\frac{1}{360}$ s, the limb-tip speed, relative to the body, would be 28·8 cm s^{-1}. This rate would be adequate to account for the reported swimming speed of over 20 cm s^{-1}.

However, it seems likely that the animal would start in low gear, with a beat of relatively long duration, and that it would then shift to higher gears, shortening the duration of each succeeding limb-beat. If acceleration were constant, swimming speeds at the ends of the five successive beats would be 4, 8, 12, 16 and 20 cm s^{-1}. Limb-tip speeds through the water would be equal to the speed relative to the body minus the appropriate swimming speed. It would be of interest to know whether the speed relative to the body increases from limb to limb in such a way that all the limbs have about the same speed through the water. If that is so, this latter speed would be about 17 cm s^{-1}, assuming acceleration to a swimming speed of 20 cm s^{-1} completed in $\frac{5}{360}$ s. Such a relatively high speed through the water would mean that the tips of the limbs are not gripping the water very effectively. Perhaps the limbs act like oars straining for maximum acceleration, or perhaps they induce jet propulsion by their sudden power-strokes towards the body (like the combplates of ctenophores, described by Sleigh and Blake in their paper).

Among the thoracic limbs, the fifth pair are not as oar-like as the others and Storch thought that they contribute nothing to propulsion. However, if they have to get the body moving, perhaps their paddle surfaces must be reduced to a size that will not overtax their muscles.

The first four pairs of thoracic limbs are fringed with swimming setae, which much expand the paddle surfaces. Nothing is known about the hydrodynamics of these very small structures. Their Reynolds number works out at 1·7, supposing that the setae are 0·001 cm thick and that they move through the water at 17 cm s^{-1}.

For the body of the animal, 0·2 cm long, moving at 20 cm s^{-1}, Reynolds number would be 400.

A more definite example of rowing is provided by the water flea Daphnia, which uses large biramous antennae as oars (Fig. 3). Sometimes the swimming is in jumps, one beat of the antennae per jump (Woltereck[25]). Sometimes the antennae beat more continuously, at rates varying in the case of D. magna from

1·66 beats per second, when only one egg is being carried, to 3·75 beats per second, when 60 eggs are carried (Fox and Mitchell[7]).

Although the swimming of *Daphnia* has often been investigated, details of the mechanism remain in dispute. Those who are interested should begin by studying the work of Woltereck,[25] even though it has been asserted that parts of this work are "all wrong" (Hutchinson[10]). Wrong or not, Woltereck presents an especially clear analysis of the problems.

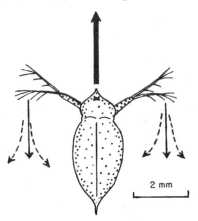

FIG. 3. *Daphnia*.

There are several species of *Daphnia* (and of related Cladocera), each with somewhat different habits. Unfortunately, it is difficult to find sets of quantitative data applicable to a single species. In addition, we have no records for speeds attained at particular limb-beat frequencies, nor is it known whether the amplitude of beat can be changed.

Despite these difficulties, it probably is worth while to make some calculations, based on the limited data available.

Records of swimming speeds are surprisingly sparse. The following figures are all for specimens swimming upward: *D. pulex*, 0·166 cm $^{-1}$ (Wautier[24]); *Daphnia* sp., 0·7 cm s^{-1} (Hardy and Bainbridge[8]); *D. longispina*, 5·8 cm s^{-1} (Cushing[6]).

In a *Daphnia magna* 0·45 cm long, the tips of the antennal endopodites perhaps have a total excursion back and forth of about 0·6 cm. At the two frequencies of beat cited above the speeds of the endopodite tips relative to the animal would then be 1·0 and 2·25 cm s^{-1}, respectively. Relative to the water these speeds during each backstroke would be 0·3 and 1·55 cm s^{-1}, assuming a swimming speed of 0·7 cm s^{-1}. For the first of these two speeds, Reynolds number for a swimming seta 0·001 cm thick would be 0·03, while for the whole body it would be 31·5.

At these very low Reynolds numbers, boundary layers would be relatively thick and viscosity would be a most important factor.

In some species of *Daphnia* the surface of the carapace is hydrophobic, whereas in other species it is hydrophilic. How these differences might affect friction drag seems not to have been investigated.

A quite different type of swimming is seen in many Crustacea which move through the water in a slow, steady glide, resulting from the rapid beating of certain appendages or parts of appendages. Usually the beat appears to be rotary, although the antero-posterior component may be more conspicuous than the transverse component. The small propulsive thrust that results could come from an oar-like action during the posterior phase of the beats, or from a propeller-like action during the transverse phases. If the beat is truly rotary, there may be a considerable saving of energy because of the absence of the repetitive acceleration and deceleration involved in a fore-stroke-backstroke cycle.

As an example of this type of swimming, the copepod *Diaptomus* may again be considered. When not making fast spurts through the water, this animal swims in a slow, steady glide, resulting from the rapid, rotary beating of the biramous antennae, the biramous mandibular palps, and the endopodites of the first maxillae (Cannon[3]). All of these appendages are provided with long swimming setae (Fig. 2).

Cannon observed the beating of the appendages with a stroboscope and he has given a detailed description of what he saw. According to him the limbs beat chiefly backwards and forwards, most rapidly during the back-strokes, each limb beginning its backstroke just ahead of the limb in front. As a result of this activity, a layer of water flows posteriorly close around the ventral half of the head. Behind the maxillae this current divides into medial and lateral eddies, rotating in opposite directions. The medial eddies, or vortices, serve to carry food to the mouth parts. The lateral vortices, which are larger, discharge a small amount of water posteriorly, contributing, Cannon thought, to the forward motion of the animal (Fig. 4). The antennae are longer than the other appendages and beat at a greater amplitude. Consequently it is their limb-tips that move fastest and do most of the work. Little energy is needed to maintain the vortices and not much water is actually transported. The appendages can therefore beat very rapidly for long periods of time.

Although Cannon describes many other details, it is evident that we know almost nothing about such things as angles of attack during different phases of the beat. Without such knowledge it is hard to understand just why the very rapid limb beats are not associated with a faster rate of swimming. The slow swimming that does occur may result from the inherent movement through the water displayed by all vortex pairs or rings (McMahon, discussion).

The actual rate of limb-beat observed by Cannon was 16·6 beats per second.

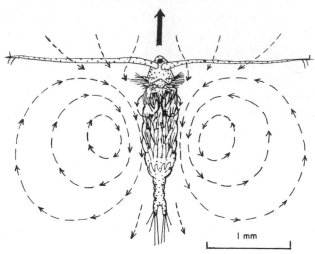

FIG. 4. *Diaptomus* gliding, ventral view.

However, Storch[22] has reported about 50 beats per second. The rate of swimming, although definitely very slow, seems not to have been recorded.

The excursion in one full beat of the longest antennal setae is perhaps about 0·1 cm. Thus, at 16·6 beats per second, the average limb-tip speed would be 1·66 cm s^{-1}. Without a correction for the forward speed of the animal, Reynolds number for the tips of these setae works out at 0·17. It would seem, then, that the propulsive mechanism is a viscous one, because of this low Reynolds number, but that the glide of the whole animal is dominated by inertia (Lighthill, discussion).

Another example of a crustacean that swims in a slow glide is the fairy shrimp, *Chirocephalus*. This is an animal about 2·5 cm long, which usually swims on its back (Fig. 5). In this position it may glide forwards, hover in one place, or drift slowly upwards, downwards, or backwards.

There are eleven pairs of trunk limbs which beat in a metachronal rhythm with a very clear fore- and backstroke action. As a result the interlimb spaces alternately enlarge and contract, so that they act as pumps, drawing

FIG. 5. *Chirocephalus*. Setae indicated only on the exopodites and on the caudal furca.

water in and then throwing it out. The water is ejected posteriorly and Cannon[4] thought that this would result in forward locomotion. However, the amount of water expelled is quite small and the same pumping continues even when the animal is going backwards or is hovering in one place (Lochhead[14]).

The ability to hover and to drift slowly in various directions seems to depend on the action of the exopodites, which beat in a rotary fashion while being at the same time carried backwards and forwards by the beating of the whole appendages. By their activity the exopodites produce a series of vortices on each side of the body, corresponding to the outer vortices described for *Diaptomus*. The direction in which the animal moves apparently is controlled by small alterations in the angle of attack or direction of beat of some or all of the expodites (Lochhead[14]). Forward locomotion is probably included in this system of control.

Lowndes[16] found that *Chirocephalus* swims forward at a speed of 1 cm s^{-1}. Probably it can swim faster, since speeds of up to 14 cm s^{-1} have been reported for another anostracan (Menner[18]).

The limb-beat frequency in *Chirocephalus* was reported by Lowndes to be 2 beats per second; $2\frac{1}{2}$ beats per second are shown on an accompanying film. Again, it seems likely that higher frequencies sometimes occur, since up to $12\frac{1}{2}$ beats per second have been reported for another anostracan (Menner[18]).

It is hard to estimate the length of swing of the exopodites because of the complexity of their movements. But perhaps a reasonable figure would be 1 cm for the total excursion in one complete beat. Thus, limb-tip speeds might range from 2 to $12 \cdot 5$ cm s^{-1}, adequate to account for most of the swimming speeds reported.

In an animal swimming forwards there will be vortex shedding from the body. If that happens in front of the limbs, it would be of interest to know whether the limb beats are synchronized with the vortices that are shed, effecting thereby a significant saving in energy (McMahon, discussion).

Turning now to a more general topic, in most Crustacea that swim the propulsive appendages are fringed with special setae, classified by Thomas[23] as "plumose setae", but more generally referred to as "swimming setae" (Fig. 6).

An enquiry into the performance characteristics of these swimming setae should yield interesting results. At present we can only speculate regarding their functional adaptations.

To begin with, we ought to know what type of sclerotin is present in these setae and how this sclerotin is intermixed with the chitin. Various mechanical properties, such as elasticity and directional resistance to bending, are doubtless adapted to functional requirements, in ways unknown to us at present.

Each swimming seta is fringed with setules of uniform size, very regularly spaced. It seems that the size and spacing of these setules is much the same for all Crustacea, but this question really should be investigated. Probably it would be helpful to make comparative studies on setae with other functional requirements, such as filter setae and mechanoreceptor setae.

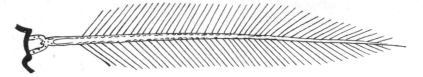

FIG. 6. A swimming seta.

Presumably a fringe of swimming setae provides a large propulsive surface with small mass. The resulting reduction in inertia could be especially important when there is repeated acceleration and deceleration of the appendage. The setae and their setules are presumably just strong enough to withstand the forces involved. The setules are presumably spaced as far apart as possible, consistent with the requirement that they provide an effective propulsive surface. For the beetle *Acilius*, Nachtigall[19] found that its swimming setae can produce up to 54% of the thrust that a solid surface would provide. Presumably the thrust would be greater if the setules were closer together, so perhaps there is some other hydrodynamic advantage that would be lost with closer spacing.

One wonders how the swimming setae of crustaceans would compare with those of *Acilius*, and how their role in propulsion would be affected by changes in Reynolds number. If the spacing of the setules is found to vary, it would be interesting to know whether this variation is correlated with such factors as speed of the setae through the water, dimensions of the propulsive surface, or viscosity of the water (correlated with temperature).

At low Reynolds numbers an aerofoil is less effective than a thin, flat or cambered plate (Alexander[1]). Perhaps one advantage of a fringe of swimming setae is that it forms a plate of adequate strength considerably thinner than the alternative possibility of two layers of cuticle plus epidermis, separated by a blood space.

Swimming setae usually diverge distally and have shorter setules there. As a result, at the edge of the fringe there are V-shaped spaces between the setae. Perhaps this feature serves to reduce turbulence by softening the edge, or perhaps it acts to prevent stalling at high angles of attack.

When appendages are being used as oars, resistance to the water must be minimized during each recovery stroke. At least in theory, there are a number of ways in which swimming setae might help to serve that need. Some of the possibilities are: the setules may slant in the direction of thrust; the setae, or the setules, may spread out like a fan during the power stroke and fold together during the recovery stroke; the setae, or the setules, may bend or hinge during the recovery stroke until parallel with the direction of motion; possibly water may flow more easily between the setules during the recovery stroke, if the rate of motion through the water is then very different from that which occurs during the power stroke.

In closing, may I emphasize once again that in the present paper I have been highly speculative, hoping thereby to stimulate enquiry into the fascinating field of crustacean locomotion.

REFERENCES

1. Alexander, R. M. "Animal Mechanics". Sidgwick and Jackson, London (1968).
2. Burrows, M. and Hoyle, G. The mechanics of rapid running in the ghost crab, *Ocypode ceratophthalma. J. exp. Biol.* **58**, 327–349 (1973).
3. Cannon, H. G. On the feeding mechanism of the copepods, *Calanus finmarchicus* and *Diaptomus gracilis. Brit. J. exp. Biol.* **6**, 131–144 (1928).
4. Cannon, H. G. On the feeding mechanism of the Branchiopoda. *Phil. Trans. Roy. Soc. Lond.* **B222**, 267–352 (1933).
5. Chadwick, L. E. The motion of the wings. Aerodynamics and flight metabolism. *In:* "Insect Physiology" (K. D. Roeder, ed.), pp. 577–636. Wiley, New York (1953).
6. Cushing, D. H. Some experiments on the vertical migration of zooplankton. *J. Animal Ecol.* **24**, 137–166 (1955).
7. Fox, H. M. and Mitchell, Y. Relation of the rate of antennal movement in *Daphnia* to the number of eggs carried in the brood pouch. *J. exp. Biol.* **30**, 238–242 (1953).
8. Hardy, A. C. and Bainbridge, R. Experimental observations on the vertical migrations of plankton animals. *J. mar. biol. Assoc. U.K.* **33**, 409–448 (1954).
9. Herrick, F. H. The American lobster, a study of its habits and development. *Bull. U.S. Fish Comm.* **15**, 1–252 (1895).
10. Hutchinson, G. E. Turbulence as random stimulation of sense organs. *In:* "Josiah Macy, Jr. Foundation, Trans. 9th Conference Cybernetics", pp. 155–158 (1953).
11. Kühl, H. Die Fortbewegung der Schwimmkrabben mit Bezug auf die Plastizität des Nervensystems. *Z. vergl. Physiol.* **19**, 489–521 (1933).
12. Kükenthal, W. and Krumbach, T. (eds) Crustacea. *In:* "Handbuch der Zoologie", Bd. 3, Hälfte 1, pp. 277–1158. de Gruyter, Berlin (1926).
13. Lindberg, R. G. Growth, population dynamics, and field behavior in the spiny lobster, *Panulirus interruptus. Univ. Calif. (Berkeley) Publ. Zool.* **59**, 157–248 (1955).
14. Lochhead, J. H. "The Feeding Mechanism of Crustacea Branchiopoda". Doctoral thesis, Cambridge University, England (1937).
15. Lochhead, J. H. Locomotion. *In:* "The Physiology of Crustacea", Vol. 2 (T. H. Waterman, ed.), pp. 313–364. Academic Press, New York (1961).
16. Lowndes, A. G. The feeding mechanism of *Chirocephalus diaphanus* Prevost, the fairy shrimp. *Proc. zool. Soc. London.* 1933, 1093–1118 (1933).

17. Macmillan, D. L. A physiological analysis of walking in the American lobster (*Homarus americanus*). *Phil. Trans. Roy. Soc. Lond.* **B270**, 1–59 (1974).
18. Menner, E. Studien über die Bewegung von *Artemia salina* und *Branchipus stagnalis. Zool. Anz.* **122**, 49–65 (1938).
19. Nachtigall, W. Locomotion: mechanics and hydrodynamics of swimming in aquatic insects. *In:* "The Physiology of Insecta", 2nd Ed., Vol. 3 (M. Rockstein, ed.), pp. 381–432. Academic Press, New York (1974).
20. Pond, C. M. The role of the "walking legs" in aquatic and terrestrial locomotion of the crayfish *Austropotamobius pallipes* (Lereboullet). *J. exp. Biol.* **72**, 447–454 (1975).
21. Smallwood, M. E. The beach flea: *Talorchestia longicornis. Cold Spring Harbor Monogr.* **1**, 1–27 (1903).
22. Storch, O. Die Schwimmbewegung der Copepoden auf Grund von Mikro-Zeitlupenaufnahmen analysiert. *Verhandl. deut. zool. Ges.* **33**, 118–129 (1929).
23. Thomas, W. J. The setae of *Austropotamobius pallipes* (Crustacea: Astacidae). *J. Zool., Lond.* **160**, 91–142 (1970).
24. Wautier, J. Réponses de quelques invertébrés aux actions combinées de la pesanteur et d'un courant d'eau vertical. *Bull. mens. Soc. linn. Lyon* **19**, 199–201 (1950).
25. Woltereck, R. Über Funktion, Herkunft und Entstehungsursachen der sogen. "Schwebe-Fortsätze" pelagischer Cladoceren. *Zoologica Stuttgart* **26**, 475–550 (1913).

17. Swimming Mechanics and Energetics of Locomotion of Variously Sized Water Beetles— Dytiscidae, Body Length 2 to 35 mm.

WERNER NACHTIGALL

Zoologisches Institut der Universität des Saarlandes, Saarbrücken, W. Germany.

ABSTRACT

Water beetles belonging to the family Dytiscidae with body length, l, between 2 mm (*Bidessus geminus*) and 35 mm (*Dytiscus marginalis*) are geometrically and dynamically compared. The following measurements were carried out on various individuals of nine species: body length l, frontal area A, geometrical shape of the frontal area, minimum and maximum swimming speed V, and beat frequency of the rowing legs f. Furthermore, the drag coefficient C_D in two species and the hydrodynamical efficiency of the rowing mechanism, η_h, and the resting metabolic rate E_r in one species were calculated.

There is a certain geometrical similarity in all beetles, the frontal areas A increasing as $l^{1.7}$. The slope of the graph of V against l is approximately 20 s^{-1}, while the slope of the graph of relative velocity against l is -0.35 (number of body lengths) $\text{s}^{-1} \text{ mm}^{-1}$). Mean Reynolds number ranged from 90 (*Bidessus*) to 14.000 (*Dytiscus*). It is shown that Newton's law of resistance is valid for values of Re at least as low as 5×10^3 in *Dytiscus marginalis*. The variation of drag coefficient C_D with body length for the whole series of beetles is discussed. Hydrodynamic drag D (N) has been calculated as a function of body length l (mm) and is shown to follow closely the law $D = 10^{-7} . l^3$. A mean hydrodynamic efficiency of the rowing apparatus $\eta_h = 0.3$ has been calculated for *Acilius sulcatus*. This gives rise to an overall efficiency of $\eta_{tot} = \eta_h . \eta_{muscle}$ of approximately 0.1. In smaller beetles η_{tot} is expected to be smaller. For a comparison of the total power needed for swimming, $P = DV/\eta_{tot}$ (under the worst possible conditions of C_D and η_h), with the power available via the physiological metabolic processes (given in multiples of the resting metabolic rate E_r), measurements for *Acilius sulcatus* were used as an experimental reference. An O_2 consumption proportional to body mass was assumed for these water beetles. The preliminary conclusion was thereby obtained that, in small beetles, pure working metabolic rate E_w

for average fast "everyday swimming" is approximately equal to the resting metabolic rate E_r, which means that only just double the metabolic activity is required in this swimming conditions. In larger beetles E_w seems to be greater. In *Acilius* $E_w \approx 10\,E_r$ for the fastest possible short bursts of swimming.

INTRODUCTION

In undulating swimmers like certain fishes and leeches the drag and thrust generating structures are integrated and not easily distinguished. In the period of a single beat in these animals certain parts of the body can switch from thrust to drag generation and back to thrust generation. Water beetles belonging to the family Dytiscidae, on the other hand, represent swimming systems whose hydrodynamically important structures scarcely interact. The rigid trunk forms a drag generating subsystem which seems to be some-what stream-lined (Fig. 1, a–e), and the third pair of legs forms a thrust generating subsystem. These hind-legs are modified to form a rowing appa-ratus by the addition of swimming hairs at the edges of the flat tibia and tarsus (Fig. 1, 1–8). In medium and larger sized Dytiscidae the first pair of legs is closely attached to the body during faster swimming. The middle pair is extended only during turning manoeuvres when it is used as a rudder or oar. In forward swimming only the third pair of legs is active as a thrust generator. During the power stroke these legs beat synchronously and are fully extended, with the swimming hairs automatically spread and the flat surface turned perpendicular to the direction of movement (Fig. 1, 1–4). Hydrodynamically the most effective part of the oar acts relatively far from the body. During the recovery stroke the separate parts of the legs are folded together, the swimming hairs are also folded, and the leg is turned in such a way that the flat side is parallel to the direction of motion (Fig. 1, 5–8).

The moving legs slightly modify the streaming characteristics of the trunk and the presence of the trunk causes a certain deviation from the free water operation of the rowing apparatus, but the interactions are negligible. There-fore water beetles form hydrodynamic systems which are relatively easy to analyse if one treats the trunk and the rowing legs separately. The hydro-dynamic characteristics of the rowing apparatus of *Dytiscus marginalis* (Hughes[8]) and of *Acilius sulcatus* (Nachtigall[12]) and of the body of *Dytiscus marginalis* (Nachtigall and Bilo[15]) have been analysed; summarised reports have been given by Nachtigall[14] and Nachtigall and Bilo.[16]

There is a range of not less than 1:20 in the body length l of water beetles in the family Dytiscidae. The smallest species, e.g. *Bidessus geminus* ($l = 2$ mm), literally never stop swimming apart from intervals of a few seconds for breath-ing at the water surface. Their relative velocity (body lengths per second) is high and does not alter very much during swimming. The largest species, e.g.

Dytiscus marginalis ($l = 35$ mm), rest often. Their average relative velocity is lower, but can be changed more markedly from low values during "everyday searching swimming" to very high values during manoeuvres, or the pursuit of prey or a female. The forms of the body and rowing legs in smaller Dytiscidae do not seem to be as well adapted as in larger species, although all the beetles live in the same environment, at the banks of small creeks or in clear, shallow, still water. Therefore a comparison of different Dytiscidae from the morphological, geometrical, hydrodynamical and energetic points of view would seem to be interesting. However, some of the measurements required for such a comparison are very difficult and in some cases impossible. Thus it is necessary to make many idealisations and estimates, so that some of the following considerations cannot be more than crude approximations. Data are taken from the papers mentioned above.

MEASUREMENTS, CALCULATIONS AND ASSUMPTIONS

The following measurements have been carried out in many individuals of the eight species discussed here (Table I): body length l, frontal area A (measured by a microprojecting method; Fig. 2), swimming speeds and beating frequencies (measured by frame analysis of hundred-frame-sections of high speed films, taken at 80 frames per second). To demonstrate the morphological-geometrical similarity, A was plotted against l in Fig. 5. Using measurements of the fastest swimming velocity which could be maintained by a beetle at a constant beating frequency, and neglecting high-speed bursts of short duration, an "average highest velocity" V_{max} was calculated for every species. The corresponding relative velocity in body lengths per second is denoted by $V_{r\,max}$. The Reynolds number is defined as $Re = lV_{max}/v$

TABLE I

	l (mm)	A (mm)^2	V_{max} (mm s^{-1})	Re (−)
1 *Dytiscus marginalis*	35·0	$1·18 \times 10^2$	400	14000
2 *Acilius sulcatus* ♂	17·0	$3·3 \times 10^1$	360	6100
3 *Ilybius fulignosus*	10·5	$1·06 \times 10^1$	220	2300
4 *Agabus bipustulatus*	9·5	$1·04 \times 10^1$	200	1900
5 *Agabus chalconatus*	8·3	$8·8 \times 10^0$	170	1400
6 *Laccophilus obscurus*	4·6	$3·17 \times 10^0$	110	500
7 *Hyphydrus ferrugineus*	4·3	$4·36 \times 10^0$	110	470
8 *Bidessus unistriatus*	3·0	$1·82 \times 10^0$	70	210
9 *Bidessus geminus*	2·0	$7·5 \times 10^{-1}$	45	90

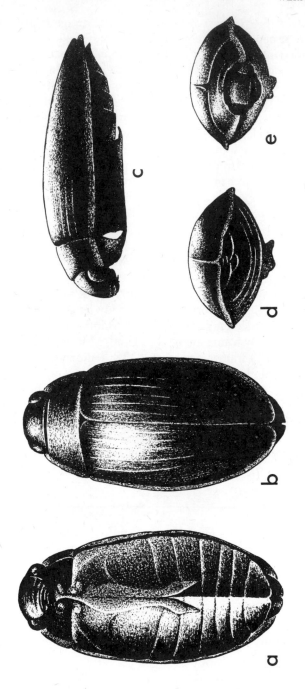

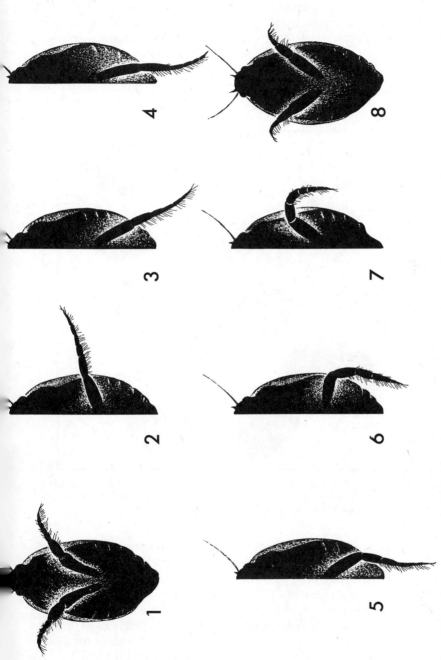

FIG. 1. Views of a male specimen of the water beetle *Dytiscus marginalis* taken from below (*a*), above (*b*), from the side (*c*), in front (*d*) and behind (*e*); limbs removed. Single phases of the power stroke and forward movement of the legs of *Acilius sulcatus*; half side view. (1–8).

(where v, the kinematic viscosity of water, is $1 \times 10^{-6} \, \text{m}^2 \, \text{s}^{-1}$). Thus the velocity, V, the relative velocity V_r and Re could be plotted as functions of l. Drag coefficients $C_D = 2D \, A^{-1} \, \rho^{-1} \, V^{-2}$, where ρ is the water density and D is the drag, could be measured for the larger species *Acilius sulcatus* and *Dytiscus marginalis* only. In the latter case C_D was determined as a function of body angle and swimming speed V. In these measurements the drag D was determined by means of a one-component balance, and the frontal area A

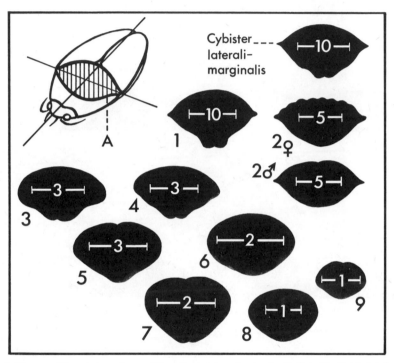

FIG. 2. Shapes of the cross-sections of the nine species of water-beetle listed in Table I, taken at that station along the body at which the cross-sectional area is maximum (and equal to A). The number within each shape gives a scale in mm.

by the projection method mentioned. In water tunnel experiments the velocity V was determined by means of a hydrodynamic propeller, in wind tunnel experiments the pressure head $\frac{1}{2}\rho V^2$ was measured by a Betz manometer. To estimate the probable range of C_D for smaller species, the values for spheres were taken as upper limits, and the values for prolate spheroids (length/width = 2) as lower limits at Reynolds numbers appropriate to each of the different smaller beetles respectively (Fig. 3). Measured values for *Acilius sulcatus* and *Dytiscus marginalis* lie well within these limits (Fig. 3). Because the bodies of smaller Dytiscidae are less slender, a tendency to the upper limits was assumed for these animals. The somewhat arbitrary curve represented in Fig. 3 by a dashed line was taken as the most probable form of the relationship between C_D and l. (Really, of course, C_D is a function of Reynolds number, Vl/v, but this can be converted into a function of l if use is made of the relationship between V and l, taken for example from Table I). Drag D could then be calculated for each of the different sized species and subsequently

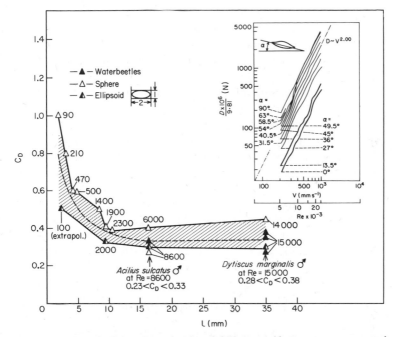

FIG. 3. Graph of drag coefficient C_D against length l. Upper and lower curves represent the values for a sphere and for a prolate spheroid (length = 2 × breadth). Also shown are measured values for two species. The numbers on the graphs are values of the Reynolds number, calculated from the peak sustained swimming speed, V_{max}, of an animal of corresponding length. Values of C_D used in subsequent calculations are taken from the broken curve. Inset: Drag D of a male *Dytiscus marginalis* (body length 35 mm) as a function of swimming velocity V and Reynolds number Re. The dashed line represents $\alpha = 63\cdot4°$ (tan $\alpha = 2\cdot0$).

plotted as a function of l (Fig. 5). First, however, it was shown that Newton's law of resistance ($D \propto V^2$) is valid for *Dytiscus* at least for Reynolds numbers above 5×10^3. In a double logarithmic scale the slope of the graph of D against Re is exactly 2 at every angle of attack (α) measured, both positive and negative (see inset in Fig. 3). The power P which the beetles must spend to move their bodies at their typical velocity through the water is equal to DV.

For a comparison of swimming energetics, measurements of the working metabolic rate E_w of the different sized animals should be made. However, this was impossible, on the one hand because the animals swim in water but breathe atmospheric air at irregular intervals, so that any measurement of E_w would represent an average over active and non-active periods, and on the other hand because the animals do not swim steadily and regularly in the respiration chamber of a Warburg apparatus. Only the resting metabolic rate E_r of *Acilius sulcatus*, individuals of which species tend to remain motionless in the respiration chamber for longer periods, could be measured and used as a reference. From this measurement and the assumption that oxygen consumption is proportional to body mass m in these insects (Kittel[9]), and the further assumption that m is proportional to l^3, the dependence of E_r on l can be deduced. The overall efficiency η_{tot} is equal to the product of hydrodynamic efficiency η_h and muscular efficiency η_{muscle} (see Wu's paper above). For *Acilius sulcatus* both η_h and η_{muscle} are taken to be roughly 0·3 (the former is the mean of values measured for this species in different swimming modes), so that η_{tot} is approximately equal to 0·1. Morphological investigations of the rowing apparatus and observations of the leg movements have led to the conclusion that η_{tot} is lower in smaller beetles. A lower limit of 0·05 was assumed for η_{tot}. Thus P/η_{tot} representing the (chemical) power spent by an animal in swimming at peak sustained velocity V_{max}, could be plotted against l. The graphs of E_r and of P/η_{tot} can now be compared. This means that the power needed for swimming can be expressed (approximately) as a multiple of the resting metabolic rate, and one can see whether this multiple changes with body size.

RESULTS AND CONCLUSIONS

Morphology and Geometrical Similarity

There is approximate geometrical similarity between all species, the beetles being dorso-ventrally flattened, with a more or less elliptical cross-section in a horizontal plane. The frontal area A also has a rather similar outline (Fig. 2), although certain of the smaller species are relatively stout and do not bear pronounced prothoracal and elytral margins (for example *Hyphydrus ferrugineus*, No. 7 in Fig. 2). A varies with body length l according to $A \propto l^{1·69}$ (Fig. 5), i.e. somewhat less than $A \propto l^2$, which would be expected if there were exact geometric similarity. This considerable morphological similarity, and the similarity in form and function of the locomotor organs, mean that the mass of the locomotory musculature may be assumed to be an approximately constant proportion of body mass. (Polke[17] showed that the mass-ratio between "chitin and the soft parts plus water" is constant in closely related

Carabids of different size). A comparison of the dynamics of locomotion of differently sized beetles would therefore seem to be of interest.

Swimming Velocity

Correlations between body length and velocity of movement are well known. Buddenbrock,[3] for example, has shown that the relative velocity V_r of terrestrial beetles of different families was not markedly lower in larger species. Hempel[6] found that the relative velocity was almost constant in Carabids, while the absolute velocity V was proportional to $m^{0.35}$ (m = body mass). In the water beetles discussed here $V = 1.04 + 2.03\,l$ and $V_r = 24.63 - 0.35\,l$ (Fig. 4); Reynolds number Re $= -1.56 \times 10^3 + 4.36 \times 10^2\,l$ (in all these equations l is taken to be measured in mm and V in mm s^{-1}). Absolute velocity approximately doubles as body length doubles, except for the larger species, *Dytiscus marginalis*, whose velocity increases more slowly. Relative velocity falls slightly as body length increases, but it does not vary greatly between the different species.

Thus Dytiscids do not exemplify part of Teissier's rule of biological similarity, which states that in differently sized animals of a systematic group

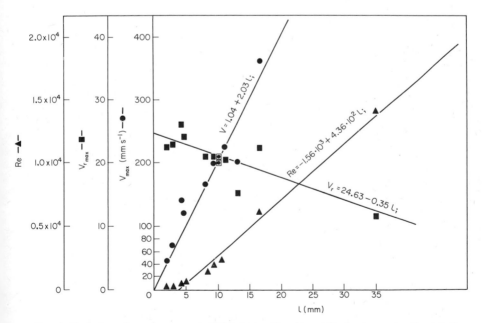

FIG. 4. Variation of maximum sustained velocity V_{max} (mm s^{-1}), the corresponding relative velocity $V_{r\,max}$ and Reynolds number Re with length l. The lines are best least-square fits to the data.

time scales change in the same way as the corresponding length scales (cf. Wilkie's paper above). They rather fulfil Buddenbrock's condition,[3] that a constant *relative* velocity is a presupposition for the existence of larger species.

The V_r values of Dytiscidae have been found to be 2 to 3 times higher than those of the closely related terrestial Carabidae (Hempel[6]). The fastest of all Carabids (*Carabus auratus*, $l = 23·5$ mm; $V = 23·3$cm s^{-1}) has the same relative velocity ($V_r = 10$) as the slowest Dytiscid (*Dytiscus marginalis*) and is surpassed 3 times by that with the highest relative velocity, *Acilius sulcatus*. The mechanics and/or energetics of rowing-swimming seem to be more favourable than of walking, in the sense that they permit higher relative

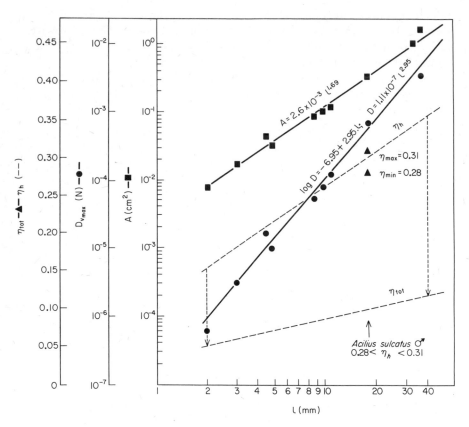

FIG. 5. Variation of frontal area A and drag D (at speed V_{max}) with length l (mm), plotted on logarithmic scales. The equations describing the best least square fits are given. Also plotted are two points representing maximum and minimum efficiency η_{tot} for *Acilius sulcatus*. The dotted lines are estimates for hydrodynamic efficiency η_h and total efficiency η_{tot}.

velocities. A higher relative velocity gives an animal a greater chance of finding prey, which is likely to be a most important factor in selection.

Body Drag

In larger species body drag D is higher as a result of both higher velocities V and greater frontal areas A, which mask a smaller drag coefficient C_D. Calculations give a correlation $D \propto l^{2 \cdot 95}$ (Fig. 5), which means that the drag increases approximately as the cube of body length l. If one assumes from the geometric similarity that there is a more or less constant fraction of muscle mass within the body, the cross-sectional area A_M of locomotor muscles, and hence the muscle force F, are proportional to the square of the body length l: $A_M \propto F \propto l^2$. This means that, other parameters being constant, smaller beetles need a smaller fraction of their muscle force to overcome the body drag when swimming at the maximum sustained speed V_{max}.

Swimming Power and Metabolic Power

Swimming power $P = D . V_{max}$ is proportional to $l^{3 \cdot 37}$ (Fig. 6). If one assumes a decline in total efficiency η_{tot} from 0·1 in large beetles to 0·05 in the smallest ones, the total power which an animal must spend in swimming at speed V_{max} is equal to P/η_{tot} and is proportional to $l^{3 \cdot 45}$ (Fig. 6). This conclusion is based on the worst possible assumptions for the variables which could not be measured directly; the total power required may be smaller than predicted but not higher. This total power is the same as the working metabolic rate E_w which the animal has to develop in order to swim at speed V_{max}; its value ranges from 6×10^{-7} W in the very smallest *Bidessus geminus* to 10^{-2} W in the large *Dytiscus marginalis*. Animal locomotion is usually a high energy-consuming process. The working metabolic rate E_w is from several times (walking in man) to more than a hundred times (flying in Diptera) higher than the resting metabolic rate E_r. This does not seem to be the case for "everyday swimming speeds" of the Dytiscidae. In *Acilius sulcatus* ♂ ($l = 17$ mm) E_r was determined. A resting beetle uses 137 mm^3 O_2 per hour at a respiratory quotient of 0·85, which means an energy budget of 2289 cal per g O_2. This coincides well with the measurements made by Hempel,[6] who determined a value of 7×10^{-2} mm^3 O_2 per hour in insect muscles. Since the specific weight of O_2 is $1 \cdot 43 \times 10^{-3}$ g cm^{-3}, *Acilius* shows a value of E_r of $6 \cdot 66 \times 10^{-1}$ cal per hour, or $7 \cdot 6 \times 10^{-4}$ W (Fig. 7). Slow everday swimming speeds of 50–100 mm s^{-1} require value of E_w of only 7×10^{-6} to 7×10^{-5} W. At $V - 256$ mm s^{-1} E equals E_r. At V_{max} (360 mm s^{-1}) E_w is approximately 2×10^{-3} W which represents only just above double the resting turnover. In bursts of very fast swimming in which $V \approx 500$ mm s^{-1} (pursuit of prey,

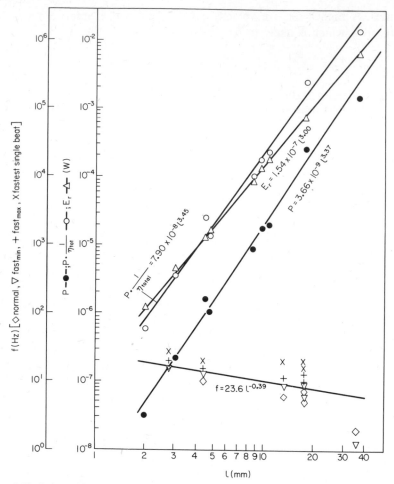

FIG. 6. Variation with length of the mechanical power used for swimming at V_{max} (P, measured in W), the metabolic power required (P/η_{tot}), and the resting metabolic rate E_r (same units). Also shown is the mean leg beating frequency f of beetles of different sizes; the line is drawn as the least squares fit to all data except that of *Dytiscus marginalis*.

flight from predators) E_w increases to approximately ten times E_r. Such manoeuvres last for only a few seconds so that under everday swimming conditions, which are a combination of slow swimming movements with many resting periods, the working metabolic rate is negligible compared with the total.

Lengthy observations of beetles kept in a huge aquarium have shown that "everyday swimming conditions" are sufficient for catching the daily requirement of prey. A typical 1-hour-protocol of a hungry beetle swimming around in an obvious searching manner was: 15 min. resting periods, 40 min. slow swimming ($V \approx$ 70 mm s^{-1}), 4·5 min. fast swimming ($V \approx$ 200 mm s^{-1}), 0·5 min. very fast swimming ($V \approx$ 400 mm s^{-1}). In this case $E_w = 0.06\ E_r$. Hence the *Acilius* swimming method seems to be a highly effective mode of locomotion which permits the animal to cover a comparatively large area when searching for prey without making too much claim on its energy

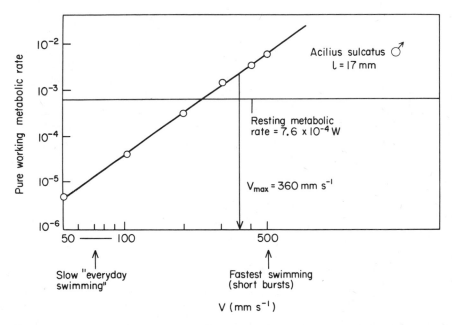

FIG. 7. Working metabolic rate E_w (W) of *Acilius sulcatus*, plotted against swimming speed V. Only for relatively high speeds, above 250 mm s^{-1}, does E_w exceed resting metabolic rate E_r.

resources. This is quite contrary to running carnivores, which spend not less than two thirds of their daily energy budget for locomotion (see Hempel[5]). However, the rowing system is very effective in accelerating the body rapidly, and uses an amount of energy which is approximately ten times higher than the resting metabolic rate. This is only possible for very short periods of time. Interestingly enough, Hempel,[5] who analyzed the energetics of the jump of locusts (Acrididae), found that the working metabolic rate for the jumping movement is approximately equal to the resting rate, 3·4 cal h^{-1} g^{-1}. This coincides with the conditions of *Acilius* swimming at approximately V_{max}.

If we calculate E_r for beetles smaller and larger than *Acilius sulcatus*, using

E_{Acilius} as a reference and assuming a respiratory activity proportional to body mass (Kittel[9]) we obtain the graph $E_r = 1\cdot54 \times 10^{-7} l^{3\cdot00}$ (Fig. 6), which crosses the graph of P/η_{tot} in the range of smaller beetles. Accordingly, $E_r = E_w$ and total power output is doubled during swimming at V_{max}. Larger beetles seem to require relatively more energy for the same purpose. Differences in the slopes of both graphs can be shown to be statistically significant, but because of the many idealisations introduced, a statistical proof seems to be of little value. The only conclusion which may be made with certainty is that, during swimming at their particular V_{max}, the differently sized water beetles have a working metabolic rate which is of the same order of magnitude as their resting metabolic rate. This is true at least for the smaller beetles which can accordingly remain in steady motion without rests at high velocities. Larger beetles such as *Acilius* and, especially, *Dytiscus* seem to reduce their velocity and/or to hold longer resting periods instead of spending relatively more energy. There may be several reasons why the swimming energetics of water insects seem to be much more favourable than the locomotion energetics of running and flying insects. Compared with flying insects, swimming beetles need no energy to stay aloft, since they are adequately suspended in the water. After rapid beating of the swimming legs, the animal covers a relatively great distance, because of its inert body mass and its rather small drag coefficient. These two methods of saving energy are not applicable to flying and running insects. Furthermore, water beetles, contrary to terrestrial insects, do not have to use energy for suspending their bodies or for pushing against gravity.

Beating Frequency of the Swimming Legs

The mean leg beating frequency f of beetles between 2 and 17 mm decreases with increasing animal size according to $f = 23\cdot6\, l^{-0\cdot39}$ (Fig. 6), with the exception of *Dytiscus marginalis*. This species has an extraordinarily low leg beating frequency combined with a low swimming speed and unusually long resting periods and thus constitutes a dynamical limiting form. A marked reduction in the difference between fastest and slowest beating frequencies is found in smaller animals; such beetles swim at a more or less constant, high frequency. Larger beetles on the other hand are able to accelerate especially quickly by executing extremely fast single beats (or a series of a few extremely fast beats). Overall, there is more variation in speed, frequency and swimming behaviour in larger beetles: the smaller ones perform more like a machine or clockwork.

Concluding Remarks

Assuming that the body length l is proportional to (body mass)$^{1/3}$ one should obtain the proportions of $A \propto m^{0.66}$, $D \propto m^{0.66}$; $P/\eta_{tot} \propto m^{0.73}$ and $f \propto m^{-0.27}$, using reduced exponents (compare Günther, Ref. 4, Table 4, p. 131). The proportions calculated with the data demonstrated here are $A \propto m^{0.56}$, $D \propto m^{0.98}$, $P/\eta_{tot} \propto m^{1.15}$ and $f \propto m^{-0.13}$. Thus there is no agreement. This may well mean that body mass cannot be used as a relevant measure of reference in these swimming beetles since their weight is compensated by the lift of the subelytral air bubble. Here frontal areas and drag coefficients seem to be more relevant for scaling considerations.

REFERENCES

1. Bauer, A. Die Muskulature von *Dytiscus marginalis*. *Z. wiss. Zool.* **95**, 594 (1910).
2. Buddenbrock, W. V. Das Schwimmen wirbelloser Tiere. *In* Bethes Handbuch der normalen und pathologischen Physiologie, Bd. **15**, 305 (1930).
3. Buddenbrock, W. V. Über kinetische und statische Leistungen grosser und kleiner Tiere und ihre Bedeutung für den Gesamtstoffwechsel. *Naturwiss.* **22**, 675 (1934).
4. Günther, B. Stoffwechsel und Körpergrösse; Dimensionsanalyse und Similaritätstheorien. *In:* "Physiologie des Menschen", Vol. 2, ed. by Gauer, Kramer und Jung. München: Urban und Schwarzenberg (1971).
5. Hempel, G. Die Energetik des Feldheuschreckensprungs *Z. vergl. Physiol.* **34**, 26 (1952).
6. Hempel, G. Laufgeschwindigkeit und Körpergrösse bei Insekten. *Z. vergl. Physiol.* **36**, 261 (1954).
7. Heumann, L. Vergleichende Untersuchungen über die Hydrostatik einiger Schwimmkäfer und ihrer Larven. *Z. vergl. Physiol.* **31**, 58 (1949).
8. Hughes, G. M. The Co-ordination of Insect Movements. III. Swimming in Dytiscus, Hydrophilus and a Dragonfly Nymph. *J. Exp. Biol.* **35**, 567–583 (1958).
9. Kittel, A. Sauerstoffverbrauch der Insekten in Abhängigkeit von der Körpergrösse. *Z. vergl. Physiol.* **28**, 533 (1941).
10. Korschelt, E. Der Gelbrand *Dytiscus marginalis* L., Bd. I. Leipzig (1923).
11. Lambert, R. and Teissier, G. Theorie de la similitude biologique. *Ann. de Physiol.* **3**, 212 (1927).
12. Nachtigall, W. Über Kinematik, Dynamik und Energetik des Schwimmens einheimischer Dytisciden *Z. vergl. Physiol.* **43**, 43–118 (1960).
13. Nachtigall, W. Hydrodynamische Messungen am Rumpf von *Dytiscus marginalis*, Vhdl. Dtsch. Zool. Gesellsch. Munchen, 317–323 (1963).
14. Nachtigall, W. Locomotion: Mechanics and Hydrodynamics of Swimming in Aquatic Insects. *In* "The Physiology of Insecta", Vol III, 2nd Edition, 381–432 (1974).
15. Nachtigall, W. and Bilo, D. Die Strömungsmechanik des Dytiscus-Rumpfes. *Z. vergl. Physiol.* **50**, 371–401 (1965).
16. Nachtigall, W. and Bilo, D. Hydrodynamics of the body of *Dytiscus marginalis* (Dytiscidae, Coleoptera), *In:* "Swimming and Flying in Nature". pp 585–595. Plenum Press, New York (1975).
17. Polke. Staatsexamensarbeit, unpublished, *Nat. Math. Fak. Mainz* (1949).

18. Flows in Organisms Induced by Movements of the External Medium

STEVEN VOGEL

Department of Zoology, Duke University, North Carolina, U.S.A.

ABSTRACT

Organisms or their dwellings commonly protrude from a solid substratum into a moving fluid medium; these situations permit energy-requiring activities to be driven by the resulting velocity gradient. One such activity is the induction of flow through organism or domicile; examples under consideration range from the ventilation of the burrows of prairie-dogs to the augmentation of pumped flow through sponges and keyhole limpets and induced airflow in the spongy mesophyll of hygrophytic leaves. Induction of flow requires asymmetry of apertures or flow field at opposite ends of internal passageways. At least two physical mechanisms are operating, one based on Bernoulli's principle and the other involving viscous entrainment. When the external flow is much greater than the internal flow, the coefficient of force drawing fluid through the passageways is nearly independent of the Reynolds number over a range of the latter from about 10 to 5000. A general discussion considers procedures for assessing the importance of induced flow in organisms in which it is suspected.

INTRODUCTION

The surface of the entire earth constitutes an interface between a solid substratum and a fluid medium. Every organism lives at or near this interface; most are attached to it, crawl over it, burrow down from it. Except, at least, for pelagic and endosymbiotic creatures, all are born and die at an interface. And furthermore, the surface of an organism itself commonly forms another such solid-fluid contact. Now a stationary fluid is exceptional, while moving fluids, driven by any of several geophysical agencies, are commonplace; thus the interfacial organism ordinarily encounters a difference in velocity as well as the difference in phase between solid and fluid.

Whether we view it as occurring at the surface or within the fluid near the surface, a velocity gradient may provide a situation permitting the extraction of energy; all that is needed is some form of transducing device. Energy is a precious commodity to organisms judging from the designs which we encounter. Let us ask, therefore, whether organisms ever capitalize on interfacial velocity gradients to drive energy-requiring processes, to enable them to occupy otherwise impractical habitats, or to simplify their own structural arrangements.

With virtually the whole bewildering array of life to survey, it becomes crucial to form some preliminary notion of just what the devices we seek will look like. The most common relevant piece of human technology is the windmill, which works not only because its rotor is in the wind, but because its feet are immovably attached to the ground. (A balloonist couldn't light his lamp with a wind-powered generator.) But a search for natural windmills would almost certainly prove quixotic. Dynamic soaring in birds employs a spatial or temporal velocity gradient in the lowest hundred or so meters of the atmosphere, but it is a specialized version of flying, itself a complex and "difficult" adaptation, and is unlikely to provide a model of wide applicability. A priori, it should be most fruitful to look for transducing arrangements of relative simplicity, in particular, those which require least modification of common anatomical designs and which serve important and general functions.

Perhaps the simplest use of a difference in velocity between two points in the environment is the generation of a pressure difference and the consequent movement of fluid. The velocity gradient in the external medium would thereby move some of that medium through the organism or a piece of plumbing made by the organism. Venturi meters and (to some extent) chimneys and carburettors function in this manner. Biological functions in which a portion of the external medium might usefully be moved through the organism include gas exchange, acquisition of food, thermoregulation, and olfaction.

At the outset it is convenient to distinguish between two schemes for inducing an internal flow and then to dismiss one with a few examples as being of less interest and generality. First, if the direction of the external flow is constant or the organism can reorient as necessary to face the flow, then the dynamic pressure of the medium may be used to drive an internal flow. Thus certain caddis-fly (Trichopteran) larvae orient their catch-nets and cases to face the oncoming water in the streams in which they live (Cummins[2]), and some scallops in unidirectional currents orient themselves so that their incurrent openings point upstream (Hartnoll[4]). But the requirements either of unidirectional external current or of reasonably rapid reorientation by the organism reduce the range of applicability of the scheme.

Another scheme, by contrast, is independent of the direction of movement of the medium, requiring only that a gradient of velocity exist, either normal to the interface or between two nearby points at the interface. Here, openings (or sets of openings) at opposite ends of an internal passage are situated normal to the plane of flow of the external medium. Induction of internal flow requires that the openings differ in some manner which results in a difference in effective velocity immediately above them. Anticipating ourselves somewhat, we recognize that two general types of arrangements

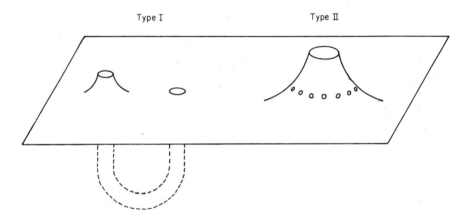

Fig. 1. Arrangements in which movement of the external medium can induce flow through a burrow-like or mound-like structure.

commonly occur (Fig. 1). In one, a U-shaped burrow has one opening even with or slightly elevated above the surrounding surface and the other protruding further above the surface, opening through a small crater (Type I). Alternatively, the geometry of the apertures or the local flow field might differ between the two openings. In the other, a conical mound has a ring of small holes around the base and these communicate with a single or a few larger apical openings (Type II). In either case, internal flow should be directed from the holes exposed to lower external velocities to those exposed to higher velocities, in short, from the low or rounded end of the burrow to the higher, cratered end (I), and from the small, basal holes to the larger, apical hole (II). It should be noted that in type II arrangements, some of the incurrent holes will be directed upstream no matter what the direction of the external flow and that therefore some portion of the internal flow will be attributable to the dynamic pressure mentioned above.

MECHANISMS

In neither the U-shaped burrow nor the conical mound can the force inducing internal flow be ascribed to a single physical mechanism. Instead, at least two mechanisms arising from rather different aspects of flowing fluids probably always participate. The first and more straightforward is a direct application of Bernoulli's principle. An elevation of the substratum, whether mound, crater, or other protuberance, leads to constriction of the streamlines above it and thus to higher flow rates in the region. The energy for this local increase in velocity must come at the expense of the pressure in the fluid. In consequence, pressure is lower just above a mound or crater than elsewhere, and flow will occur in a pipe connecting the apex of mound or crater to any other point on the substratum. The rate of flow in the pipe or burrow should be approximately given by setting the pressure generated according to Bernoulli's equation equal to the pressure resisting flow in a pipe according to Poiseuille's law. (Internal flow will be laminar in all applications to be considered.)

Thus

$$\Delta P = \rho(V_{ext}'^2 - V_{ext}^2)/2 = 8\mu l V_{int}/a^2 \tag{1}$$

and

$$V_{int} = \rho a^2 (V_{ext}'^2 - V_{ext}^2)/16\,\mu l \tag{2}$$

where ρ and μ are the density and viscosity respectively of the fluid, l and a are the length and radius of the pipe, and V_{int}, V_{ext}, and V_{ext}' are the average internal and two external velocities. V_{int} must, of course, be much smaller than V_{ext}', or else the internal flow would itself appreciably relieve the predicted pressure difference; the mechanism would still be effective, but this theory would severely overestimate the internal velocity.

In practice, however, this simple formula is difficult to apply. In the form given, it provides no information on the effect of the shape of the exit aperture on induced flow, yet the shape proves quite important: in most circumstances sharp-edged apertures form better exit holes than wider and rounder craters. And in any case, the formula holds strictly only for frictionless flow along a particular streamline. At the relatively low Reynolds numbers involved here, boundary layers are thick and the actual pressure distributions are not obvious; thus Bernoulli's equation is at best only a locally applicable approximation.

The second mechanism depends upon rather than ignores the viscosity of the fluid. As a result of the resistance of real fluids to rapid shear rates, stationary (or slower) fluid in a pipe may be "pulled" out of an aperture normal to a moving stream, a phenomenon termed "viscous entrainment".

The latter has received little attention except in connection with the location and size of the static holes of Pitot tubes. As a result of viscous entrainment, fluid will in general leave an aperture exposed to higher external velocities and enter an aperture at the other end of a pipe which is in a region of lower external flow.

Viscous entrainment entails several additional complications. Due to the increased area of contact between internal and external fluid, a larger hole provides a better exit than does a small hole; thus fluid will enter a small aperture and leave a pipe through a larger one even with no difference in external velocity between the ends of the pipe. And the mechanism is sensitive to the gradient of the velocity component directed away from the surface of the substratum (C. P. Ellington, Jr, personal communication). Thus location near the leading edge of a flat plate or atop a sharp-edged crater improves the performance of an aperture as an exit for fluid.

The general characteristics of a good exit hole—large size, upstream location, sharp edge, elevation above substratum—have been verified with a variety of models in flow tanks and wind tunnels. At the same time, it has become evident that the detailed geometrical optimization of an exit depends on the details of the particular situation. The results of ordering a set of nine apertures from best to worst exits provides an extreme example—the hole which performed best at a Reynolds number (see below) of 900 was midway between best and worst at a Reynolds number of 600 (Vogel[7]).

SCALING AND INDUCED FLOW

Just as consideration of mechanisms refines our ability to predict the appearance of devices suitable for inducing internal flow, consideration of scaling can aid prediction of the range of size, speed, and internal resistance over which these devices are likely to prove advantageous. For problems of low speed fluid flow without heat transfer, elasticity, or liquid–gas interfaces, the basic parameter for considerations of scaling is the Reynolds number. Investigating variations of a force (F) and the physical characteristics of flow with changes in velocity, size, and fluid properties requires only that one determine how a dimensionless coefficient of force, C_F, varies with the Reynolds number, Re, also dimensionless:

$$F = C_F \rho S V^2 / 2$$

and

$$\mathrm{Re} - \rho l V / \mu. \tag{3}$$

For present purposes, the characteristic length, l, in the Reynolds number will be taken as the diameter of the pipe, S as the cross-sectional area of the apertures, and V as the free-stream velocity, V_{ext}. Because of the dual physical mechanisms inducing flow in our burrow or mound as well as the considerable complexity of the mechanism of viscous entrainment, it has not proven possible as yet to predict the manner in which C_F will vary with Re from the basic equations of fluid flow. Conversely, it is a relatively simple matter to determine the relationship empirically, and the results, although they provide little insight into mechanisms, are just as useful for questions of scaling.

Several similar models of a basic U-shaped burrow were constructed for testing in a flow tank. Using the diameter of the apertures as the unit of length, each model had one aperture elevated one unit above the substratum (a metal plate), with a sharp rim and walls tapering 30° from its axis and had the other aperture even with the substratum. Apertures were five units from the sharpened leading edge of the metal plate and ten units from each other. Flexible "Tygon" transparent tubing connected the two apertures on the underside of the plate; a wide range of lengths of tubing permitted adjustment of the resistance of the pipe. Flow in the tubing was measured simply by timing the movement of the front of an injected slug of dye, taking advantage of the laminar, fully developed flow and the rule that the velocity on the axis of the pipe is twice the average velocity in such circumstances. Poiseuille's law enabled conversion of internal flow rates to forces. The use of models varying eight-fold in linear dimensions together with a fifty-fold variation in velocity permitted equivalent measurements over a 400-fold range of Reynolds numbers; except, perhaps, at the low end, this spans most of the range expected to be of biological interest. To minimize relief of induced pressure by internal flow, tubing was kept long enough so that average internal flow did not exceed about 3% of external flow.

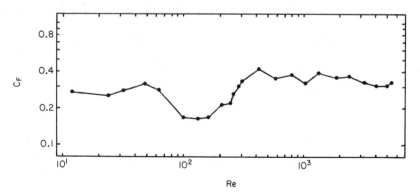

FIG. 2. The variation of force coefficient with Reynolds number for models similar to Type I of Fig. 1.

Figure 2 shows the results of these measurements. Over the entire 400-fold Reynolds number range, the force coefficient varies only about two-fold! As a first approximation, C_F may be considered constant, with a value of about 0·3. (The rather non-random-looking variation about the average has not yet been properly investigated.) Thus

$$F = 0·15 \, \rho S V_{ext}^2 \qquad \text{or} \qquad P = 0·15 \, \rho V_{ext}^2. \tag{4}$$

Combining the expression above with Poiseuille's Law, one gets

$$V_{int}/V_{ext} = 6 \times 10^{-3} \, \rho S V_{ext}/\mu l \tag{5}$$

which permits some predictions concerning the scaling of induced flow. Pressure, which is of only indirect interest to the organism, has been eliminated, and the remaining parameters are all measurable.

Assuming geometrical similarity among a series of cases of induced flow (so that length of pipe squared is proportional to cross-section) it is clear from (5) that

$$V_{int}/V_{ext} \propto \text{Re}. \tag{6}$$

Thus the mechanisms generating internal flow in the models are less effective at low Reynolds numbers: for a given fluid, small size and low speed militate against induction of flow, but without serious discontinuities as far as is known.

Equation (5) also predicts that for a given fluid and a given geometry of pipes and apertures

$$V_{int} \propto V_{ext}^2. \tag{7}$$

It is again clear why it was necessary to keep V_{int} much below V_{ext}, in practice by lengthening the tubing as the external velocity was increased: at high external velocities, the internal flow becomes disproportionately great. Since internal flow relieves the induced pressure, the relationship must break down as V_{ext} is increased. This result is evident upon re-examination of measurements on models of the burrows of prairie-dogs (Vogel et al., Ref. 9, Fig. 4). In the latter case, with relatively high Reynolds numbers (thousands) and low internal resistance, the expected behaviour according to (7) is seen up to a V_{int}/V_{ext} ratio of about 0·1: above this value (with a surprisingly sharp inflection) the curve straightens toward

$$V_{int} \propto V_{ext} + C. \tag{8}$$

It is not expected that this linear range would be encountered in many other biological systems.

What happens if the length of the pipe (or the internal resistance, proportional to the length) varies? Equation (5) implies that, if other parameters are held constant, the velocity ratio, V_{int}/V_{ext}, will be inversely proportional to pipe length. However, as noted above, the relationship should break down for short pipes and high Reynolds numbers. And this is, in fact, what one finds using models similar to those described earlier. Figure 3 shows the direct proportionality between the diameter to length ratio of the pipe and the velocity ratio at constant external velocity and diameter for long pipes, with

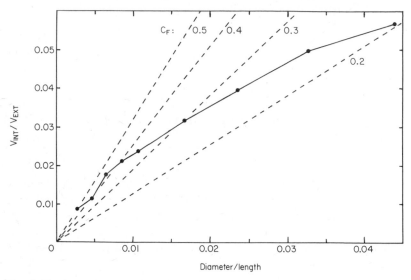

FIG. 3. The increasing deviation from a linear relation between the velocity ratio and the diameter-to-length ratio as the latter increases.

increasing deviations as length is reduced. Since pipe length will not always be an isolated variable when one compares diverse biological systems, a more useful relationship for the linear range might be

$$\frac{V_{int}}{V_{ext}} = 4.7 \times 10^{-3} \, Re\left(\frac{d}{l}\right). \tag{9}$$

Finally, it ought to be noted that, if other parameters are the same, induction of internal flow will be more effective (about fifteen times so at ordinary temperatures) in water than in air, due to the lower kinematic viscosity of water. In Nature, however, airflows are typically more rapid than flows of water; $1 \, m \, s^{-1}$ in air is barely noticeable, while the same velocity in water is torrential. Thus the more rapid natural flows of air compensate roughly for the less effective induction, and we can predict more-or-less equivalent applicability of the model to organisms in the two media.

APPLICATIONS TO ORGANISMS

How can organisms be identified which make use of the motion of the medium to move fluid through themselves or their dwellings? We have, at this point, a model or hypothesis in search of an organism; a general explanation in need of specific problems. And we have some information on mechanisms and scaling rules to help predict what appropriate devices would look like and where they might prove useful. The search thus reverses the usual approach in which a biologist asks about the structure or function of a particular organism, organ, or organelle. It therefore requires a breadth of exposure to natural history and morphology out of fashion at present as well as some physiological and physical insight. The practical alternative is extensive interaction with a diversity of biologists.

Table I summarizes a series of biological examples of induced flow for which there is either experimental support or strong grounds for suspicion. These examples will provide illustrative material in the discussion which follows, written as both guidance and encouragement for anyone who chooses to pursue such problems. While testing a possible case of induced flow is in principle straightforward, in practice it often proves awkward. A general procedure can be outlined, although it should be noted that not all elements are necessary or practical in all suspected cases and that the order given is not especially significant.

1. Peruse the writings of the ecologists and natural historians—is induced flow reasonable given the habits and habitats of the organism? All of the organisms tabulated are indeed exposed to moving fluids. Limpets and marine sponges are not described from "still" water. Prairie-dogs inhabit the windy plains of North America. Turret spiders burrow in open fields and coastal dunes. Archeocyathid fossils come from coarse sediments reflecting "high energy" environments. The tropical termites of interest live in open savannah.

2. Deduce, as far as possible, the benefit to be obtained on some reasonable assumption of the function of the induced flow. It is easy to calculate that prairie-dogs risk asphyxiation in their burrows without some system of forced ventilation; a similar situation has been noted for termite mounds (Luscher[6]). For filter-feeders such as sponges, the cost of filtration has been estimated as approaching the yield (Jørgensen[5]), so any mechanism reducing this cost should be beneficial. Plants (as a result of their own prior activities) live in a world in which a crucial input, CO_2, is exceedingly dilute; artificially elevating its concentration above atmospheric levels typically increases productivity. Thus a mechanism for supplying CO_2 without the concentration gradient required for diffusion should (if desiccation can be avoided) be of use.

3. Consider the manner in which the organism's morphology varies among

TABLE I. Induced flows—biological examples.

Organism	Structure	Type	Function	Status, references
Prairie-dog, *Cynomys ludovicianus*	Burrow	I	Gas exchange	Vogel *et al.*[9]
Bony fish (various)	Nostrils and passages	I	Olfaction	Inference from habit and structure
Burrowing shrimp *Upogebia, Callianassa*	Main passage of burrow	I	Uncertain	Inference from burrow arrangement
Hygrophytic plants (various)	Cratered leaf stomata	I	CO_2 supply	Flow measurements, model studies
Turret spiders *Geolycosa*	Cratered burrow, porous soil	I	humidification	Observation of flow in burrow
Sponges *Halidhondria*, others	Entire hydraulic system	II	filter-feeding	Vogel,[7] field measurements
Keyhole limpets Fissurellidae	Mantle cavity, gills, keyhole	II	uncertain	Flow observations on shells in tank
Tropical termites *Macrotermes*	Porous mound walls, openings	II	gas exchange	Weir,[10] Vogel and Bretz[8]
Archeocyathids Model of fossil	Hydraulic system	II	filtration?	Balsam and Vogel[1]

habitats which differ with respect to relevant environmental factors such as wind or current. Such variations, if present, should be consistent with maintenance of induced flow under natural conditions or the substitution of other devices where necessary. Thus the exit apertures of sponges have been described as being located on higher oscular chimneys where a given species inhabits calmer water (DeLaubenfels[3]), a situation consistent with the thicker boundary layers as well as the lower currents in such habitats. Similarly, the turrets on spider burrows in coastal (and presumably windier) localities are apparently lower than those of inland fields. Termite mounds in rain forest have different ventilatory arrangements (probably using a lung-like arrangement and internal convection) than conspecific mounds in open country (Luscher[6]).

4. Determine the manner in which the internal velocity varies with changes in the velocity of the medium under reasonably representative field conditions. Results of such measurements may be compared with the predictions made earlier if the winds or currents provide a sufficient range of speeds. In addition, the relative contributions of active (metabolically driven) and passive induced) components of internal flow may be assessed where (as in sponges) the former is substantial. For prairie-dog and turret spider burrows, semi-quantitative observations were made with smoke generators in the absence (at the time) of adequate microanemometers for field use. For sponges, it was possible to record the velocity adjacent to the animal simultaneously with that just beneath the aperture, but it was impractical to make such measurements on days when wave-induced local currents were high. Instead resort was made to the less desirable alternative of artificially augmenting the currents around the animals.

5. The notoriously irregular variation of winds and currents and the difficulties of measurement of flow in the field make it useful, either instead of or in addition to (4), to determine the relationship between internal and external flow in a flow tank or wind tunnel. Furthermore, other relevant variables such as internal resistance may often be more conveniently determined in the laboratory. Often, however, moving the organism to the laboratory is impractical (for example, prairie-dog burrows) or the burrows constructed in the laboratory are of questionable quality (turret spiders and burrowing shrimp). The measurements on sponges actively pumping in a flow tank have been found, by comparison with subsequent field measurements, to reflect, in part, the relatively low active pumping rates of disturbed animals.

The possibility of shifting from one fluid to another should not be ignored. As long as the Reynolds number is kept unchanged, the patterns of flow and force coefficients should be normal; a shift, for example, from air to water permits a reduction in velocity of about 15-fold and increase forces about

L

4-fold. Dye movements in water are more easily observed than smoke movements in air. Thus hygrophytic leaves were filled with water and fluid entry into the stomata measured by exposing them to moving dye solutions for brief periods and then extracting the leaves for photometric measurements. The procedure also reduced the influence of diffusion to a manageable level and permitted equivalent estimates of diffusion from exposures to still dye.

6. Physical models may be built and tested in wind tunnel or flow tank, the patterns of flow observed, and V_{int} and V_{ext} measured. Some morphological study or measurements of internal resistance of real organisms may be a prerequisite to constructing adequate models; the models are very handy, however, for eliminating active responses, morphological variation, and other inconveniences. (The original impetus for the entire project was provided by two simple models—a U-shaped length of plastic tubing attached to the bottom of a wash basin and a perspex sponge-like object mounted in a length of rain gutter). Models may be of considerable sophistication and, for questions about mechanisms, may be of widely varying degrees of abstraction from the organism. Models of archeocyathids, for example, were as close to reality as could be inferred from the fossils since the organisms were obviously unavailable, while other models explored only the influence of the shape and size of apertures and resembled no particular organism. The advantages of shifting size, velocity and medium are great; only the Reynolds number need be kept constant, as mentioned earlier. Thus the models of archeocyathids were tested in air since anemometry was at that point more convenient—the velocity was merely increased fifteen-fold. And models of the burrows of prairie-dogs fitted conveniently into the local wind tunnel when built one-tenth natural size; the velocity was increased ten-fold in compensation.

7. Consider, finally, alternative functions of the structural arrangement in question. If alternative functions are evident, does one have a truly multifunctional structure, or is induced flow merely an accidental consequence of an arrangement which satisfies some other need? The mounds on prairie-dog burrows have been claimed to prevent flooding and the elevated oscular apertures of sponges to prevent reingestion of previously filtered water; these are probably multifunctional. But the asymmetrical, U-shaped burrows of many intertidal worms may be accidents reflecting little more than that the excavation began at one end. In the latter, induced flow is strong when the burrow is empty, but reduced or absent when the worm fills the tube.

8. It is generally useful to do correlative physiological studies on the organism itself to assess further the importance of induced flow in its functioning; since such studies are both diverse and essentially conventional, they will not be considered further.

REFERENCES

1. Balsam, W. L. and Vogel, S. Water movement in archaeocyathids: evidence and implications of passive flow in models. *J. Paleont.* **47**, 979–984 (1973).
2. Cummins, K. W. Factors limiting the midrodistribution of larvae of the caddis-flies *Pycnopsyche lepida* (Hagen) and *P. guttifer* (Walker) in a Michigan stream (Trichoptera: Limnephilidae). *Ecol. Monogr.* **34**, 271–295 (1964).
3. DeLaubenfels, M. W. The sponges of Woods Hole and adjacent waters. *Bull. Mus. Comp. Zool.* **103**, 1–55 (1949).
4. Hartnoll, R. G. An investigation of the movement of the scallop, *Pecten maximus*. *Helgolander Wiss. Meersunter.* **15**, 523–533 (1967).
5. Jørgensen, C. B. "Biology of Suspension Feeding." Pergamon Press, Oxford (1966).
6. Luscher, M. Der Sauerstoffverbrauch bei Termiten und die Ventilation des Nestes bei *Macrotermes natalensis* (Haviland). *Acta Tropica* **12**, 289–307 (1955).
7. Vogel, S. Current-induced flow through the sponge, *Halichondria*. *Biol. Bull.* **147**, 443–456 (1974).
8. Vogel, S. and Bretz, W. L. Interfacial organisms: passive ventilation in the velocity gradients near surfaces. *Science* **175**, 210–211 (1972).
9. Vogel, S., Ellington, C. P., Jr. and Kilgore, D. L., Jr. Wind induced ventilation of the burrow of the prairie-dog, *Cynomys ludovicianus*. *J. Comp. Physiol.* **85**, 1–14 (1973).
10. Weir, J. S. Air flow, evaporation and mineral accumulation in mounds of *Macrotermes subhyalinus* (Rambur). *J. Anim. Ecol.* **42**, 509–520 (1973).

19. Effects of Size on the Swimming Speeds of Fish.

C. S. WARDLE

Marine Laboratory, P.O. Box 101, Aberdeen, Scotland.

ABSTRACT

The cruising (aerobic) and the maximum (anaerobic) swimming speeds of fish are limited by two quite distinct factors related to the length of the fish's body.

It has been shown that the fish's maximum sustained cruising speed is matched by a maximum sustainable rate of passage of oxygen from the water to the aerobic muscle. A series of fish types is presented showing a progression in physiological and anatomical adaptations designed both to elevate this turnover of oxygen and increase sustained swimming speed. Evidence observed in anatomical features from species spanning a wide size range suggests that this series of fish types has evolved to avoid critical Reynolds numbers which may dictate ceilings in cruising speed. The significant iso-Reynolds number lines become apparent when the maximum cruising speeds for the different sizes of fish are examined within this series of fish types.

Common to all types and sizes of fish are the anaerobic lateral swimming muscles. New measurements have shown the contraction time of these muscles to increase with the length of the fish. One complete tail beat cycle requires time for two contractions of the lateral muscles, one on each side, and moves the fish forward one "stride", equal to about 0·7 times the length of the fish. The maximum burst swimming speed is thus limited by the maximum stride frequency which in small fish (length, L = 0·10 m) at 14°C results in speeds up to 25 body lengths per second ($L s^{-1}$) but in larger fish (L = 1·0 m) allows only 6 $L s^{-1}$. An increase in temperature raises the speed and a decrease lowers it.

INTRODUCTION

Among the diverse teleost fish can be found species which grow to all sizes up to and occasionally longer than 3 m. On the basis of a small range of sizes, rules of thumb have been established, which suggest maximum speeds

of 3 body lengths per second (3 L s⁻¹) for sustained cruising swimming, and
10 L s⁻¹ for burst swimming. Recent observations of the behaviour of fish
in fishing gears (Hemmings[16] and Wardle, unpublished) have indicated that
the capabilities of "wild" fish cannot be explained by such simple rules and
neither of these figures is tenable over a large size range. Valid records of
maximum swimming speeds and sustained swimming speeds of fish spanning
a large size range are difficult to obtain, because of the inherent problems
either in measuring true speed in the field, or of damaging, inhibiting or
restricting the fish by transferring them to the laboratory.

Fish show an enormous variety of adaptations to their environment yet
apart from a very few highly specialised species it is always an advantage for a
fish to be able to accelerate rapidly and move briefly at a high speed to seek
shelter from a predator. On the other hand, a fast sustained cruising speed is
only an advantage in certain special life styles such as pelagic feeding species.
Among those species of teleosts that have had their swimming speeds measured
there can be seen a series of types showing different degrees of achievement
in burst speed. The elevation of sustained swimming performance of fish is
seen to be closely linked both to those adaptations of anatomy which help
to increase the sustained metabolic level and to those features of body shape
that help to overcome or reduce the restricting effects of the water on the
movement of the fish.

The present account examines the existing measurements of swimming
speed in relation to features of the physiology and anatomy and finds two
distinct limits to swimming speeds of fish related to length.

THE TYPES OF SWIMMING MUSCLE

Teleost fish swim by alternately contracting the left and right lateral
muscles causing cyclic oscillations of the caudal (tail) fin. A functional distinction can be made between two types of swimming muscle (Hudson,[20]
Webb[36]). The dark or red muscle—perhaps better termed *aerobic* muscle—
contracts only when oxygen is available to the cells and any restriction in
oxygen availability limits its rate of performance. The white or *anaerobic*
muscle can contract rapidly and powerfully in the absence of oxygen and
becomes exhausted when all the glycogen stored in the muscle cells has been
converted to lactic acid. Rebuilding of the glycogen store uses oxygen and can
take relatively long periods of up to 24 hours.

By careful placement of electrodes Hudson[20] was able to demonstrate
electrical activity in the red muscle of rainbow trout, *Salmo gairdneri*, at
slow swimming speeds and noted the addition of electrical activity in the
white muscle as swimming speed increased. The recruitment of white muscle

contractions occurred as the tail beat frequency increased to values between 3·05 and 3·60 cycles per second. The fish were 26–34 cm long and the maximum cruising speed was estimated to occur at 3·75 cycles per second.

A group of 9 cod, *Gadus morhua*; 9 saithe, *Gadus virens*, and 2 haddock, *Melanogrammus aeglefinus* with lengths between 27 and 45 cm swam at 0·367 m s^{-1} (0·81 to 1·47 L s^{-1}) for 102 hours in a moving light pattern in the gantry tank at Aberdeen. At the end of the experiment the fish were stunned electrically and resting levels of lactate and glycogen were found in the white muscle (Wardle[35]). An increased use of the glycogen in the white muscle was measured in groups of saithe (17–21 cm) made to swim for six hours at up to 4·62 L s^{-1} (Johnston and Goldspink[22]). Rainbow trout forced to swim at maximum burst speed convert 50% of the glycogen store to lactic acid in 2 minutes (Black *et al.*[5]).

There is a threshold swimming speed above which endurance suddenly becomes limited because a greater share of the energy for muscle contraction is derived from the anaerobic fuel store. Measurements of this threshold speed, or maximum sustained cruising speed, have shown that different speeds are achieved by different types of teleost fish.

Increasing the Levels of Aerobic Metabolism

Evidence from anatomical studies has shown that those teleosts which attain higher sustained cruising speeds to suit a particular way of life have elevated that set of apparatus assigned to transferring oxygen from the water to the aerobic swimming muscles. These adaptations include increases in the proportion of pigmented, dark or red muscle (Blaxter *et al.*,[6] Boddeke,[7] Boddeke *et al.*,[8] Love,[25] Webb[38]), increases in the circulatory system including heart size, heart rate, haematocrit (Blaxter *et al.*,[6] Klawe,[24] Wardle,[34] Wardle and Kanwisher[37]) and increases in the gill area (Gray,[14] Hughes,[21] Muir,[28] Muir and Hughes,[29] Muir and Kendall[30]).

EXAMPLES RELATING SUSTAINED SWIMMING SPEED
AND AEROBIC METABOLISM

Three teleost groups with distinct levels of aerobic metabolism are listed in order of increasing sustained swimming performance; the Gadidae, represented by the relatively unstreamlined cod, haddock and saithe; the Salmonidae, the most systematically studied group both for measurements of sustained swimming and metabolic rate, and including the sockeye salmon, *Oncorhynchus nerka*; the Scombroid families as defined by Zharov,[39] including the superior streamlined mackerel and tuna types.

The Gadidae have only small proportions of separate red aerobic swimming muscle. Their lateral muscles are dominated by white muscle which acts through fins of large surface area to give rapid acceleration to burst swimming speeds using anaerobic energy. The maximum rate of taking oxygen from water in haddock (24·8 cm: 10°C) approached $300 \text{ mg O}_2 \text{ kg}^{-1} \text{h}^{-1}$ ($1·22 \text{ J kg}^{-1} \text{s}^{-1}$) at a maximum sustained speed of just over 2 L s^{-1} (Tytler[32]). Other records of maximum sustained cruising speeds in Gadidae are given in Table Ia.

TABLE I. Maximum sustained (aerobic) cruising speeds recorded for (a) Gadidae and (b) Salmonidae

Species	Length (m)	Temperature (°C)	$L \text{ s}^{-1}$	$\text{mg O}_2 \text{ kg}^{-1} \text{h}^{-1}$*	Ref.
(a) saithe	0·163	10·4	3	—	15
haddock	0·248	10	2	300	32
cod	0·353	8	2·6	—	4
gadidae	0·27 to 0·45	10	0·81 to 1·5	—	unpublished
haddock (in sea)	0·30 to 0·50	8	<2	—	16
cod	0·70 to 0·80	10	<1	—	unpublished
(b) sockeye salmon	0·10	10	6	700	9
sockeye salmon	0·20	10	4	>600	9
sockeye salmon	0·42	10	3	>600	9
sockeye salmon	0·54	10	2·6	600	9

*   *Footnote*: $1 \text{ mg O}_2 \text{ kg}^{-1} \text{h}^{-1} \approx 4·08 \times 10^{-3} \text{ Jkg}^{-1} \text{s}^{-1}$.

For the Gadidae the impression is that there is a progressive reduction in the maximum sustained cruising speed as the size increases; $3–4 \text{ L s}^{-1}$ for fish up to 20 cm; 2 L s^{-1} for fish up to 30 cm reducing to no more than $0·5–1 \text{ L s}^{-1}$ for fish above 100 cm. The maximum length reached by cod is 169 cm, saithe 120 cm and haddock 100 cm.

The performance and oxygen consumption of the Salmonidae have been carefully studied by J. R. Brett and a recent summary of Brett's work (Brett and Glass[9]) indicates again a decrease in speed measured in lengths per second as size increases (see Table Ib).

The figures for sockeye salmon in Table I show nearly twice the sustained speed range of the haddock and just twice the maximum oxygen consumption recorded for the haddock by Tytler.[32]

The fish in the Suborder Scombroidei are noticeably streamlined in shape and have a style of swimming which is termed carangiform with lunate tail (Webb[38]). They have a pelagic life and are committed to continuous swimming with a *minimum* sustained cruising speed associated with maintenance of lift by the pectoral fins (Magnuson[26]), and ram ventilation of their extended gill areas (Muir and Kendall,[30] Brown and Muir[10]). Records of *Euthynnus affinis* (37–42 cm) swimming voluntarily at 1·3–2·2 L s^{-1} (Cahn,[11] Magnuson[26]), indicate that these *minimum* swimming speeds are as high and often higher than the sustained maximum cruising speeds of the Gadidae and Salmonidae considered above. This species of tuna increases its voluntary swimming speed when stimulated by food. Fish (40 cm) were recorded feeding at speeds between 3·8 and 10 L s^{-1} and non-feeding speeds ranged from 2·9–12·5 L s^{-1} (Walters[33]). No systematic measurements have been made of the maximum sustained cruising speed in the Scombroidei but it is generally understood that many species of the Scombroidei can cruise for long periods at speeds of up to 3 m s^{-1}. Bluefin tuna *Thunnus thynnus* grows up to 300 cm in length, and migrates long distances. Two bluefin tuna tagged in the Gulf of Mexico swam to waters off Norway, a distance of 4,200 nautical miles in 118 and 119 days (Mather,[27] misquoted by Carey and Teal[12]). These fish were approximately 2·5 and 2·8 m long and show an average migration speed of 0·75 m s^{-1}, no more than 0·3 L s^{-1}.

No measurements of oxygen turnover have been successfully made in the Scombroidei but there are many pronounced features of their anatomy and physiology that suggest evolution of the apparatus necessary to increase oxygen turnover to values well above those of the Salmonidae. Particularly noticeable are increased proportions of red aerobic swimming muscle (Rayner and Keenan[31]), special adaptation of the blood circulation to maintain a relatively high temperature in the red muscle (Kishinouye,[23] Carey and Teal,[12] Carey et al.[13]), increase of the surface area and streamlining of the gill filaments (Gray,[14] Muir and Hughes[29]), and very high levels of blood haemoglobin (Klawe[24]). All these features and others are directly related to elevation of the metabolic rate of the aerobic swimming muscle to peak levels. The commitment of a larger proportion of their tissues to the continuous presence of high levels of oxygen makes these fish particularly sensitive to capture and experimental handling.

THE SIGNIFICANCE OF BODY SHAPE

The elevation of metabolism as recorded through the three groups, Gadidae, Salmonidae and Scombroidei is directly linked with the observed increase in maximum speed of sustained swimming. There is also a change in body shape towards the exaggerated streamlined shapes of mackerel and tuna associated with adaptation to cruising at high optimum speed. The body of the cod (Fig. 1(a)) is cluttered with many adjustable dorsal and ventral fins with large surface areas. The caudal fin is large, flexible and membranous, suitable for creating a maximum acceleration and achieving a high burst speed. The sockeye salmon (Fig. 1(b)) has only a single dorsal fin, a ventral fin on the trailing edge ahead of the caudal peduncle and a reduced adipose fin strategically placed on the dorsal trailing edge, perhaps to help position and control turbulence generation in cross flow while cruising at its slightly

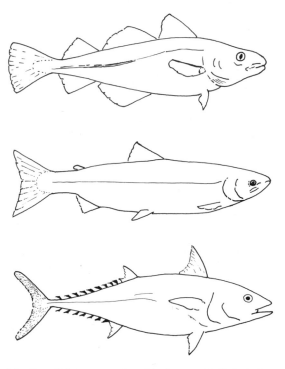

FIG. 1. Outlines of the three fish types mentioned in the text. (a) Cod, *Gadus morhua* (L.), family Gadidae. (b) Sockeye salmon, *Oncorhynchus nerka*, family Salmonidae. (c) Bluefin tuna, *Thunnus thynnus* (L.), Suborder Scombroidei.

higher speeds. The Scrombroidei (Fig. 1(c)) are clearly designed to cope with the continuous problems created by water flow over their bodies at high sustained swimming speeds. Their shape seems to approach an optimum profile for a particular speed of swimming. Yet as their swimming involves continuous sinusoidal movement of the tail direct comparative measurements of drag on the swimming body in order to determine this optimum are unpractical. Hertel (Ref. 17, p. 128) considers tuna, like dolphins, to have profiles similar to those designed by aeronautical engineers to assist maintenance of laminar flow at higher Reynolds numbers. Magnuson (Ref. 26 in Fig. 13) demonstrates the slicing of cross flow by the finlets on the dorsal and ventral trailing edges of *Euthynnus affinis* suggesting microturbulence generation. Such effects are used in aircraft design where the generated vortices correspond to the size of the elements and are discussed by Hoerner[19] under the headings of slotted flaps (p. 6–14) and pervious flaps (p. 12–13) which are positioned to slice up the flow and thus avoid heavy turbulence.

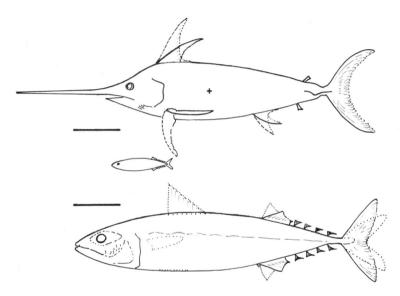

FIG. 2. Scale drawings of the swordfish, *Xiphias gladius* L (upper; scale marker 25 cm long), and Atlantic mackerel, *Scomber scombrus* L . (lower; scale marker 5 cm long). (For discussion see text.) The extremes of movement of the fins are shown by dotted and continuous lines. The swordfish pectoral fin does not fold flat against the body wall but locks in a position pointing posteriorly but standing out from the body. In the swordfish there are narrow sensory grooves (containing remnants of fins present in the younger stages) between the 1st and 2nd dorsal fin and between the anal and 2nd ventral fin. The sword in the specimen drawn was broken.

Two pelagic cruising species recently examined by the author are shown as scale drawings in Fig. 2. The swordfish grows occasionally longer than 400 cm, the mackerel to 50 cm (note the two scales in Fig. 2). The first dorsal fin of the mackerel folds into a socket flush to the body and the pectoral and pelvic fins fold flat against the body wall. The second dorsal fin of the mackerel and the anal fin fold only partially and have no sockets. The nonfolding dorsal and anal fin together with the finlets remain to some degree permanently extended and suggest by their presence a breakdown in laminar water flow two-thirds back along the body where the profile begins to narrow. None of the fins of the swordfish examined could be folded and each had the restricted range of movements indicated in Fig. 2. The leading edge of all the large fins of the swordfish are thick and fleshy; the trailing edges tend to be broken or serrated and very thin. It would seem that all the fins are permanently extended to control and stabilise the effects of turbulent water flow over the larger body. The small sketch of the mackerel drawn to the same scale beneath the head of the swordfish, Fig. 2, illustrates that laminar flow may break down at a similar distance back along the body of both fish. The first nonfolding fin on the mackerel is at 20 cm from the anterior tip of the head. The first dorsal fin of the swordfish is 27·5 cm behind the anterior tip of the lower jaw.

In describing the features of the Suborder Scombroidei, Zharov[39] discusses the variable development of a collar of rough scales, the pectoral corselet, posterior to the gill opercula. The corselet is generally smooth in smaller sized species and covered in rough scales in larger species and must generate microturbulence immediately behind the head. Zharov also discusses the relative migration of dorsal fins forwards along the body as size increases both during development and among species achieving different sizes.

Further measurements and accurate observations are clearly required of the positions of folding and non-folding fins and related structures in both a wide range of sizes of the same species and in species reaching different sizes. The pelagic cruising species about which we know so little seem to hold the clues to many aspects of the critical effects of body size on movement through water.

SUSTAINED SWIMMING SPEEDS AND REYNOLDS NUMBER

The sustained swimming speeds discussed are outlined on a graph of velocity against length (Fig. 3). The speeds do not correspond, except for a few species near 20 cm in length, to a 3 body length per second line but cruising speeds are significantly reduced as size increases. Some clues to the value of the critical Reynolds number (i.e. velocity × length ÷ kinematic viscosity of water), where laminar flow becomes turbulent, are given in studies of flow over cylinders and flat plates (Hertel, Ref. 17, Fig. 254, and Hoerner, Ref. 19. Fig. 2-6.5). The critical Reynolds numbers are found between 1×10^6 and 3×10^6. Iso-Reynolds number lines for appropriate fixed temperatures are plotted on the velocity length graph, Fig. 3. Note that the kinematic viscosity of sea water varies considerably with temperature but not with salinity; the values used were taken from Hoerner.[19]

In Fig. 3, the maximum speed of sustained swimming can immediately be related to the Reynolds number. The sustained swimming speeds attained by the larger Gadidae and Salmonidae at 10°C are divided by a line a little less than Re $= 1 \times 10^6$ at 10°C. The Scombroidei, with exaggerated stream-lined profiles and elevated aerobic metabolism, are not affected at this level but are apparently limited by a Reynolds number near 3×10^6. In all cases the maximum aerobic metabolic rate may be the limit to sustained swimming speeds achieved by smaller sizes and the increase in drag associated with the critical Reynolds number becomes important in restricting performance as the size of the fish's body increases.

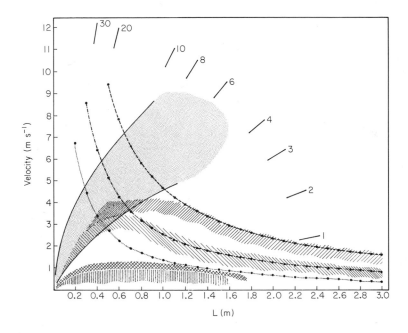

FIG. 3. The limits of swimming speed plotted as velocity against length, related to Reynolds number. Iso-Reynolds number lines $----$ 3×10^6, 5°C; $-\cdot-\cdot-$ 3×10^6, 30°C; $\cdots\cdots$ 1×10^6, 10°C. The cross hatched areas represent estimated (see text) sustained swimming speeds of Salmonidae ※, Gadidae ‖‖ and Scombroidei, warm body in cold water /// and Scombroidei, warm body in warm water ◈. For comparison, the dotted area includes maximum burst swimming speeds for all temperatures between 0°C and 20°C and stride lengths between 0·6 and 0·8 (from Wardle[36]). The third curved scale allows reading of the speed of any point on the graph in body lengths s^{-1} by lining up that scale with the origin of the graph.

Surface structures have been developed by fish, for example the small adipose fin in salmon, to cope with turbulence generation once a Reynolds number of 1×10^6 has been exceeded. The evidence suggests that these surface features have been developed or selected in order to control or lessen the effects of the substantially increased drag at higher cruising speeds and thus allow a worthwhile gain from raising the aerobic metabolic level and not as is often claimed to increase maximum burst swimming speed.

An Interesting Effect of Temperature

The larger bluefin tuna have been shown to maintain a temperature of near 30°C in water temperatures down to 6°C (Carey and Teal[12]). Drawn in Fig. 3 are two iso-Reynolds number lines for values of 3×10^6, one at 5°C, the temperature of the coldest water entered by bluefin tuna, and one at 30°C. It seems significant that bluefin tuna tagged with miniature internal and external temperature transmitters were shown to choose to swim in the cooler layers of a thermocline for long periods (Carey et al.[13]). In the water of lower temperature there will be significantly less drag for the same speed and size (Fig. 3). Note that, by elevating sustained metabolism and improving the hydrodynamic structure still further, the large warm blooded aquatic mammals can sustain speeds in the region of 6–7 m s^{-1} and deal with turbulence at Re $= 1 \times 10^7$ (Hill[18]).

MAXIMUM ANAEROBIC BURST SWIMMING SPEEDS

Wardle[36] developed a technique whereby the contraction time of pieces of lateral anaerobic swimming muscle could be measured. It was found that contraction time (T) increased with the length (L) of the fish and with decrease in temperature. Bainbridge[1] had pointed out that swimming speed was closely related to tail beat frequency. The fish moves forwards a constant proportion of L, ($A \times L$), for each completed cycle. Hudson[20] re-examined and confirmed this relationship. Using accumulated measurements of contraction time, Wardle[36] was able to predict the maximum swimming speed at any temperature, for all fish within the size range measured. The prediction was based on the estimation of maximum tail beat frequency where one cycle requires time for one contraction of the muscle on the left side followed by one contraction on the right side of the body, and maximum speed $U_{max} = A \times L/2T$. Figure 4 shows the relationship between length and muscle contraction time with the original data from Wardle[36] together with new measurements. The derived lines marked by temperature are those from Wardle[36] and the 14°C line is used to calculate the speed curve (in Fig. 5) for $A = 0.6$ to 0.8. Existing measurements of burst swimming speeds such as those given by Wardle[36] and Bainbridge[3] lie reasonably close to this estimated speed curve and videotape records of fish swimming at maximum speed have shown the predicted tail beat frequencies (Wardle[36]). In con-

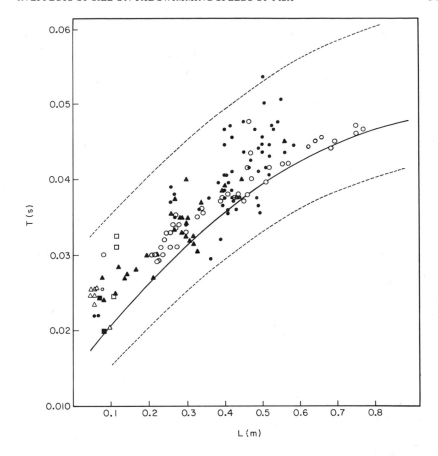

FIG. 4. Contraction time of anaerobic swimming muscle plotted against fish length. The symbols refer to *Salmo salar* ●; *Pleuronectes platessa* ▲; *Trachinus vipera* □; *Gadus morhua* ○; *Clupea harengus* △; *Melanogrammus aeglefinus* ■. The minimum contraction time at 14°C (continuous line) was derived from the data in Wardle[36] and is used to construct Fig. 5 (see text). The dotted line (upper) is minimum contraction time at 0°C and (lower) at 20°C based on a measured change in contraction time of 0·01 s per 1°C.

trast to the speed range outlined for sustained swimming speed (Fig. 3) the burst speed curves penetrate directly through the iso-Reynolds lines (Fig. 5). However, the effect of the increasing drag on the body is reflected in the displacement of the maximum speed curve from near 25 L s^{-1} for a 10 cm fish to only 6 L s^{-1} at 100 cm. This result suggests that the effect of the increased drag on the larger body has moulded or restricted the evolution of the muscle properties and the matching strength of the supporting structures so

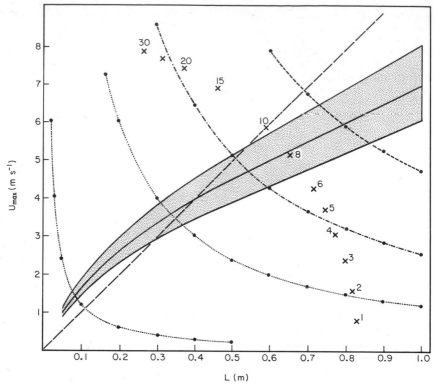

Fig. 5. Maximum burst swimming velocity plotted against fish length and related to Reynolds number. Iso-Reynolds number lines ········ 1×10^5 and 1×10^6 at 14°C; —·—·— 3×10^6, 30°C (as Fig. 3); — — — 3×10^6 5°C (as Fig. 3). The dotted area includes the predicted burst swimming speeds of fish calculated from the contraction time of muscle at 14°C shown in Fig. 4. The three firm lines indicate speeds attained by a stride length of, from top to bottom, 0.8 L, 0·7 L and 0·6 L. The crosses when connected to the origin give body lengths s^{-1}, a line is marked in at 10 L s^{-1}, for discussion see text.

that the body of the larger animal cannot be unduly strained, as discussed by Hill.[18]

Preliminary examination of the geometric scaling of cod as they grow from 20 cm to 100 cm show that their mass remains close to L^3, their fin dimensions remain closely related to L and their fin areas to L^2. The muscle mass of cod was measured using techniques described by Bainbridge[2] and was found to reduce from 47·4 % of total fresh mass at 20 cm to 37·8 % for fish 100 cm long, the muscle mass being related to 3·75 $L^{2·86}$. Bainbridge (Ref. 2, Fig. 8) found an increase in the proportion of muscle mass as length increased in fish between 5–20 cm, the muscle mass being related in dace to $L^{3·04}$, trout to

$L^{3.06}$ and goldfish to $L^{3.40}$. The change in the trend from an increase to a decrease in the proportion of muscle with increasing length occurs as the slope of the speed curve (Fig. 5) decreases between $Re = 1 \times 10^5$ and $Re = 1 \times 10^6$ at 14°C.

The Power and Endurance of Maximum Speed Swimming

Bainbridge[3] estimated the power required to achieve the measured maximum swimming speeds. His velocity-length curve, for a laminar flow and a power value of $4 \times 10^5 \, erg \, s^{-1} \, g^{-1}$ of muscle (40 W kg^{-1}), follows very closely the position of the maximum swimming speed curve at 14°C (0·8 L, Fig. 5). Bainbridge's figure of 40 W kg^{-1} of muscle can be related to the total store of 5,580 J kg^{-1} potential energy normally present as glycogen in resting anaerobic muscle (Wardle[36]) and released by anaerobic conversion to lactic acid. If completely depleted at 40 J s^{-1} this store would allow 140 seconds of laminar burst swimming. However, it is unlikely that burst swimming is laminar at Reynolds numbers above 1×10^6 and a reduction in endurance might be expected. This energy is more likely to be released in short bursts of diminishing strength (Wardle[36]). The total distance moved forwards before the anaerobic fuel is depleted will be directly related to the number of strides and thus to the fish length.

ACKNOWLEDGEMENTS

I acknowledge constructive discussions on the subject of swimming fish with colleagues R. S. Batty, R. S. T. Ferro, J. J. Foster, C. C. Hemmings and A. Reid. I thank R. S. Batty and W. Mojsiewicz for help with measurement and interpretation of cod isometrics.

REFERENCES

1. Bainbridge, R. The speed of swimming of fish as related to size and to the frequency and amplitude of the tail beat. *J. exp. Biol.* **35**, 109–133 (1958).
2. Bainbridge, R. Speed and stamina in three fish. *J. exp. Biol.* **37**, 129–153 (1960).
3. Bainbridge, R. Problems of fish locomotion. *Symp. zool. Soc. Lond.* **5**, 13–32 (1961).
4. Beamish, F. W. Swimming endurance of some northwest Atlantic fishes. *J. Fish. Res. Bd. Can.* **23**, 341–347 (1966).
5. Black, E. C., Connor, A. R., Lam, K. and Chiu, W. Changes in glycogen, pyruvate and lactate in rainbow trout (*Salmo gairdneri*) during and following muscular exercise. *J. Fish. Res. Bd. Can.* **19**, 409–435 (1962).
6. Blaxter, J. H. S., Wardle, C. S. and Roberts, B. C. Aspects of the circulatory

physiology and muscle systems of deep-sea fish. *J. mar. biol. Ass. U.K.* **51**, 991–1006 (1971).

7. Boddeke, R. Size and feeding of different types of fish. *Nature, Lond.* **197**, 714–715 (1963).

8. Boddeke, R., Slijper, E. J. and Van der Stelt, A. Histological characters of the body musculature of fishes in connection with their mode of life. *Proc. K. Ned. Akad. Wet., Ser C., Biol. Med. Sci.* **62**, 576–588 (1959).

9. Brett, J. R. and Glass, N. R. Metabolic rates and critical swimming speeds of sockeye salmon (*Oncorhynchus nerka*) in relation to size and temperature. *J. Fish. Res. Bd Can.* **30**, 379–387 (1973).

10. Brown, C. E. and Muir, B. S. Analysis of ram ventilation of fish gills with application to skipjack tuna. *J. Fish. Res. Bd Can.* **27**, 1637–1652 (1970).

11. Cahn, P. H. Sensory factors in the side-to-side spacing and positional orientation of the tuna *Euthynnus affinis* during schooling. *Fishery Bull., Fish. Wildl Serv. U.S.* **70**, 197–204 (1972).

12. Carey, F. G. and Teal, J. M. Regulation of body temperature by the blue fin tuna. *Comp. Biochem. Physiol.* **28**, 205–213 (1969).

13. Carey, F. G., Teal, J. M., Kanwisher, J. W., Lawson, K. D. and Beckett, J. S. Warm-bodied fish. *Am. Zool.* **11**, 137–145 (1971).

14. Gray, I. E. Comparative study of the gill area of marine fishes. *Biol. Bull. mar. biol. Lab., Woods Hole,* **107**, 219–225 (1954).

15. Greer Walker, M. and Pull, G. Skeletal muscle function and sustained swimming speeds in the coalfish *Gadus virens* L. *Comp. Biochem. Physiol.,* **44A**, 495–501 (1973).

16. Hemmings, C. C. Direct observation of the behaviour of fish in relation to fishing gear. *Helgolander wiss. Meeresunters,* **24**, 348–360 (1973).

17. Hertel, H. "Structure, Form and Movement". Reinhold, New York, 251 pp. (1966).

18. Hill, A. V. The dimensions of animals and their muscular dynamics. *Sci. Prog. Lond.* **38**, 209–230 (1950).

19. Hoerner, S. F. "Fluid Dynamic Drag". 2nd ed. Published by author, Brick Town, N.J., U.S.A. (1965).

20. Hudson. R. C. L. On the function of the white muscles in teleosts at intermediate swimming speeds. *J. exp. Biol.* **58**, 509–522 (1973).

21. Hughes, G. M. The dimensions of fish gills in relation to their function. *J. exp. Biol.* **45**, 177–195 (1966).

22. Johnston, I. A. and Goldspink, G. A study of glycogen and lactate in the myotomal muscle and liver of the coalfish *Gadus virens* L. during sustained swimming. *J. mar. biol. Ass. U.K.* **53**, 17–26 (1973).

23. Kishinouye, K. Contributions to the comparative study of the so-called scombroid fishes. *J. Coll. Agric. imp. Unu. Tokyo,* **8**, 294–475 (1923).

24. Klawe, W. L., Barrett, I. and Klawe, B. M. H. Haemoglobin content of the blood of six species of scombroid fishes. *Nature, Lond.* **198**, 96 (1963).

25. Love, R. M. Dark muscle. *In* "The Chemical Biology of Fishes". Academic Press, London. pp. 17–35 (1970).

26. Magnuson, J. J. Hydrostatic equilibrium of *Euthynnus affinis*, a pelagic teleost without a gas bladder. *Copeia,* 56–85 (1970).

27. Mather, F. J. Transatlantic migration of two large bluefin tuna. *J. Cons. perm. int. Explor. Mer.* **27**, 325–327 (1962).

28. Muir, B. S. Gill dimensions as a function of fish size. *J. Fish. Res. Bd Can.* **26**, 165–170 (1969).

29. Muir, B. S. and Hughes, G. M. Gill dimensions for three species of tunny. *J. exp. Biol.* **51**, 271–285 (1969).
30. Muir, B. S. and Kendall, J. E. Structural modifications in the gills of tunas and some other oceanic fishes. *Copeia*, 1968, 388–398 (1968).
31. Rayner, M. D. and Keenan, M. J. Role of red and white muscle in the swimming of the skipjack tuna. *Nature, Lond.* **214**, 392–393 (1967).
32. Tytler, P. Relationship between oxygen consumption and swimming speed in the haddock, *Melanogrammus aeglefinus. Nature, Lond.* **221**, 274–275 (1969).
33. Walters, V. On the dynamics of filter-feeding by the wavyback skipjack (*Euthynnus affinis*). *Bull. mar. Sci.* **16**, 209–221 (1966).
34. Wardle, C. S. New Observations on the lymph system of the plaice *Pleuronectes platessa* and other teleosts. *J. mar. biol. Ass. U.K.* **51**, 977–990 (1971).
35. Wardle, C. S. An assessment of the role of lactic acid as a limit to fish performance during capture. ICES C.M. 1971 Gear & Behaviour Cttee, B:18, 10 pp (1971).
36. Wardle, C. S. Limit of fish swimming speed. *Nature, Lond.* **255**, 725–727 (1975).
37. Wardle, C. S. and Kanwisher, J. W. The significance of heart rate in free swimming cod. *Gadus morhua:* Some observations with ultrasonic tags. *Mar. Behav. Physiol.* **2**, 311–324 (1974).
38. Webb, P. W. Hydrodynamics and energetics of fish propulsion. *Bull. Fish. Res. Bd Can.* **190**, 158 pp (1975).
39. Zharov, V. L. Classification of the scombroid fishes (Suborder Scombroidei, order Perciformes). *Voprosy Ikhtiologii*, **7**, 209–224 (1967). [Translation (1968) available at U.S. Bur. Comm. Fish. Biol. Lab., P.O. Box 3830, Honolulu, Hawaii 96812.]

20. Effects of Size on Performance and Energetics of Fish.

P. W. WEBB

School of Natural Resources, The University of Michigan, Ann Arbor, Michigan 48104, U.S.A.

ABSTRACT

The effects of size (related to the characteristic dimension length) on performance and propulsive energetics of fish are discussed. Discussion is based on dimensional theory in the absence of sufficiently comprehensive data for any given species. Simple dimensional analysis (Bainbridge,[4] Gray[14]) provides explanation for the facts that specific speed (in body length s^{-1}) and tail beat frequency decrease with length, but the relationships obtained do not describe the observations well. More detailed analysis of the power expended by the propulsive wave, equated to aerobic metabolic power for a hypothetical salmonid-type fish, shows that the specific propulsive wavelength (wavelength/length) should decrease with length. This results in improved efficiency of the caudal propeller in larger fish for small amplitude motions. Possible effects of large amplitude motions off-setting these efficiency increases are found to be small. Thrust coefficients decrease with Reynolds number faster than the theoretical drag coefficients and imply that swimming drag is relatively lower for larger than for smaller fish. The drag of larger fish would be expected to approach theoretical drag values.

INTRODUCTION

The literature on fish energetics and mechanics is now extensive (see Webb,[30] Hoar and Randall[16]). Much of the information available includes data on the effects of size, particularly for performance and metabolic rate (Brett,[7, 8] Beamish,[5] Brett and Glass[9]). Fewer studies have examined propulsive kinematics with respect to size (Bainbridge,[3] Hunter and Zweifel[17]). Unfortunately, many studies do not include sufficient data to interconvert the measurements to a common parameter related to size. As a result, most analyses of the effects of size on fish locomotion are still theoretical (Hill,[15] Bainbridge,[4] Gray[14] and Webb[30]). However, such analyses can provide some explanation for observed size effects and they can predict other probable effects.

In addition, previous theoretical analyses have shown that larger fish should use their available metabolic or muscle power more economically than smaller fish if paradoxical situations are to be avoided at their observed performance levels (Bainbridge,[4] Webb[30]). More economical use of these power sources is likely to be reflected in propulsion efficiency changes related to kinematic changes with size.

This paper reviews earlier approaches to the problems of energetics, performance and size and compares results with observations. A simple dimensional approach is found to be inadequate to explain the size effects, which are subsequently evaluated for a hypothetical salmonid-type fish to predict likely kinematic changes with size and their energetic and mechanical consequences.

A SIMPLE DIMENSIONAL APPROACH

Size will be characteristized by total length, L. Then the simplest approach to size effects on locomotory performance and energetics is to attempt to predict a relationship between swimming speed and length. This approach has been used analytically by Hill,[15] Bainbridge[4] and Gray[14]. It was developed in greatest detail by Bainbridge[4] in evaluating "Gray's Paradox", the apparent discrepancy between muscle power available and mechanical power required to overcome drag (Gray[13]).

Bainbridge[4] calculated the thrust (=drag) power required by swimming fish and cetaceans, assuming that the drag was equal to that of an equivalent stretched-straight rigid body. This theoretical drag is calculated from the standard Newtonian equation:

$$P_T = \tfrac{1}{2}.\rho.S.U^3.C_D \qquad (1)$$

where
$\qquad P_T =$ thrust (or drag) power

$\qquad \rho =$ density of water

$\qquad S =$ wetted surface area

$\qquad U =$ swimming speed

$\qquad C_D =$ drag coefficient.

C_D is related to Reynolds Number and boundary layer flow conditions. For a flat plate moving at zero angle of incidence to the free stream and with laminar flow in the boundary layer:

$$C_D = 1.32 . R_L^{-0.5} \tag{2}$$

and for turbulent flow in the boundary layer (when the surface is smooth):

$$C_D = 0.072 . R_L^{-0.2} \tag{3}$$

when

$$R_L = L . U . /v \tag{4}$$

and

$$R_L = \text{Reynolds Number}$$

$$v = \text{kinematic viscosity of water.}$$

Equations (1) to (4) determine the minimum theoretical drag for the flat plates with the same length and wetted surface area as a hypothetical fish. Theoretical drag for an equivalent rigid body is commonly determined by increasing C_D by a factor of 1·2 (Gero,[12] Bainbridge,[4] Webb[28, 30]).

The power required to overcome P_T is generated by the muscle system. Bainbridge[4] assumed that muscle mass was half the body mass, and that half the muscle worked at any instant. The power developed was taken as proportional to muscle mass. Bainbridge[4] took the relationship between muscle mass, M_m and length to be:

$$M_m = 0.005 . L^{2.9} \tag{5}$$

so that muscle power output, P_m is related to length by:

$$P_m \propto L^{2.9}. \tag{6}$$

The efficiency of converting muscle power to thrust power (the caudal propeller efficiency, η_p) was assumed to be 0·75, and independent of size. Therefore, P_T will be related to L according to:

$$P_T = P_m . \eta_p \propto L^{2.9}. \tag{7}$$

Bainbridge[4] considered that S was proportional to L^2 so that for laminar boundary layer flow we can write, from Eqns (1)–(4) and (7):

$$U \propto L^{0.56}, \qquad U/L \propto L^{-0.44} \tag{8}$$

and for turbulent boundary layer flow:

$$U \propto L^{0.39}, \qquad U/L \propto L^{-0.61}. \tag{9}$$

Gray[14] also points out that tail beat frequency, f, is related to swimming speed and length by

$$U \propto f.L. \tag{10}$$

Then, from Eqns (8) and (9), for laminar boundary layer flow:

$$f \propto L^{-0.44} \tag{11}$$

and for turbulent boundary layer flow:

$$f \propto L^{-0.61}. \tag{12}$$

These estimates differ from the predictions of Wu above ($f \propto L^{-0.88}$ in laminar flow), mainly because they follow Gray[14] in assuming that fL/U is independent of size (as found by Hunter and Zweifel[17] for sufficiently large swimming speeds). Wu's estimate, on the other hand, was obtained by equating the drag ($\frac{1}{2}\rho SU^2 C_D$) to the thrust, assumed to be proportional to $\rho d_T^2 f^2 A^2$, where d_T is the depth of the caudal fin and A is the amplitude of its beat (cf Eqn 16 et sq. below); both d_T and A are taken to be proportional to L. Only more complete data can really tell which relation is the better at which swimming speeds.

Hunter and Zweifel[17] have made a detailed analysis of relations between U, f, and L for several species which show that Eqn (10) holds in practice only for values of f above some minimum intercept at zero speed. The intercept value of f is approximately proportional to $L^{0.67}$. The intercept value of f is small compared to the value for sprint speeds so Eqn (10) is then reasonably valid as shown by Bainbridge.[3]

These generalizations show that specific swimming speeds (U/L) and f decrease with increasing L, as generally observed, but it is necessary to determine how accurately Eqns (8) and (9), and (11) and (12) describe observed performance and tail beat frequency.

Most of the muscle mass consists of white twitch muscle, so the ideal speed for the purpose of comparison is the maximum sprint speed, which can be maintained for only about one second. Bainbridge[3] presents data from which maximum (sprint) tail beat frequencies f_{max}, and speeds, U_{max}, can be obtained as functions of length (Table I). Although the sample size is small, it is clear

TABLE I. Relations with length, L, for maximum tail beat frequency, f_{max} and swimming speed, U_{max}, in sprint activity. Calculations based on data in Bainbridge[3] L measured in cm.

	Leuciscus leuciscus	Carassius auratus	Salmo irideus
Length range (cm)	5.3–24.1	7.4–22.6	4.1–29.4
$f_{max}(s^{-1})$	$42\,L^{-0.32}$	$92\,L^{-0.88}$	$31\,L^{-0.18}$
$U_{max}(cm\,s^{-1})$	$31\,L^{0.65}$	$89\,L^{-0.05}$	$22\,L^{0.81}$
$n.$	6	4	4

that there is great variability in the value of the various exponents in Eqns (8), (9), (11), and (12), few of which approach the order expected.

The same relationships between swimming speed and length would be expected for any specified performance level, as long as the swimming speed is great enough for Eqn (10) to hold. In the case of cruising speeds this requires the additional assumption that red (slow) muscle occupies a constant proportion of the muscle mass, for Eqn (6) to hold.

TABLE II. Morphometric parameters and predicted relations between swimming speed and length for two species of salmonids. Data from Webb (Ref. 30 and unpublished).

	Oncorhynchus nerka	*Salmo gairdneri*
Length range (cm)	11·4 to 35·0	12·8 to 39·4
n	19	33
Wetted surface area, S (cm^2)	$0·23\ L^{2·14}$	$0·28\ L^{2·11}$
R	0·99	0·99
Muscle mass M_m (g)	$0·0015\ L^{3·38}$	$0·0020\ L^{3·29}$
U (laminar boundary layer flow)	$\propto L^{0·70}$	$\propto L^{0·67}$
U (turbulent boundary layer flow)	$\propto L^{0·51}$	$\propto L^{0·49}$

Jones et al.[18] provide data on the maximum speed which could be maintained for 10 minutes (10-min. U_{crit}) for ten species of MacKenzie River fish. They found that:

$$U \propto K . L^b \tag{13}$$

where the overall mean value of $b = 0·41$, with standard deviation 0·27 and a range of $-0·17$ to 0·75, again illustrating the interspecific variability in performance/length relations.

Wardle[27] has shown that f_{max}, and hence U_{max}, will be limited by muscle contraction times t. f_{max} is given by $1/(2t)$ and for four species of marine fish at 14°C, Wardle's data gives f_{max} proportional to $L^{-0·40}$.

It is clear that no valid interspecific generalizations of the form of Eqns (8), (9), (11), and (12) are possible on the basis of the assumptions made. However, this does not preclude the possibility that such descriptions will apply for a given species when data are more complete. Such data are presented in Table II for two species of salmonids. These can be compared with observations for the maximum 60-min swimming speed (60-min U_{crit}) from Brett and Glass[9] as shown in Table III. Data for higher performance levels are not available. Relations between 60-min U_{crit} and length are shown for three temperatures. The exponents may be considered as different determinations of a single value, but subject to experimental error, so that:

$$60\text{-min } U_{crit} \propto L^{0.63} \qquad (14)$$

and is independent of temperature. A similar relation would be expected for trout.

TABLE III. Empirical relations between length and the 60-min U_{crit} for sockeye salmon at three temperatures. Data from Brett and Glass.[9] U measured in cm s^{-1}, L in cm.

Temperature °C	Swimming Speed/Length Equation
5·0	60 min $U_{crit} = 9.03\ L^{0.62}$
15·0	60 min $U_{crit} = 13.46\ L^{0.63}$
20·0	60 min $U_{crit} = 12.68\ L^{0.63}$

In comparing the equations in Table II with Eqn (14), it is still assumed that red muscle is a constant percentage of the muscle mass. It is not known if this is correct. Gatz[11] has re-interpreted Bainbridge's[3] observation of swimming performance on the basis of measurements of red muscle mass for the different species. Although Gatz reports interspecific differences in red muscle mass, he does not report any size effects.

The speed/length relation in Eqn (14) falls between those predicted from the theoretical drag with either fully laminar or fully turbulent boundary layer flow (Table II). This might be taken to imply that the boundary layer flow is transitional for swimming fish, as observed by Aleyev and Ovcharov.[1, 2] In contrast, more detailed observations and calculations from hydromechanical models suggest that P_T is proportional to $U^{2.8}$, implying fully turbulent boundary layer flow (Webb[28, 30, 31]). Furthermore, P_T is commonly three to five times greater than the theoretical drag for small fish (Kliashtorin,[19] Webb[30]), implying additional drag components which may have quite different relations with swimming speed and size.

A simple dimensional approach to energetics and mechanics may give a qualitative description of performance, but cannot give much explanation of the factors affecting locomotion. Consequently, a more comprehensive approach is required to elucidate size relations for locomotory performance.

APPLICATION OF HYDROMECHANICAL MODELS

A more useful approach to the problems of size in locomotion mechanics and performance is to try and relate empirical observations on metabolism and kinematics. In order to analyze these relations, discussion will be restricted to salmonids, particularly sockeye salmon and rainbow trout. Complete rigorous data are not available for either species. Metabolic data are fairly complete for sockeye salmon (Brett and Glass[9]), while kinematic data are mainly available for rainbow trout (Bainbridge,[3, 4] Hunter and Zweifel,[17] Webb[28]). In the following analysis it is assumed that locomotory performance is reasonably similar among salmonids. Therefore, while the general conclusions are probably valid, specific results are subject to error. Unless otherwise stated, empirical equations and deductions from them are given for a temperature of 15°C, the optimum temperature for prolonged swimming at the 60-min U_{crit} speed of sockeye salmon (Brett[6]).

Metabolism

Brett and Glass[9] summarize ten years of oxygen consumption data for sockeye salmon. Relationships between standard and active metabolic rate are shown in Table IV. Metabolic scope was calculated as a measure of the energy available for propulsion after satisfying maintenance requirements (Table IV). Metabolic scope (active − standard oxygen consumption) was

TABLE IV. Relationships of size with metabolic rate and metabolic scope for sockeye salmon. Data from Brett and Glass.[9] M is given in g and L in cm.

Temperature °C	Standard Metabolism mg O_2 h^{-1}	Active Metabolism mg O_2 h^{-1}	Metabolic Scope J s^{-1}
5·3	$0·057\,M^{0·914}$	$0·467\,M^{0·999}$	$1·06 \times 10^{-5}\,L^{3·12}$
15	$0·156\,M^{0·846}$	$0·951\,M^{0·963}$	$2·37 \times 10^{-5}\,L^{3·02}$
20	$0·184\,M^{0·885}$	$0·824\,M^{0·995}$	$1·56 \times 10^{-5}\,L^{3·14}$

converted to J s^{-1} assuming 1 mg O_2 = 14·7 J. Scope is shown as a function of length, calculated from the empirical relation between length and mass, M:

$$M = 0·0066 \cdot L^{3·09} \tag{15}$$

(the regression coefficient for this equation was 0·99, from a sample of 19 specimens).

Again, differences in the values of the exponents relating metabolic scope and length are attributed to experimental error, the mean exponent then being 3·09. The length relations for metabolic scope and those for muscle mass (Table II) show a different exponent, implying that the metabolic rate of muscle will be relatively lower per unit mass for larger than for smaller fish. This is commonly observed for total metabolism (Fry[10]).

Thrust Power

Thrust power, P_T, can be calculated from hydromechanical models. Lighthill's[21] simplified bulk momentum model is most suitable for our present purposes and will initially be used without correction for large amplitude motions. Use of the simplified model rather than a more detailed one results in only a small error (Lighthill,[22] Wu[33]), while other factors such as body thickness, upstream vorticity/trailing edge interactions and virtual mass/size corrections are negligible for salmonid-type fish (Weihs,[32] Wu and Newman,[34] Newman,[25] Newman and Wu,[26] Webb[31]).

Lighthill's bulk momentum model states for the small amplitude case:

$$P_T = P - P_K \tag{16}$$

where P = total rate of working of the propulsive wave

     P_K = kinetic energy lost to the wake.

$$P = m.w.W.U. \tag{17}$$

and $P_K = \tfrac{1}{2}.m.w^2.U. \tag{18}$

and $w = W(V - U)/V \tag{19}$

where m = virtual mass per unit length

     W = trailing edge lateral velocity

     V = backward velocity of propulsive wave

     U = swimming speed.

Furthermore

$$m = (\rho \pi d_T^2)/4, \tag{20}$$

the root-mean-square value of W is given by:

$$W = \pi f A/\sqrt{2}, \tag{21}$$

and $V = f\lambda, \tag{22}$

where $\qquad d_T$ = trailing edge depth

$\qquad\qquad f$ = tail beat frequency

$\qquad\qquad A$ = maximum side-to-side displacement of the trailing edge

$\qquad\qquad \lambda$ = wavelength of propulsive wave.

λ is a mean value for the propulsive wave, which in practice varies along the length of the body of a swimming fish (Lindsey[24]). Hunter and Zweifel[17] have assembled data for both f and A for several species. For rainbow trout f is related to length by:

$$f = (U + 1{\cdot}28\,L^{0{\cdot}67})/0{\cdot}64\,L, \qquad (23)$$

when L is measured in cm, f in Hz, and U in cm s^{-1}. At the 60-min U_{crit}, the data of Table III suggest $U \propto L^{0{\cdot}63}$ (Eqn 14), so that Eqn (23) gives, to fair approximation, $f \propto L^{-0{\cdot}35}$. A is commonly of the order of $0{\cdot}2\,L$ for different species. However, observations on sockeye salmon and rainbow trout suggest that specific amplitude, A/L, decreases with size (these data are summarized by Webb[30] from Bainbridge[4] and Webb[28]). Using all these data, we obtain:

$$W = 14\,L^{0{\cdot}51}. \qquad (24)$$

Observations show that:

$$d_T = 0{\cdot}23\,L \qquad (25)$$

so that

$$m = 0{\cdot}04\,L^2 \qquad (26)$$

(measured in g cm^{-1}).

  Determination of λ is more problematic. Webb[30] attempted to calculate λ assuming $U/f\lambda \to 0{\cdot}7$ at the 60-min U_{crit} since this value approaches the optimum for good efficiency (Lighthill,[20, 21, 22] Wu[33]). This assumption implies that mechanical caudal propeller efficiency is constant, as might be expected when fish are stimulated to maximum activity. For the salmonid data, this leads to λ being directly proportional to length, a solution that may be a fortuitous result and not necessarily applicable to all species.

  It is thus initially assumed that $\lambda \propto L$, in which case $(V - U)/V$ is independent of size;

then $\qquad\qquad\qquad\qquad w \propto L^{0{\cdot}51}. \qquad (27)$

Equations (16), (17) and (18), with m substituted from (26), w from (27), W from (24), and U again from (14), give

$$P_T \propto L^{3{\cdot}65}. \qquad (28)$$

Comparison with Observation

The relation (28) can be compared with a value of P_T inferred from measurements of the metabolic energy expended. Thus:

$$P_T = P_{aerob} \cdot \eta_m \cdot \eta_p \tag{29}$$

where P_{aerob} = metabolic power available (metabolic scope)

η_m = muscle efficiency

η_p = caudal propeller efficiency.

From Table IV, the relation between P_{aerob} and L has a mean exponent of 3·09. η_p is assumed above to be independent of size and η_m will similarly be assumed size independent. Then:

$$P_T \propto P_{aerob} \propto L^{3.09}. \tag{30}$$

Evidently the theoretical prediction (28) does not describe well the observed variation with length. Therefore, it is necessary to reconsider the assumptions made. These can be considered under two headings: metabolic energy expenditure and efficiency.

Metabolic Energy

P_{aerob} does not account for all the metabolic energy made available to the propulsive system. It is known that anaerobic energy is also released at prolonged speeds including the 60-min U_{crit} (see Webb[30]). However, in order to explain discrepancies in size relations, for example between Eqns (28) and (30), anaerobic energy would have to increase with $L^{0.56}$, with total metabolic power increasing with $L^{3.65}$. This seems unlikely. At best anaerobic metabolism might increase with muscle mass, proportional to $L^{3.39}$ for sockeye salmon (Table II). However, this would not rectify the discrepancies. Webb[29] discussed likely anaerobic contributions to total metabolism and concluded they were small at the activity level considered here. In general, it is unlikely that the apparent discrepancies between prediction and observation can be attributed to anaerobic energy contributions.

Efficiency

In Eqn (29) both muscle efficiency, η_m, and caudal propeller efficiency, η_p, are assumed constant for all sizes. This is reasonable for η_m, but η_p could well vary with length if certain kinematic factors vary with size. So far, the data used and assumptions made take the specific propulsive wavelength, λ/L, to be a constant. However, changes in λ/L with size and the concomitant size dependence of λ would affect η_p. If η_p is P_T/P, then from Eqns (16), (17), and (18):

$$\eta_p = 1 - 0\cdot5\,[(V - U)/V]$$
$$= 1 - 0\cdot5\,[(f\lambda - U)/f\lambda].$$

(31)

To explain the discrepancy between Eqns (28) and (30), η_p must increase with size and, from Eqn (31), this would be obtained if λ/L decreased with increasing length. However, if λ/L decreases with size, P, P_K, and P_T will not vary in the same way with L, and λ will vary in a complex fashion with L.

In order to evaluate the possibility that λ/L is size dependent, P and P_{aerob} can be related according to:

$$P = P_{aerob} \cdot \eta_m.$$

(32)

Substituting for P from equations (17), and (19):

$$m \cdot W^2 \cdot U \cdot [(f\lambda - U)/f\lambda] = \eta_m \cdot P_{aerob}.$$

(33)

In order to make numerical estimates of $f\lambda$, the relationships given in Table IV for P_{aerob} and in Table III for U can be used, together with Eqns (20)

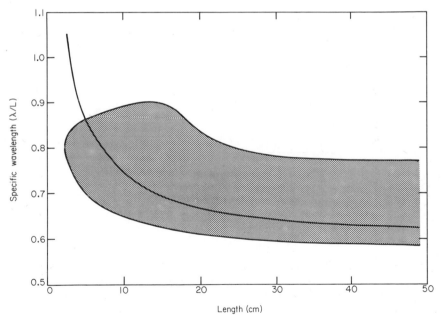

FIG. 1. Results of calculations of specific wavelength, λ/L, shown as a function of length. The stippled area shows the range of values measured by C. C. Lindsey (1975, personal communication).

and (21) for m and W respectively. A value of 0·3 is reasonable for η_m and is assumed here. λ can then be determined from Eqn (33).

λ is not related to length in any simple fashion and is most conveniently shown graphically as λ/L, plotted against L (Fig. 1). λ/L is predicted to be large for small fish, decreasing rapidly at first with increasing size but tending

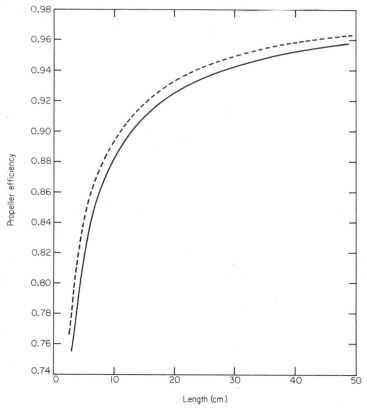

FIG. 2. Results of calculations of caudal propeller efficiency as a function of length for the small amplitude bulk momentum model (dotted line) and large amplitude model (solid line).

to approach length independence for larger fish. C. C. Lindsey (1975, personal communication) has made preliminary measurements of λ/L for several species. These are of the same order, and show qualitatively the same relations with length as predicted by the present analysis (Fig. 1).

The above solution for λ can be inserted into Eqn (31) to obtain η_p, which is shown in relation to length in Fig. 2. η_p increases rapidly with size but tends to approach a maximum value of around 0·95 for larger fish. Caudal propeller efficiencies of this magnitude are high, but not impossible (Wu[33]).

The discussion so far is for small amplitude propulsive movements and does not take into account the large amplitude movements characteristic of fish. Lighthill[23] shows that for large amplitude motions it is necessary to take into account changes in the orientation of the plane normal to the trailing edge across which momentum is shed to the wake. The relevant angle, θ, can be calculated as that subtended by the trailing edge and the axis of forward motion (Lighthill[23]). Energy loss is increased according to:

$$P_K = (\tfrac{1}{2} . m . w^2)/\cos \theta. \tag{35}$$

Thus increasing θ will increase $1/\cos \theta$, increasing energy loss and hence reducing η_p. This is important in the present context because decreasing λ/L for a given A/L increases $1/\cos \theta$ as θ increases. Consequently, gains in η_p through decreasing λ/L with L might be offset by the increases in $1/\cos \theta$. However, for a given λ/L, $1/\cos \theta$ is reduced when A/L decreases, corresponding to the situation considered here for larger salmonids. Therefore, the decrease in A/L with L might be explained as a mechanism to maintain high η_p as λ/L decreases.

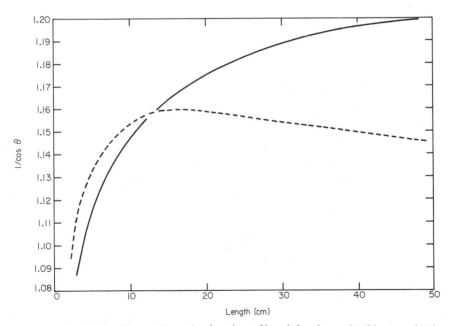

FIG. 3. Results of calculations of $1/\cos \theta$ as functions of length for observed tail beat amplitudes (dotted line) and constant tail beat amplitude of $0·2 L$ (solid line).

In order to evaluate this possibility, mean values of $1/\cos\theta$ were calculated for standing waves of constant amplitude, first with A/L constant at $0\cdot2$ and then for the amplitudes used here, taken from Webb[30]. Wavelengths were taken from Fig. 1. The results of those calculations are shown in Fig. 3. When A/L is constant, $1/\cos\theta$ increases continuously with size. However, when A/L decreases with size, $1/\cos\theta$ first increases rapidly to a maximum and then decreases for larger fish; the total reduction in $1/\cos\theta$ in this case is relatively large.

However, since P_K is small relative to P, the effect on η_p of reducing A/L with L can be shown to be negligible (Fig. 2). Consequently, it would be difficult to explain the reduction in A/L with L on the basis of improving η_p. However, the possibility is one that warrants experimental evaluation.

One further problem remains, and this is in the solution to P_T. Lighthill[21] emphasizes that although w must be small compared to W for good efficiency, it must not be so small that the thrust becomes too low to overcome drag. The problem is pertinent because decreases in λ/L with L lead to decreases in w (Eqns (19) and (22)). Consequently w might be too small to generate sufficient power to overcome drag.

This problem can be evaluated using Eqn (29) to calculate P_T with values for η_m as above and η_p from Fig. 3. Then a thrust (drag) coefficient, C_D, can be calculated:

$$C_D = (P_{\text{aerob}} \cdot \eta_m \cdot \eta_p)/(\tfrac{1}{2} \cdot \rho \cdot S \cdot U^3). \tag{36}$$

This can be compared with the theoretical drag coefficients from Eqns (2) and (3), multiplied by $1\cdot2$ to represent values for a thick body. These are shown as a function of Reynolds number in Fig. 4. The drag coefficient exceeds the theoretical value for turbulent boundary layer flow for smaller fish, but for larger fish, partly laminar boundary layer flow would be required. This is consistent with observation (Aleyev and Ovcharov[1, 2]).

CONCLUSIONS

For salmonid type fish, the flow in the boundary layer is expected to be largely turbulent. Then, with the observed metabolic power available, larger fish would be placed in paradoxical situations in which the power required would exceed the power available (Webb[39]), unless there were efficiency changes positively correlated with size and/or drag coefficients decreasing with size. Large increases in propeller efficiency are possible with moderate decreases in specific wavelength which also result in concomitant decreases in thrust (=drag) coefficients. Both requirements appear to be biologically and mechanically reasonable within the size range considered here.

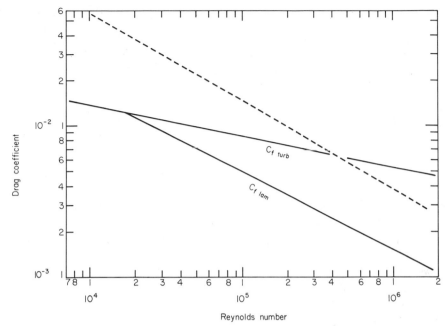

FIG. 4. The broken line represents the drag coefficients calculated from Eqn (36), plotted as a function of Reynolds Number. Theoretical drag coefficients are also shown for an equivalent rigid body with laminar ($C_{f\,\text{lam}}$) and turbulent ($C_{f\,\text{turb}}$) boundary layer flow (solid lines).

REFERENCES

1. Aleyev, Y. C. and Ovcharov, O. P. Development of vortex forming processes and nature of the boundary layer with movements of fish. *Zool. Zh.* **48**, 781–790 (1969).
2. Aleyev, Y. C. and Ovcharov, O. P. The role of vortex formation on locomotion of fish, and the influence of the boundary between two media on the flow pattern. *Zool. Zh.* **50**, 228–234 (1971).
3. Bainbridge, R. The speed of swimming of fish as related to size and to frequency of the tail beat. *J. Exp. Biol.* **35**, 109–133 (1958).
4. Bainbridge, R. Problems of fish locomotion. *Symp. Zool. Soc. Lond.* **5**, 13–32 (1961).
5. Beamish, F. W. H. The capacity of fish to exercise. *In* Hoar, W. S. and Randall, D. J. (eds). "Fish Physiology", Vol. 7. Academic Press, New York and London (in press, 1976).
6. Brett, J. R. The respiratory metabolism and swimming performance of young sockeye salmon. *J. Fish. Res. Board Can.* **21**, 1183–1226 (1964).
7. Brett, J. R. The relation of size to the rate of oxygen consumption and sustained swimming speeds of sockeye salmon (*Oncorhynchus nerka*). *J. Fish. Res. Board Can.* **22**, 1491–1501 (1965).
8. Brett, J. R. Swimming performance of sockeye salmon in relation to fatigue time and temperature. *J. Fish. Res. Board Can.* **24**, 1731–1741 (1967).

9. Brett, J. R. and Glass, N. R. Metabolic rates and critical swimming speeds of sockeye salmon (*Oncorhynchus nerka*) in relation to size and temperature. *J. Fish. Res. Board Can.* **30**, 379–387 (1973).

10. Fry, F. E. J. The effect of environmental factors on the physiology of fish. pp. 1–98. *In* Hoar, W. S. and Randall, D. J. (eds). "Fish Physiology", Vol. 6. Academic Press, New York and London (1971).

11. Gatz, A. J. Speed, stamina, and muscles in fishes. *J. Fish. Res. Board Can.* **30**, 325–328 (1973).

12. Gero, D. R. The hydrodynamic aspects of fish propulsion. *Am. Mus. Novit.* **1601**, 1–32 (1952).

13. Gray, J. Studies in animal locomotion. VI. The propulsive powers of the dolphin. *J. Exp. Biol.* **13**, 192–199 (1936).

14. Gray, J. "Animal Locomotion". World Naturalist Series, Weidenfeld and Nicolson, London (1968).

15. Hill, A. V. The dimensions of animals and their muscular dynamics. *Sci. Prog.* **38**, 209–230 (1950).

16. Hoar, W. S. and Randall, D. J. (eds.) "Fish Physiology", Vol. 7, Locomotion. Academic Press, New York, N.Y. (in press 1976).

17. Hunter, J. R. and Zweifel, J. R. Swimming speed, tail beat frequency, tail beat amplitude and size in jack mackerel, *Trachurus symmetricus*, and other fishes. *Fish. Bull.* **69**, 253–266 (1971).

18. Jones, D. R., Kiceniuk, J. W. and Bamford, O. S. Evaluation of the swimming performance of several fish species from the MacKenzie River. *J. Fish. Res. Board Can.* **31**, 1641–1647 (1974).

19. Kliashtorin, L. B. The swimming energetics and hydrodynamic characteristics of actively swimming fish. *Express Information* (*1973*), 1–19 (1973).

20. Lighthill, M. J. Note on the swimming of slender fish. *J. Fluid Mech.* **9**, 305–317 (1960).

21. Lighthill, M. J. Hydromechanics of aquatic animal propulsion. *Annual Rev. Fluid Mech.* **1**, 413–446 (1969).

22. Lighthill, M. J. Aquatic animal propulsion of high hydromechanical efficiency. *J. Fluid Mech.* **44**, 265–301 (1970).

23. Lighthill, M. J. Large-amplitude elongated-body theory of fish locomotion. *Proc. R. Soc. Lond.* **B 179**, 125–138 (1971).

24. Lindsey, C. C. Form, Dimension, and locomotion in fishes. *In* Hoar, W. S. and Randall, D. J. "Fish Physiology" Volume 7. Academic Press, New York and London (in press, 1976).

25. Newman, J. N. The force on a slender fish-like body. *J. Fluid Mech.* **58**, 689–702 (1973).

26. Newman, J. N. and Wu, T. Y. A generalized slender-body theory for fish-like forms. *J. Fluid Mech.* **57**, 673–693 (1973).

27. Wardle, C. S. Limit of fish swimming speed. *Nature* (*Lond.*) **225**, 725–727 (1975).

28. Webb, P. W. The swimming energetics of trout. I. Thrust and power output at cruising speeds. *J. Exp. Biol.* **55**, 489–520 (1971a).

29. Webb, P. W. The swimming energetics of trout. II. Oxygen consumption and swimming efficiency. *J. Exp. Biol.* **55**, 521–540 (1971b).

30. Webb, P. W. Hydrodynamics and energetics of fish propulsion. *Bull. Fish. Res. Board Can.* **190**, 1–158 (1975).

31. Webb, P. W. Hydrodynamics: non-scombroid fish. *In* Hoar, W. S. and Randall, D. J. (eds.). "Fish Physiology", Vol. 7. Academic Press, New York and London (in press, 1976).

32. Weihs, D. Large amplitude motions of finned slender bodies. Workshop on slender body theory, University of Michigan, June 12–15 (1973).
33. Wu, T. Y. Hydrodynamics of swimming fish and cetaceans. *Adv. Appl. Math.* **11**, 1–63 (1971).
34. Wu, T. Y. and Newman, J. N. Unsteady flow around a slender fish-like body. *In* "Proc. Int. Symp. Directional Stability and Control of bodies Moving in Water", London p. 33–42 (1972).

21. Effects of Size on Sustained Swimming Speeds of Aquatic Organisms.

D. WEIHS

Dept. of Aeronautical Engineering, Technion-Israel Institute of Technology, Haifa, Israel.

ABSTRACT

The influence of body size on the preferred crusing speeds of aquatic creatures is examined by a theoretical approach. The criterion for finding these preferred speeds is based on energy balances. The assumption made is that fish will tend to swim at speeds for which the energy required per unit distance crossed, or per unit energy collected by feeding, is minimal. Regulating the swimming speed to conform with these criteria can lead to large savings in the energy required for specific tasks such as migration for climatic or spawning reasons.

It is shown that the optimum speeds found by applying the principles mentioned vary rather slowly with fish size. This suggests a possible explanation for the fact that the swimming speeds of whales and large sharks are relatively smaller than those of smaller fish—in terms of animal lengths per second.

INTRODUCTION

One of the basic parameters affecting the mode of life of aquatic creatures is their swimming speed. This is especially significant for migratory species as the speed, coupled with climatic and other similar factors, will determine the range of migration, the times required for such movements, etc.

Swimming speeds have been studied extensively for many different aquatic species. These results have been reviewed in various contexts (see Blaxter[1] for an excellent example) and will not be discussed in detail here except when certain data are required specifically.

Three typical ranges of swimming speeds have been defined (Webb[7])

333

that of sustained (i.e. long term), prolonged and burst swimming. The present paper concerns itself with the first type only. Sustained swimming speeds would be observed during migrations, or routine motion when there is no specific disturbance to cause higher swimming speeds (proximity of prey, a predator, currents, etc.). One can therefore expect that the speed will be determined by efficiency considerations, which would be secondary in the presence of the influences mentioned above.

Previous analyses of this problem by the author show that optimal sustained speeds are obtained at metabolic rates of approximately twice the resting value. The optimality condition was defined in terms of minimum energy required per unit distance crossed, (Weihs[8]) or of maximum food intake per unit energy required for foraging (Weihs[9]).

This result is now applied to obtain the size-dependence of the optimum speeds.

ANALYSIS

Take the case of a neutrally buoyant fish, which requires negligible energy for keeping at a constant depth, and which is swimming at a constant speed. At the normal range of swimming speeds of fish the Reynolds number is much larger than unity. As a result, the drag on the animal is written as

$$D = \tfrac{1}{2}\rho V^{2/3} C_D U^2 = K C_D U^2 \tag{1}$$

where ρ is the water density, V the animal volume, U the forward speed and C_D the drag coefficient which varies with the Reynolds number. The second form of (1) groups together in K the quantities independent of speed. For constant speed swimming the thrust (T) has to balance the drag exactly. The rate of energy expenditure per unit time, for swimming at uniform speed, is

$$P = \frac{1}{\eta} TU = \frac{1}{\eta} K C_D U^3 = \frac{1}{\eta} K C_{D_0} U^{3-\delta} \tag{2}$$

where η is the propulsive efficiency and $C_D(\mathrm{Re})$ is written as $C_{D_0} . U^{-\delta}$, separating out the empirical velocity relationship (Klyashtorin[4]) and leaving C_{D_0} independent of U. The dependence of the efficiency on the swimming speed has been measured within recent years for some fish species, including trout, salmon and goldfish (Klyashtorin,[4] Webb[7]). The results are described by

$$\eta = \beta U^\alpha \tag{3}$$

where β and α are constants. For the species examined $0.7 < \alpha < 1.0$. The total rate of energy expenditure of the animal is, after substituting (3),

$$W = P + M = \frac{1}{\beta} KC_{D_0} U^{3-\alpha-\delta} + M \qquad (4)$$

where M is the "standard" rate, which is measured when the fish is at rest and includes all energy consuming processes unrelated to swimming.

Having obtained the expression for the rate of working required for constant speed swimming, one can now look for velocity ranges where this energy is most efficiently utilized. This approach has been used by the author to find the most efficient speed for migration (Weihs[8]) where the criterion set was maximisation of the distance crossed per unit energy used. A different criterion can be established for foraging—where one would require the ratio of energy intake to expenditure to be maximal (Weihs[9]). Both of these analyses led to the same value of the optimal speed. The former model will be briefly reviewed here.

The distance crossed per unit time (i) is numerically equal to the swimming speed. Thus, from (4), the distance crossed per unit energy used is

$$\frac{i}{W} = \frac{U}{(1/\beta)KC_{D_0}U^{3-\alpha-\delta} + M}. \qquad (5)$$

The optimal speed U_0 fulfils the condition

$$\frac{d}{dU}\left(\frac{i}{W}\right) = 0 \qquad (6)$$

which, after some manipulation, leads to

$$U_0 = \left(\frac{\beta M}{KC_{D_0}(2 - \alpha - \delta)}\right)^{1/(3-\alpha-\delta)}. \qquad (7)$$

The optimum velocity (7) is described also by the ratio

$$\left(\frac{M}{P}\right)_0 = 2 - \alpha - \delta \qquad (8)$$

Effects of Scaling on the Optimal Speed

The influence of scaling on the optimum swimming speed can now be found by analysing the size-dependence of the various quantities appearing in Eqn (7). This is achieved by a combination of dimensional reasoning and use of empirical data, when such is available.

No direct data was found for the effect of size on the coefficients β and α in Eqn (3). Webb[1] attempted calculating the overall efficiency η as a function of length at the highest sustained speed from data on sockeye salmon but obtained a curve, of which he said "...the shape of this curve is highly improbable." These quantities are therefore assumed size independent, i.e., α, $\beta \propto L^0$, where L is the fish length.

Brett and Glass[2] found that the rate of energy expenditure at rest, M, was a function of fish mass (for salmon-*Oncorhynchus nerka*), as well as temperature. The average result was $M \propto (\text{mass})^{0.88}$, for the range of temperatures 5°–20°C. For these fish the length was proportional to the $(\text{mass})^{0.32}$, i.e. only slightly different from the theoretical relationship, resulting from geometrical similarity, of length $\propto (\text{mass})^{1/3}$. From these findings

$$M = \text{const. } L^{0.88/0.32} = \text{const. } L^{2.7}. \tag{9}$$

These results are obviously not valid for aquatic species in general. However the fact that the metabolic rate increases less rapidly than the volume (i.e. the standard rate per unit mass decreases with size) is well established. For mammals, this is expressed roughly by $M = \text{const. } L^{2.3}$ (Schmidt-Nielsen[5]).

In the denominator of (7) K is proportional to the length squared and C_{D_0} changes as the $(\text{length})^{-\delta}$. The latter result is obtained when one recalls that the drag coefficient C_D in Eqn (2) is a function of the Reynolds number— which in turn is directly proportional to the length.

The remaining quantity in (7), δ, is indirectly influenced by fish size. The information on the drag of actively swimming fish is sparse, so that the drag coefficients found for stiff slender bodies or plates is used in most studies of this kind. For the relevant range of Reynolds numbers $0 < \delta < 0.3$. The upper value stands for lower Reynolds numbers, i.e. small fish moving slowly. A typical value of δ, which will be used here, can be deduced from Webb's[6] data on trout, from which $\delta = 0.21$.

Applying the above length-dependences in (7), we find that

$$U_0 \simeq \text{const. } L^{0.43} \tag{10}$$

with possible variations in the exponent between 0·35 and 0·50 due to the large scatter in the size relations of the quantities involved. Thus, while one cannot give accurate extrapolations of optimum swimming speeds as a function of animal size, some important qualitative information may be obtained.

This argument is complicated by the influence of other factors, such as excitement or the temperature, on the resting rate. Webb[7] has shown that the optimum speed of yearling sockeye salmon can change by a factor of 2 when the ambient temperature is changed from 5°C to 24°C. Similar effects can be expected for quantities such as β and δ (as the Reynolds number will change).

Most of the relevant data is for fish of the 25–40 cm size range. The calculated optimum speed for 30 cm trout is approximately 1 body length per second i.e. 30 cm s^{-1}. Using this value as a starting point Fig. 1 shows the variation of optimum cruising speed as a function of body length. The solid line represents Eqn (10) while the crosshatched area covers the range of uncertainty described above. For example, a 3 m shark is expected to forage at about 80 cm s^{-1} i.e. much less than the smaller fish, in terms of body length.

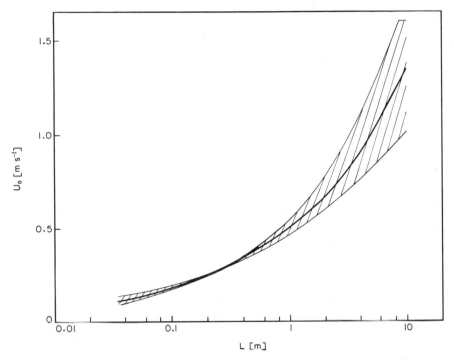

FIG. 1. Optimum swimming speed, in terms of energy expenditure, as a function of fish size. The thick line describes the size dependence and the shaded area is the range of uncertainty.

While no data were found for the actual speed during free, sustained swimming, several experimental observations tend to support the theoretical predictions of the present paper. First, from Eqn (8) it appears that for fish moving in the wild the average energy expenditure rate will be approximately twice the standard rate. This has been found (Kerr[3]) for the various species in a lake.

Also, the maximum specific swimming speeds (per unit length) are known to decrease with growing fish size, so that it is reasonable to expect the result

of the present analysis for the optimum speed, which will be the state in which
the fish in the wild will tend to be.

REFERENCES

1. Blaxter, J. H. S. Swimming speeds of fish. FAO Fish. Rep. 62, Vol. 2, 69–100 (1969).
2. Brett, J. R. and Glass, N. R. Metabolic rates and critical swimming speeds of Sockeye Salmon (*Oncorhynchus nerka*) in relation to size and temperature. *J. Fish.
   Res. Bd. Canada*, **30**, 379–387 (1973).
3. Kerr, S. R. A simulation model of lake trout growth. *J. Fish. Res. Bd. Canada*, **28**,
   815–819 (1971)
4. Klyashtorin, L. B. The swimming energetics and hydrodynamic characteristics
   of actively swimming fish. *Commercial oceanology and underwater technology* **6**,
   1–19 (1973) (English translation).
5. Schmidt-Nielsen, K. "Animal Physiology," Prentice-Hall, Englewood Cliffs, N.J.
   (1964).
6. Webb, P. W. The swimming energetics of trout (Pts. I and II) *J. Exp. Biol.* **55**,
   489–520 and 521–540 (1971).
7. Webb, P. W. "Hydrodynamics and Energetics of Fish Propulsion." Bull. Fish
   Res. Bd. Canada No. 190, Ottawa (1975).
8. Weihs, D. Optimal fish swimming speed. *Nature* **245**, 48–50 (1973).
9. Weihs, D. An optimum swimming speed of fish based on feeding efficiency,
   Israel J. Tech. **13**, 163–169 (1975).

Note added in proof. Recent sonar observations of motions of a basking shark
off the south-west coast of England (F. R. Harden-Jones, pers. comm.) showed
a cruising speed of slightly over 90 cm s^{-1} for a 9 m long fish—i.e. within the
lower band of Fig. 1.

22. Froude Propulsive Efficiency of a Small Fish, Measured by Wake Visualisation.

CHARLES W. McCUTCHEN

Laboratory of Experimental Pathology, National Institute of Arthritis, Metabolism and Digestive Diseases, National Institutes of Health, Bethesda, Maryland 20014, U.S.A.

ABSTRACT

If a fish swims in thermally stratified water, the surface of confluence formed by rejoining of the waters that have flowed round his two sides is marked by a discontinuity in temperature and refractive index, and is visible in shadowgraphic projection. By analysing motion pictures of the wake of a Zebra Danio (*Brachydanio Rerio*), I find an upper limit of 0·56 for the Froude propulsive efficiency of a 3·15 cm fish swimming in the push and coast mode. The low Froude efficiency is not caused by pushing on too small a mass of water, but by pushing in far from the best direction on large masses.

In continuous swimming the fish also pushes in wrong directions, but pushes on much more water, so the Froude efficiency is higher.

I suggest that the fish pushes in hydronamically wrong directions because his muscles are too irreversible to use as springs. The energy thus wasted was once kinetic energy of lateral motion of the aft end of the fish's body and the nearby water, energy that he has no way to recover and store.

INTRODUCTION

Considering the man-hours spent studying fish propulsion, we know precious little about what the fish does to the fluid.[1, 2, 4, 12, 15]. Conventional flow visualization methods (Merzkirch[13]), such as dye streamers, need too much cooperation from the fish to be successful in open water. If the fish is confined to a narrow passage, neither he nor his wake will behave naturally (Fig. 1).

339

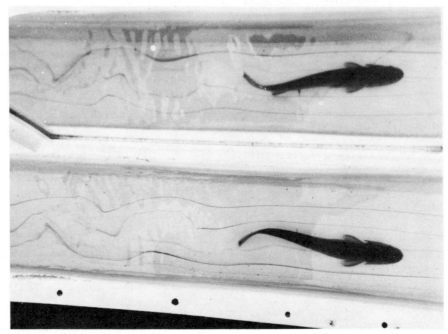

Fig. 1. A 10 cm trout swimming in water marked with ink streamers in a 7·5 cm wide channel that restricts the motion of the wake. (In Ref. 12 the dimensions of fish and channel were erroneously given as 14 and 10 cm respectively.) To view this pair in stereo stand a front-surfaced mirror on the dividing line, which should be parallel to one's nose, and look at the clear view directly with one eye and the blurry view via the mirror with the other eye.

Wake Visualisation by Stereo Shadowgraphs of Stratified Fluid.

We need to sensitize the fish's whole wet world to disturbances. This can be done. One stratifies the water thermally, which has no visible effect until the water is rearranged by passage of the fish. Then, the jump in temperature and consequently in refractive index across the surface of confluence where the waters from opposite sides of the fish rejoin, makes the wake visible in shadowgraphic projection (Fig. 2).

Two shadowgraphs in stereo display the wake in three dimensions. The most convenient apparatus is the magic cap that makes water visible (Fig. 3). Because the projection lights move with the hat, the synthesized wake has ordinary parallax and looks natural except that one sees it, in effect, from the position of the lights.

Away from surfaces of confluence, flow is shown only by the motion of preexisting inhomogeneities in the water. If needed these can be created by judicious stirring.

Stratification can strongly influence the flow pattern if the hydrodynamic event lasts longer than a small fraction of the Väisälä–Brunt free oscillation period, $2\pi\,[(\text{acceleration of gravity})\,d\,(\log_e \text{density})/d(\text{depth})]^{-1/2}$ (Hill[5]), 10 or 15 s in my apparatus, where the temperature gradient is $1\text{–}2°C\,cm^{-1}$. This return to equilibrium, aided by thermal diffusion, causes the markings in the water to fade away after 15 seconds or so, and prepares it for the next experiment.

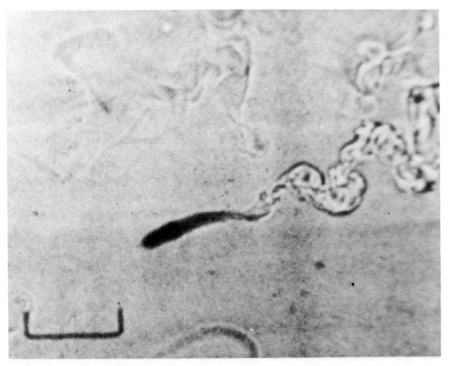

FIG. 2. A shadowgraph shows where stratified water has been disturbed. The marker in this and other shadowgraphs is 2·54 cm long.

Movies, in Mono, of Fish Wakes

I have observed many species of tropical fish. The magic cap reveals much, but fish small enough to swim naturally in a laundry sink can move too abruptly for the eye to follow. One misses the early history of the wake. So I took shadowgraph movies, in mono, at 44 ± 2 frames per second of a Zebra Danio (*Brachydanio Rerio*), losing the third dimension but gaining speed and a permanent record. The apparatus is shown in Fig. 4.

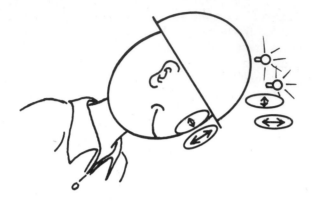

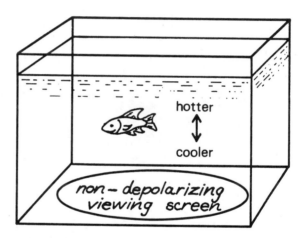

Fig. 3. A useful stereo shadowgraph apparatus. Microscope illuminator bulbs are adequate point sources. Polaroid filters make the shadowgraph from the left bulb invisible to the right eye, and vice versa.

Light from the point source is rendered parallel by a 30 cm diameter, 90 cm focal length, plano-convex, plastic field lens, and passed vertically through the aquarium, whence an identical lens focuses it on the camera lens.

Focusing the camera much closer than the plane of the fish, or much further away, makes the shadowgraphs very sensitive to refractive inhomogeneities, at the expense of a little distortion. The movies were made with the camera focussed half way from itself to the nearer field lens. A schlieren system could eliminate the distortion, but it requires field lenses with less spherical aberration than mine.

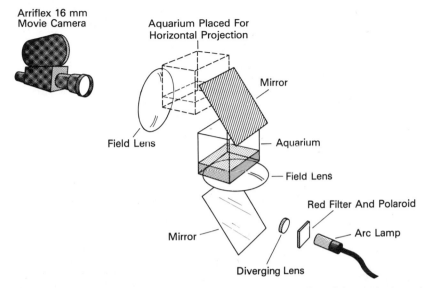

FIG. 4. Shadowgraphic movie apparatus. The red filter reduces the effect of chromatic aberration of the field lenses. The Polaroid is aligned in the plane of incidence of the mirrors, which are rear silvered, so as to reduce the intensity of the unwanted front-surface reflections.

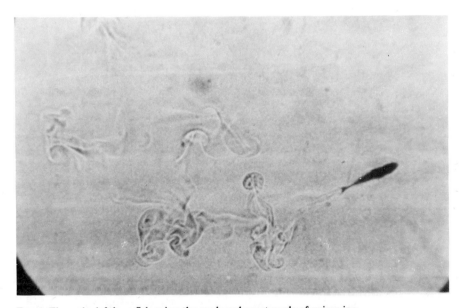

FIG. 5. The wake left by a fish using the push and coast mode of swimming.

Because waves on the water surface would distort the shadowgraphs, a 3 mm thick sheet of clear plastic was floated on the surface with styrofoam blocks.

FIG. 6. Dorsal, anal, and tail fins are spread during the tail's first swing. Stratification causes some vertical distortion of the shadowgraph. When the tail span was measured the fluid was at uniform temperature.

Water depth varied between 6 and 9 cm. The fish swam near mid-level most of the time and was only 3·15 cm long, so the presence of the top and bottom should have had little effect on either fish or wake. The temperature of the water at mid-level was 26·7°C so its viscosity was 0·0086 poise.

[At this stage the movie was shown, at 6/11 of true speed.]

THE "PUSH-AND-COAST" MODE OF SWIMMING

The movies show that most often the fish travels in the "push-and-coast" mode, a vigorous left-right (or right-left) tail flick followed by a straight coast, followed by another tail flick and so on (Fig. 5). All course changes occur at the tail flicks. Shadowgraphs taken horizontally through the tank show the dorsal, anal and tail fins to be spread during the tail's first motion (Fig. 6). During the return stroke the tail fin is fully spread but the dorsal and anal fins are partly collapsed (Fig. 7). In the first part of the coast, the fins and tail are collapsed (Fig. 8). They spread as the fish slows down (Fig. 9).

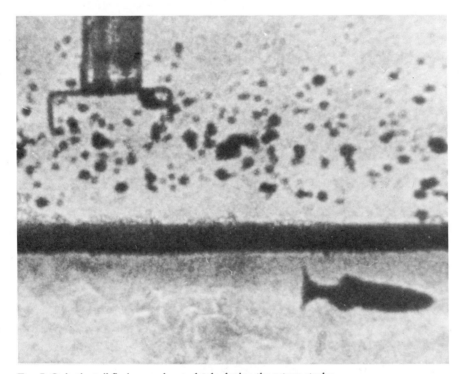

FIG. 7. Only the tail fin is spread completely during the return stroke.

The left-right sequence of tail strokes involves a turn to the left, or vice versa, from as little as 22° (Fig. 10), up to 169° (Fig. 11). The fish can avoid the change of course by flicking his tail to the left, to the right, and then slowly leftward until his body is straightened out. This leaves him going in the same direction as before, but displaced to the left of his original path (Fig. 12). The highest speed recorded, 70 cm s^{-1}, occurred at the end of such an acceleration (Fig. 13). More strokes may be used, sometimes enough for the fish's speed to reach equilibrium, and successive cycles to repeat each other. In this steady swimming the dorsal and anal fins are at least half-collapsed. The tail fin is spread, though perhaps not quite so fully as in the tail flicks of the push and coast mode. The tail swings sideways a distance of 0·24 fish lengths peak to peak, and the fish advances 0·6 fish lengths through the water with every cycle.

Occasionally the fish brakes with his pectoral fins, sometimes with the aid of puffs of water out of his mouth (Fig. 14). When not moving he hovers with his pectoral fins fanning the water downward, and his tail fin moving it downward with a downward travelling wave (Fig. 15). Zebra Danios keep themselves a little denser than water. The downward motion of the recoiling water is seen clearly with the magic cap.

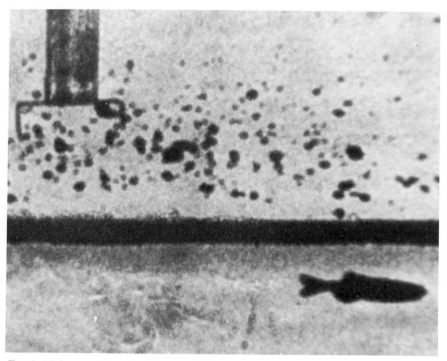

FIG. 8. All fins are collapsed at the start of the coast.

The pectoral fins are sometimes held out as the fish moves forward (Fig. 16). I think they are used like airplane wings to generate lift. The magic cap cannot settle whether the wakes go downward, as they ought if the explanation is true, because the events are too fast for the eye to follow.

The Fish does not Leave a Ladder Vortex Propulsive Wake in Steady Swimming

A prominent feature revealed by the movie is the amount of sideways motion in the recoiling water left by the fish. The observed motion of the wake of a steadily swimming fish is a lateral expansion of its initial sinuous shape (Fig. 17). The sideways gusts that form the wake are launched from positions centered near the maximum excursions of the tail rather than where it crosses the centerline and has its greatest lateral velocity. The wake is not the zig-zag ladder propulsive vortex trail, with its meandering aft-flowing central jet that was proposed by Kárman and Burgers for the self-propelled flapping wing.[7]

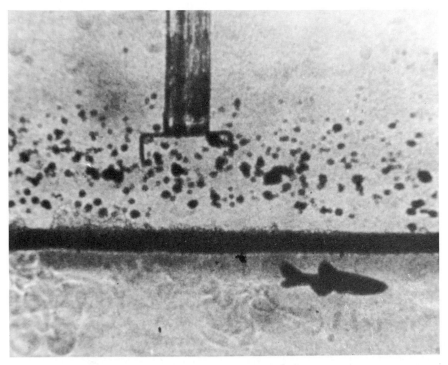

FIG. 9. Dorsal, anal and tail fins spread as the coasting fish slows down. In fact, the photo shows the end of the previous coast.

Fish Schooling

Weihs[17] has suggested that fish in schools station themselves in a diamond pattern array so that each avoids the jet from the fish in front. The absence of the jet seems to rule out this explanation. Instead, the diamond patter (which I have never seen for myself) may serve to put each fish into the path of the lateral gusts from the fishes in front, so it can use the gust energy to help in its own propulsion.

The "Push" Puts as Much Energy into the Water as into the Fish

In the push and coast mode, propulsion and drag wakes are mostly in different places (Fig. 18). The gusts of water forming the propulsive wake move in almost opposite directions. On average, the first gust travels at 92·9° to the direction of motion of the fish before the acceleration. Because

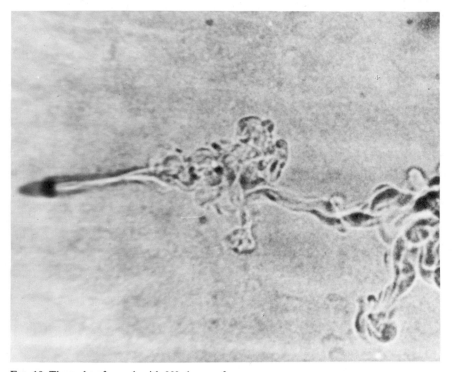

FIG. 10. The wake of a push with 22° change of course.

the fish turns during the acceleration, the second gust has a substantial backward component relative to the fish's eventual course, the first gust a substantial forward component. Relative to the fish's *change* in velocity the second gust is even more favourably directed, the first gust even more adversely. If the turn is 90° and the fish's initial speed is low, the first and second gusts are aimed about dead ahead and dead aft respectively relative to the fish's final course (Fig. 19).

The fish propels himself by creating gusts; as the gust is formed, momentum is transferred to the fluid and to the container. The gust is a historical record of this transfer. It travels in the direction of the transferred momentum, and its "impulse", the fluid density times half the first moment of the vorticity in the gust, is a vector quantity with dimensions of momentum that exactly equal the momentum transferred (Lamb[8]). This is true even if the container is closed and infinitely massive, so the total momentum of the fluid itself is zero.[10] Dividing the impulse of a gust by the velocity with which the gust propagates gives its effective mass, about but not exactly equal to the mass of the fluid that travels along with the gust. For example, the effective mass of Hill's spherical vortex is 3/2 that of the travelling sphere of fluid (Lamb[9]).

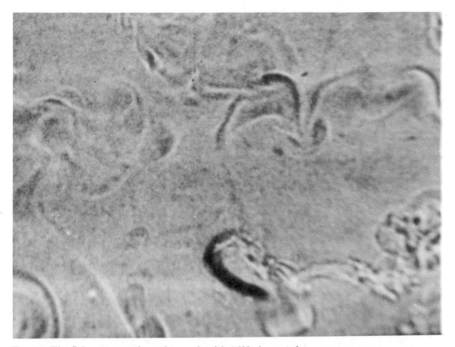

FIG. 11. The fish part way through a push with 169° change of course.

For a given impulse, the kinetic energy of a gust is large if the vorticity is concentrated, less if it is spread out. The energy of Hill's spherical vortex is 5/7 the product of its impulse and its velocity of propagation. The peak velocity, at the center, is 5/2 the propagation velocity.

If the gust shape is stretched out in the propagation direction, and the velocity within it made more uniform, the propagation velocity approaches the peak velocity, and the ratio of energy to the product of propagation

TABLE I. Mechanical parameters of impulsive accelerations. The angles are defined in Fig. 18.

θ change of course (degrees)	α angle between first gust and initial velocity of fish (degrees)	β angle between first gust and final velocity of fish (degrees)	γ angle between second gust and final velocity of fish (degrees)	final speed of fish (cm s⁻¹)	(initial speed)/(final speed)	speed of first gust (cm s⁻¹)	speed of second gust (cm s⁻¹)	(impulse of first gust)/(final impulse of fish)	(impulse of second gust)/(final impulse of fish)	(mass of first gust)/(mass of fish)	(mass of second gust)/(mass of fish)	(energy of first gust)/(change in fish energy)	(energy of second gust)/(change in fish energy)	Froude efficiency
22	86.5	64.5	132	14.4	0.200	3.7	4.52	1.95	2.48	7.59	7.90	0.501	0.811	0.433
23.5	98	74.5	135.5	14.8	0.278	4.52	5.34	0.857	1.34	2.81	3.71	0.283	0.525	0.553
24	100	76	134.5	10.3	0.160	3.90	4.11	1.11	1.60	2.93	4.01	0.432	0.655	0.479
32	99.5	67.5	150	6.16	0.167	3.70	3.29	0.580	1.25	0.965	2.34	0.359	0.687	0.489
32	100	68	122	4.9	0.417	3.57	2.67	2.48	2.98	3.40	5.47	2.19	1.97	0.194
35	80	45	145	22.6	0.582	3.70	5.34	0.155	0.772	0.947	3.27	0.038	0.276	0.761

35	97	62	146	8·22	0·250	3·70	3·49	0·694	1·36	1·44	3·20	0·357	0·617	0·507
36	96	60	128·5	20·7	0·420	5·59	6·62	2·46	3·03	9·11	9·47	0·807	1·18	0·335
38	88	50	138	4·93	0·250	3·70	2·47	3·04	3·71	4·05	7·40	2·43	1·98	0·185
38	99	61	152·5	10·3	0·200	4·11	4·52	0·507	1·23	1·27	2·80	0·211	0·563	0·564
42·5	102	59·5	146·5	3·29	0·250	3·49	2·47	0·706	1·41	0·666	1·88	0·798	1·13	0·342
43	98·5	55·5	145	3·29	0·438	3·49	1·85	0·416	1·12	0·392	1·99	0·546	0·780	0·430
43	92	49	144·5	24·8	0·417	4·76	7·04	0·739	1·45	3·85	5·11	0·172	0·498	0·599
44	102·5	58·5	147	3·29	0·189	3·49	2·26	0·841	1·56	0·793	2·27	0·925	1·11	0·329
46	97	51	143	6·16	0·200	3·90	3·49	1·67	2·39	2·64	4·22	1·10	1·41	0·285
47	86	39	158	8·22	0·300	4·11	3·29	0·323	1·13	0·646	2·82	0·177	0·498	0·597
123	80	−43	190	18·6	0·056	4·55	6·21	0·413	1·35	1·69	4·04	0·101	0·453	0·644
163	79	−84	206	28·8	0·429	11·0	7·40	0·862	1·67	2·26	6·87	0·403	0·498	0·526
34	85	51	127	12·3	0·333	3·70	4·73			2·64	4·38	0·269	0·729	0·501
35	91	56	125	13·6	0·455	5·34	3·90			2·64	4·38	0·513	0·454	0·508
52	98	46	163	6·16	0·367	3·90	3·49			2·64	4·38	1·22	1·62	0·260
56	94	38	160	7·60	0·432	3·90	3·90			2·64	4·38	0·855	1·42	0·306
67	97·5	30·5	158	18·5	0·333	5·34	5·34			2·64	4·38	0·247	0·410	0·603
92	94	2	182	6·16	0·000	3·29	3·08			2·64	4·38	0·753	1·10	0·353
98·5	82	−16·5	183·5	18·6	0·222	4·55	6·62			2·64	4·38	0·166	0·584	0·572
100	92·5	−7·5	184·5	12·3	0·333	4·11	4·93			2·64	4·38	0·332	0·791	0·471
101	93	−8	183·5	10·3	0·240	3·70	4·52			2·64	4·38	0·361	0·895	0·443

velocity and impulse approaches its lower limit, 1/2, the value for a solid body moving in a vaccum.

Because no external force acts on the system of fish and water, impulse is conserved. In a tail-flick acceleration, the vector sum of the impulses of the gusts must equal minus the change in impulse of the fish, which we know because we know his mass, 0·27 g, and can measure the change in his velocity. A thin, streamlined body like a fish moving lengthwise has little hydrodynamic added mass. The impulse and energy of the fish and of the surrounding water are little higher than the momentum and energy of the fish alone.

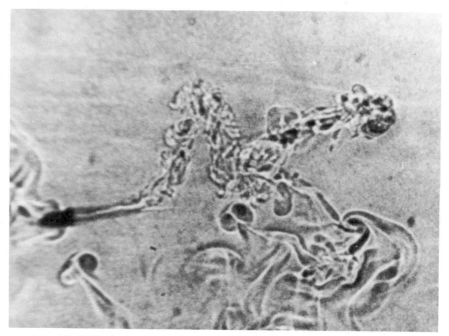

FIG. 12. A left right left sequence that leaves the fish pointing in the same direction as before but displaced to the left of his original path.

We know less about the gusts. We can measure their velocities, but not their equivalent masses. We therefore know the direction but not the magnitude of their impulses. However, if there are only two gusts, and they do not travel in exactly opposite directions, we can solve for these magnitudes. Each gust impulse forms one side of a triangle whose third side is the change in the fish's impulse. The data is in the first part of Table I.

Multiplying the impulse of each gust by half its velocity gives a lower limit on its energy. The sum of the energies of the two gusts, divided by the change in energy of the fish, averages 1·53 for 18 different accelerations. The fish puts more energy into the water than into himself.

This is so wasteful that one looks for errors in the calculation. The data on two accelerations were impossible: The observed change in fish momentum could not be the consequence of gusts travelling in the observed directions. The likely reason is water motion caused by previous passages of the fish,

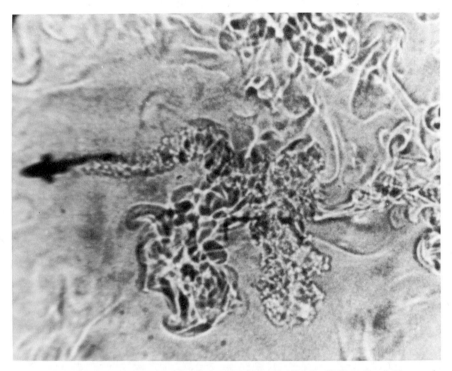

FIG. 13. A much faster three flick sequence, right left right, leaving the fish moving at 70 cm s^{-1}. The wake gusts of the three flicks have finer grained turbulence than the older gusts in the picture.

but this, or any other cause, could no doubt also make errors that would not force observations to be rejected. Here could be a source of bias.

There cannot be much wrong with the gust velocity measurements. Trouble, if any, must enter via the impulse values, which are sensitive to errors in measuring gust direction. Do the measured gust impulses correspond to reasonable gust masses? Dividing the gust impulses by the gust velocities we find considerable scatter, but the average is 2·64 and 4·38 fish masses respectively for the first and second gust.

The ratio between the masses is reasonable. The tail travels further in making the second gust, and the dorsal and anal fins may add contributions. The straight-line distance from start to finish of the tail's double flick measures from 1·4 to 2·6cm. Taking the total arc length of tail travel as 1·5 times the mid-value, 2 cm, the volume of fluid influenced by the tail, according to conventional fluid dynamics, is a cylinder 3 cm long whose diameter is 0·92 cm, the vertical span of the tail. This is 2 cm³ or 7·4 fish volumes. Considering that the force tapers off at the beginning and end of each gust, the agreement with the 7·02 fish masses inferred from the measured impulses is excellent.

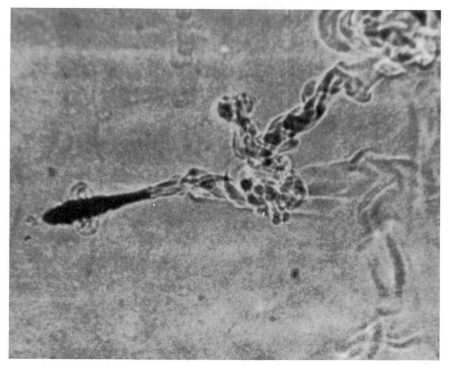

FIG. 14. Wake caused by braking with pectoral fins and a puff of water out of the mouth, and also earlier tail flicks.

The measured gust masses are about what we would have assumed, had there been no way to measure the gust impulses directly. So in the second part of Table I we use these values to calculate gust impulses and energies for the impossible acceleration, and for several accelerations where the gusts were too nearly oppositely aimed for a direct solution to mean anything. On average, for 9 of these accelerations, the energy added to the water is at least 1·41 times that added to the fish.

Froude propulsive efficiency can be defined as useful work divided by total work. For an instantaneous impulsive acceleration, it is the energy added to the fish divided by the energy added to the fish and to the water. It averages 0·458 for the fully solved accelerations, 0·446 for those where gust masses were assumed, not very good at all, especially as these values are upper limits.

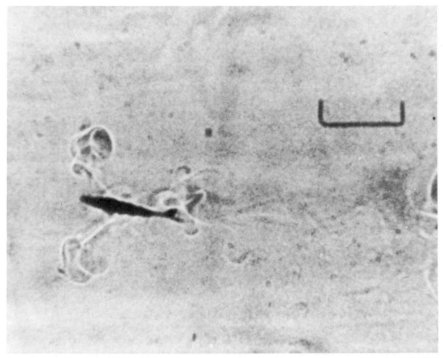

FIG. 15. Hovering, fanning the water downward with pectoral fins and massaging it downward with downward travelling waves in the tail fin.

But the acceleration takes between 0·45 and 0·84 fish lengths, which approaches half the length of the subsequent coasts. To the kinetic energy added to the fish and subsequently dissipated during the coast should be added a quantity of energy up to half as large which is dissipated by drag during the acceleration. Thus the Froude efficiency might be as high as 0·556, which is still not very high. Considering the peculiar directions that the fish pushes on the water, a low value should not surprise us. If he pushed directly aft on 7 fish masses of water his Froude efficiency would be 7/8.

[In the discussion, Lighthill observed that virtually every "push" of the "push-and-coast" swimming shown in the movie constituted a turn. Turning manoeuvres were analysed by Weihs,[18] and a Froude efficiency of over 40% does not seem too bad for such manoeuvres. Since turns are not necessary for intermittent swimming, it would not be surprising if their Froude efficiency were lower than that of other possible intermittent swimming modes. Repeated turns might be performed for reasons other than minimum energy expenditure.]

TABLE II. Speed in intervals of steady swimming, and the angle between the outline of the gusts on one side of the wake and the outline of the gusts on the other.

Speed (cm s^{-1})	Full angle of wake (degrees)
13·2	14·5
18·0	19
18·0	15
19·3	16·5
19·6	21
19·8	18
21·3	19
25·1	17
30·6	14
31·9	18

Steady Swimming has High Froude Efficiency

Is this push-and-coast cycle the most efficient mode of swimming? Lighthill[11] says that intermittent swimming is likely to be more efficient than steady swimming, since in the coast phase there are not lateral motions to thin the boundary layer and raise the drag. But during the push phases of push-and-coast the lateral motions are much more violent than in steady swimming, and the extra drag may balance, or overbalance, the saving during the coasts.

For push-and-coast to be more efficient the drag, averaged over distance, would have to be much less than that in steady swimming because the latter has much better Froude efficiency. Zebra Danios do not often swim steadily, but Table II gives the data for 10 sequences. The full angle of the envelope of the expanding sinuous wake pattern lay between 14 and 21° with an average of 17·2°. The speeds ranged from 13 to 32 cm s^{-1}, averaging 21·7 cm s^{-1}. There was no obvious correlation between speed and wake angle. The velocity of the wake gusts averages 0·151 times the speed of the fish. From the movie, the tail span in continuous swimming is perhaps 0·84 cm instead of 0·92 cm as in the pushes of push-and-coast, so a gust will have an effective mass of 6·15 rather than 7·37 times the fish mass per fish length of tail travel measured in push-and-coast. This is 8·72 fish masses per fish length of fish travel. It then takes 5·03 fish lengths of wake to contain kinetic energy equal to the kinetic energy of the fish. This is the amount of energy that the fish would put into the water in a single push that brought him to the same speed from a standing start.

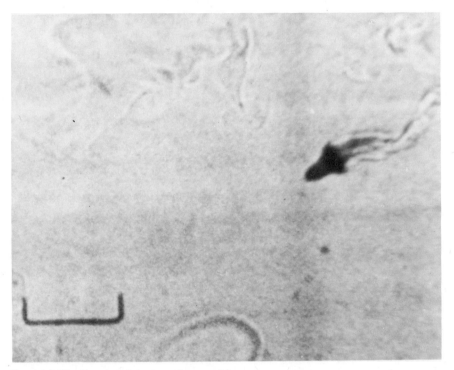

FIG. 16. Lifting(?) wake left by spread pectoral fins.

The distance vs time curves for 5 separate coasts are fitted by: drag $= k$ (speed)$^{3/2}$ dynes, where $k = 0·139$, $0·149$, $0·154$, $0·165$, and $0·180$, with an average of 0·157. This is the drag that would act on a thin sheet the same length as the fish, 3·15 cm, and 0·72 cm wide (Hoerner[6]). The fish's body is 0·62 cm deep at its deepest, and 0·31 cm deep at the shallowest part of the caudal peduncle, giving him less wetted area than the sheet when his fins are collapsed.

The work done against drag in 1 cm of travel at $21 \cdot 7 \, \text{cm s}^{-1}$ is $15 \cdot 9$ ergs, whereas the energy in 1 cm of wake is $4 \cdot 01$ ergs, making the Froude efficiency $0 \cdot 798$, or somewhat less because all wake energy values are lower limits.

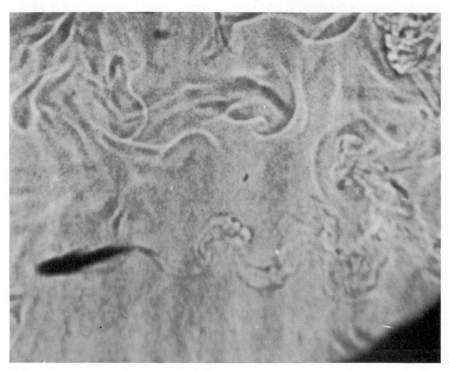

FIG. 17. The wake gusts from a steadily swimming fish spring from nearer the extremes of tail travel than from the points of maximum lateral velocity, and travel laterally at nearly constant speed.

Lateral Recoil of Wake Water is Larger than Propulsion Alone Would Account For

To cause the lateral motion of the wake the tail must apply a lateral force whose absolute value averages $53 \cdot 1$ dynes, aimed first one way and then the other. The propulsive force required is $15 \cdot 9$ dynes, so the fish pushes much harder sideways than longitudinally, the resultant push being directed at $73 \cdot 4°$ to his line of travel. Pushing laterally hurts the Froude efficiency less in steady swimming than it does in push and coast because the fish pushes, in effect, on much more water.

The shape of the tail's path would allow this angle to be much smaller.

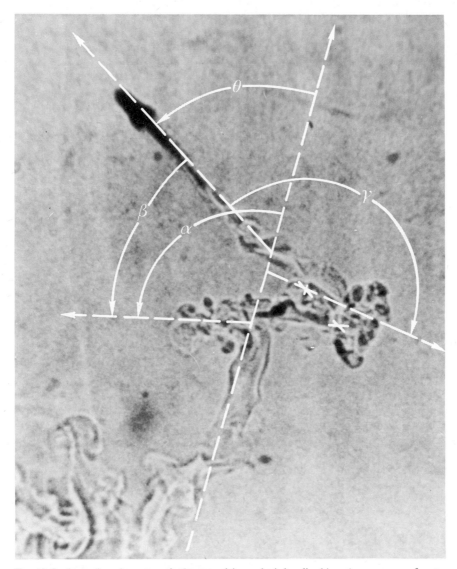

FIG. 18. In the push and coast mode the propulsive wake is localized in pairs or groups of gusts, and most of the drag wake lies between the gusts. Final direction of motion makes angle θ with initial direction. First gust makes angle α with initial direction of motion, β with final. Second gust makes angle γ with final direction of motion.

The fish oscillates his tail with an amplitude of 0·12 fish lengths once per 0·6 fish length of travel. It crosses the midline at about 52°. If the propulsive force is sinusoidal with time, starting and finishing at the extremes of motion, the average direction of each thrust would be 45° to the fish's line of travel if we neglected drag forces. Including induced drag, but not skin drag, which is accounted for in the skin drag of the fish, the lateral component of the fish's thrust can hardly be more than 1·2 times the longitudinal component.

FIG. 19. Right angle turn from slow speed. First gust goes dead forward, second dead aft relative to final direction of fish.

Were this the only lateral force the Froude efficiency would be 94·8 %. Yet the deduced lateral force is 3·35 times the measured drag. Unless the skin drag of the swimming fish is 2·79 times that of a fish coasting at the same speed, which seems most unlikely [but which is less than the figure of about 4 quoted by Lighthill[11]], there is too much lateral recoil to be accounted for by propulsion.

In push-and-coast, also, the path of the fish's tail would permit him to aim his propulsive gusts much more nearly aft than he does.

Muscular Irreversibility May Prevent the Fish from Using Hydrodynamically Ideal Strokes

So why does the fish push so hard sideways, or even forward, as happens in push-and-coast? Screw propellers push much more directly aft. And why do not the gusts originate at the peaks of the tail's lateral velocity, as they would if propulsion, or the boundary layer shed with lateral momentum by the laterally moving tail, were solely responsible?

The super high pitch propeller of the racing boat "Miss Budweiser", with 58 cm pitch and 33 cm diameter,[14] pushes only 37° from dead aft at 3/4 the way out along each blade. Obviously some engineering constraint foreign to propeller design acts on the fish, whose equivalent angle is 73·4°.

Propeller blades do not fly off at a tangent to their circle because the hub applies the large centripetal forces needed to keep them in their circular paths. The forces do no work because each is at right angles to the motion of its particular blade.

The fish tail oscillates instead of orbiting. The force that reverses its momentum at the end of each swing acts parallel to the tail's velocity. During deceleration the tail does work on the force, losing kinetic energy as it goes. In the subsequent acceleration the force does work on the tail, resupplying it with kinetic energy.

The force could come from a mighty spring, making the fish a one-tined tuning fork. But a tuning-fork fish would always have to wiggle at about the same frequency,[12] whereas real fish wiggle at different frequencies to swim at different speeds. In any case, a newly-dead fish is limp. There is no spring.

Muscles can provide the same forces as a spring, but their efficiency for absorbing and returning mechanical work over a complete cycle of extension and contraction is zero or negative (Tucker[16]).

I suggest that rather than lose all the tail's energy twice per cycle by making his muscles imitate a spring, the fish saves some of it by using hydrodynamic lift to reverse the tail's motion. If the consequent gusts are heavy enough compared to the tail and associated afterbody, most of the energy is saved.

The proposed tail-reversing force is a quarter cycle later in phase than the lateral component of the propulsive force. For our fish, the sum of the two forces lags the propulsive component by 68·8°, which accounts for the observed gust pattern. The absolute value of the tail-reversing force averages 49·6 dynes, and its peak is $49·6\ \pi/2 = 77·9$ dynes if the force is sinusoidal with time.

At a swimming speed of $21·7\ \mathrm{cm\ s}^{-1}$ the peak acceleration of the tail is

1966 cm s^{-2}, so the tail's effective mass is 0·0396 g, or about 1/7 fish masses, a reasonable value because it comprises the tail fin, some of the afterbody of the fish, and hydrodynamic added mass contributed by the nearby water. The tail's peak lateral velocity is 27 cm s^{-1}, whereas the recoil velocity of the wake associated with tail reversing is 3·1 cm s^{-1}. The tail, in effect, is turned around by an elastic collision with about 2·4 fish masses of water, in the process losing 0·23 of its kinetic energy, which is expensive, but better than losing it all, and maybe more besides. Were all the tail's energy lost at each turn around, the effect would be equivalent to a Froude efficiency of 0·52.

As the fish swims faster and faster, or grows bigger and bigger while remaining the same shape, his skin drag increases as Reynolds number $R^{3/2}$ for laminar flow or as $R^{9/5}$ for turbulent flow.[6] The tail-reversing forces go as R^2, as does the induced drag they cause. As speed or size rises, recoil from the tail-reversing forces accounts for a larger fraction of the fish's drag, which gets more nearly proportional to R^2. The fish is worse off than a torpedo, whose drag goes as $R^{13/7}$ at the worst.[6] It is easily shown that the Froude efficiency of our Zebra Danio should be highest at the speed where the propulsive and tail-reversing wake energies are equal. The efficiency is then 82·7% and the speed is 6·04 cm s^{-1}.

Above this speed the Froude efficiency gets worse. There is nothing an individual fish can do about this. But evolution over the years has reduced the relative mass of the tail and afterbody in many fast swimming fish like the mackerel, and especially in fish like the tuna that are large as well as fast. The tail's area is reduced, but its span is kept large so it still interacts with a large mass of water. The afterbody tapers so sharply, and to such a small diameter, that finlets seem to be used as vortex generators to keep the flow from separating as the afterbody swings laterally through the water.

The relatively small tail does not stall in steady swimming because high Reynolds number reduces the fish's skin drag coefficient, but it might well stall during accelerations, when it acts against the inertial reaction of the fish's body. The fast accelerators, even large ones like the barracuda, have tails of large area.

ACKNOWLEDGEMENTS

I thank Professors Lighthill and Weis-Fogh for the opportunity to present this work. Robert Finney and Paul Mortillaro provided the camera, advice and technical aid that made the movie possible. James Wilkins told me to polarize the light to avoid unwanted reflections at the front surfaces of the mirrors. Tim Pedley helped reduce the number of computational errors.

REFERENCES

1. Aleev, Io. G. and Ovcharov, O. P. On the role of vortex formation in the loco-motion of fish and the influence of the boundary between two media on the flow pattern. *Zoologicheskii Zhurnal*, 50, 228–234 (1971).
2. Allan, W. H. "Underwater Flow Visualization Techniques". N.O.T.S. Technical Publication, 2759 (1961).
3. Debler, W. and Fitzgerald, P. "Shadowgraphic Observations of the Flow Past a Sphere and a Vertical Cylinder in a Density-Stratified Liquid." University of Michigan Technical Report EM-71-3 (1971).
4. Gero, D. R. The hydrodynamic aspects of fish propulsion. *American Museum Novitiates* No. 1601, 1–31 (1952).
5. Hill, M. N. "The Sea," Vol. 1. Interscience, New York, p. 37 (1962).
6. Hoerner, S. F. "Fluid Dynamic Drag." Published by the author. 148 Busteed Drive, Midland Park, New Jersey 07432. Page 2.4ff (1965).
7. von Karman, T. and Burgers, J. M. General aerodynamic theory: Perfect fluids. Vol. 2 (367 pp) of "Aerodynamic Theory", Durand, W. F., (ed.) Springer Verlag, Leipzig, 6 vols. pp. 306–308 (1934).
8. Lamb, Sir Horace. "Hydrodynamics." Dover, New York, p. 214 ff (1945).
9. Lamb, Sir Horace. "Hydrodynamics." Dover, New York, p. 246 (1945).
10. Lanchester, F. W. "Aerodynamics." D. van Nostrand, New York p. 5 ff (1908).
11. Lighthill, M. J. Large amplitude elongated-body theory of fish locomotion. *Proc. Roy. Soc.* **B179**, 125–138 (1971).
12. McCutchen, C. W. The trout tail fin: A self-cambering hydrofoil. *J. Biomechanics*, 3, 271–281 (1970).
13. Merzkirch, W. "Flow Visulatization." Academic Press, New York and London (1974).
14. "*Miss Budweiser:* Milestones and memories 1975." Anheuser-Busch, St. Louis (1975).
15. Rosen, M. W. "Water Flow about a Swimming Fish." N.O.T.S. Technical publication 2298 (1959).
16. Tucker, V. A. The energetic cost of moving about. *Amer. Sci.* **63**, 413–419.
17. Weihs, D. Hydrodynamics of fish schooling. *Nature*, **241**, 290–291 (1973).
18. Weihs, D. A hydrodynamical analysis of fish turning manoeuvres. *Proc. Roy. Soc.* **B182**, 59–72 (1972).

PART IV

AERIAL LOCOMOTION

23. Introduction to the Scaling of Aerial Locomotion.

JAMES LIGHTHILL

Department of Applied Mathematics and Theoretical Physics, University of Cambridge, England.

1. PREFACE

My first duty in introducing the scaling of aerial locomotion is to do what Professors Alexander and Wu did when they introduced (respectively) the scaling of terrestrial and aquatic locomotion; namely, to emphasize the extreme complexity of the problems concerned with the scaling of animal flight. Their complexity derives in part from the fact that flight has many different functions: flight for escape, for food seeking, for food taking, for migration, for reproductive purposes, etc. Again, many different modes of flight are extensively used: horizontal straight flight, zigzagging, climbing, diving, bounding, soaring, take-off, landing and hovering. Analysis of scaling must take into account the different relative importance of different flight functions, and of different flight modes, for different groups of animals. These types of complication are familiar to us already from the lectures on terrestrial and on aquatic locomotion.

By contrast, a unique source of complexity in the study of animal flight derives from the interaction (a highly nonlinear interaction, intricate also in other ways) between a flying animal's activities aimed at (i) aerodynamic weight support and (ii) the overcoming of air resistance. Yesterday's papers on aquatic locomotion reminded us both of the difficulties in scaling the resistance to motion offered by a fluid (because of the subtle dependence on Reynolds number) and of further complexities in the generation of the thrust needed to overcome such resistance; however, the fact that an aquatic animal's weight is largely supported by hydrostatic buoyancy does simplify the investigation of its movements. Terrestrial locomotion, on the other hand, is complicated by the continued need for weight support, but simplified by the virtual insignificance of fluid resistance (which, for example, permits the energy required to use a particular mode of locomotion over a given distance to be independent of speed). In the study of flight there is a twofold complexity, because of the intricate interaction between aerodynamic weight support and aerodynamic thrust generation.

We may note also that a unique type of organ, the wing, *combines* these two functions in all animals capable of sustained flight. This intensifies the interaction problems (somewhat as in terrestrial animals where, usually, a single type of organ, the leg, combines the functions of weight support and of locomotion). The tremendous economy resulting from a wing which combines the airspeed-maintenance and weight-supporting functions has put a premium on the development of many refinements of wing structure and wing motion which still further complicate the analysis of animal flight.

For all these reasons, it is essential that the scaling of aerial locomotion should be treated from the combined points of view of specialists in several different disciplines. In an introductory lecture one can only scratch the surface of such a complicated subject. Nevertheless, because the forces supporting the weight of a flying animal are aerodynamic forces, and the main resistance to movement which he must overcome is aerodynamic resistance, it may be quite appropriate that an introduction to the scaling of aerial locomotion is here given by an aerodynamicist.

I have elsewhere (Lighthill[7]) given an extended general survey "Aerodynamic Aspects of Animal Flight". From that rather elementary survey, which will perhaps be found a useful general synopsis by those unfamiliar with its subject matter, I here extract only those parts directly relevant to the scaling of flapping flight (Sections 2.1 and 2.2) and of soaring (Section 2.3). This material all appears in Section 2, the first main section of this lecture, concerned with flight at airspeeds well above the stalling speed; that is, in conditions where, in the strong relative wind, aerodynamic weight support is achieved relatively more easily, but airspeed maintenance against resistance raises power-consumption problems and is achieved relatively less easily.

The remainder of the lecture consists of new material. First the scale-dependent advantages of "bounding flight" (with intermittent flapping; still at airspeeds well above the stalling speed) are analyzed in Section 2.4. Here I present concepts derived from the work of my students S. B. Furber and J. M. V. Rayner; see also Mr. Rayner's discussion contribution later in the Proceedings. Next, several considerations related to thrust generation in flapping flight are put forward. After a preliminary crude analysis (Section 2.5) limited to small amplitudes of flapping, and relevant at most to fast forward flight, a considerably more refined analysis is made (Section 3.1) and applied especially to slow forward flight (Section 3.2). Finally, the last main section, 4, is concerned with the problems of scaling the hovering of animals in still air. It gives preliminary indications of how a large number of different constraints may act to limit the range of wing-beat frequencies possible for a given size, and also to set upper and lower limits to the sizes of hovering animals.

2. FLIGHT AT AIRSPEEDS WELL ABOVE STALLING SPEED

2.1. Size-Speed Relationships

In the light of the extremely variable balance of relative importance between the different functions of flight in different groups of animals, as well as in different flight modes, it is hardly surprising that every collection of data relating important flight parameters (Fig. 1, for example, showing size-speed relationships in insects) shows a wide scatter of points. At the same time, a general statistical tendency for some characteristic airspeed (that is, flight speed relative to the air) to increase as some measure of animal size increases is common to all such collections of data.

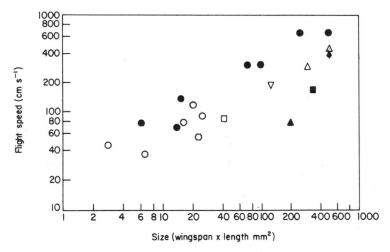

FIG. 1. Airspeed plotted against a measure of size for various insects (Lewis and Taylor[6]). ○ Homoptera, △ Lepidoptera, ▽ Coleoptera, □ Psocoptera, ● Diptera, ▲ Neuoptera, ◆ Apoidea, ■ Ichneumonoidea.

The classical interpretation of such a trend by aerodynamicists is expressed in terms of *wing loading* (mg/S, where mg is total weight and S is wing area; we shall see below why characteristic airspeeds tend to increase with wing loading). Geometrical similarity would make wing area S proportional to

$m^{2/3}$ so that wing-loading would increase with size in proportion to $m^{1/3}$. Lighthill (Ref. 7, Fig. 39) reproduced a famous log–log plot of wing-loading versus mass due to von Karman,[5] giving data for many bird species and for the Wright brothers' aircraft! The band of data points was thick but lay approximately along a straight line of slope 1/3. Furthermore, the thickness of the band was partly due to the fact that species near the bottom were birds of prey which, relative to other birds of similar size, need lower wing loadings so that they are able to seize prey and fly away successfully carrying that additional "payload".

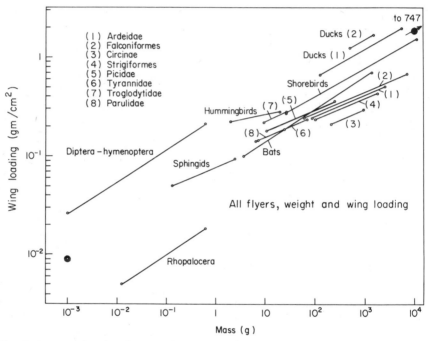

Fig. 2. A comprehensive diagram of wing loading versus total mass (Greenewalt[3]). Wing loadings given in g cm^{-2} may be multiplied by 98 to yield values in Nm^{-2}.

Lighthill (Ref. 7, Fig. 40) also reproduced more comprehensive data for birds from the massive survey of Greenewalt.[2] This data, when expressed in terms of the *ratio* of wing-loading mg/S in Nm^{-2} to $m^{1/3}$ in kg$^{1/3}$, assigns to birds from a wide range of families typical "intermediate" values between 50 and 100; it assigns a lower range of values, between 30 and 60, to another group of families (typically birds of prey) and a generally higher range of values (100 to 200) to bird families which avoid low-speed flight (particularly aquatic birds which fly only after an extended runway-type take-off).

The appearance in July, 1975 of a major new analysis of animal-flight data by Greenewalt[3] allows me here to show a diagram (Fig. 2) embodying results on a still wider range of flying animals and, this time, the "Boeing 747"! In this important paper, Greenewalt draws attention to significant departures, within particular families or orders of flying animals, from the scaling laws appropriate to conditions of geometrical similarity. Nevertheless, he emphasizes through this diagram that deviations from a line representing an average scaling like $m^{1/3}$ (the line going through the two closed circles, and the "Boeing 747" point, and corresponding to a value of 90 Nm^{-2} kg$^{-1/3}$ for the above ratio) are by modest factors only. Furthermore, reasonable interpretations of most of those deviations can be discerned.

Thus, both the birds of prey in the strict sense (Falconiformes), and other birds that fly carrying prey (the owls Strigiformes and herons Ardeidae), have *reduced* values (30 to 60) of the above ratio. The lines marked "Ducks (1)" (representing not only the Anatinae proper but also the coots Fulicinae, divers Gaviidae and grebes Podicipediformes) and "Ducks (2)" (goosanders Merginae and pochards Aythyinae) represent water birds with *augmented* values (110 to 160 Nm^{-2} kg$^{-1/3}$). On the other hand, the large group lumped together as 'shorebirds' (including plovers Charadriinae, sandpipers and woodcocks Scolopacidae, pigeons Columbidae, parrots Psittacidae, kingfishers Alcedinidae, rails Rallinae, geese Anserinae, swans Cygninae and bustards Otididae) yields intermediate values from 80 to 100, while the tits Parulidae, flycatchers Tyrannidae and woodpeckers Picidae yield values from 60 to 80 (as, incidentally, do all of the restricted number of bat species considered by Greenewalt). These two latter groups may be regarded as having "typical" values, 60 to 100, for wing-loading divided by the cube root of body mass. Indeed, other groups' deviations from from them are by at most a factor of 2 either way, and, apparently, associated with particular modes of life.

For insects, what seem to be extreme values of the same ratio are somewhat farther apart: from about 250 for an average line, consistent with geometrical scaling, representing Diptera and some Hymenoptera (mainly Apoidea), to about 20 for another average line (also consistent with geometrical scaling) representing butterflies Rhopalocera. Many of the butterflies (alternatively called Papilionoidea) are exceptional fliers in several respects: they pursue their nectar-seeking mode of life not through true hovering but through a capability for slow flight unusual in relation to their size; they escape predators through a specialised rapid take-off using the Weis-Fogh "fling" mechanism (twice), as well as through a specialised ability to give their flight a "randomized" appearance; and they achieve all this with an abnormally low ratio of muscle mass to total body mass (Greenewalt, Ref. 3, p. 7). By contrast many Diptera and Hymenoptera are exceptionally active hoverers as well as being fast fliers in relation to their size. (Note that the humming-birds Trochilidae, with analogous capabilities, also have generally above-average values, 100 to 180, of the wing loading over cube root of mass; on the other hand, their wing-loading is almost scale-independent (Fig. 2); indications of why this should be so for such *large* hovering animals are given in Section 4.1.) Studies of these two extreme insect groups must seek in the future to define not only qualitative reasons for reduced or enhanced wing-loadings but also quantitative explanations of the reduction and enhance-ment factors. In the meantime we note that ratios for the hawk-moths Sphingidae, with modes of life intermediate between the above extremes, take "normal" values between 70 and 100.

The trends in wing-loading (mg/S) discussed above are particularly im-portant because of their influence on characteristic airspeeds U for hori-zontal flight, in which the animal's weight mg is supported by some sort of average aerodynamic lift on wings of total area S. That influence is classically expressed by an equation

$$(mg/S) = \tfrac{1}{2}\rho U^2 C_L \qquad (1)$$

(where ρ is the air density) which does little more than define an average lift coefficient C_L; but the equation's importance lies in the fact that aerodynamic experiments over a very wide range of wing shapes have indicated certain broadly applicable characteristic properties for the lift coefficient.

On the one hand, the tendency for C_L to increase with the wing's angle of attack to the relative wind comes to an abrupt end when C_L reaches a certain maximum value, $C_{L\max}$. Physically, this is because at a certain angle of attack the airflow separates abruptly from the leading edge of the wing; then the air pressures on the upper surface rise to values much *less* sub-atmospheric than before, producing large loss of lift. This is the phenomenon described as "stalling" in the case of fixed-wing aircraft, and we can say that equation (1) defines a "stalling speed" for a flying animal when C_L is given the maximum value ($C_{L\max}$) that its wings are capable of developing. This is typically a little over 1 for insect wings, or a little less than 2 for bird wings under the aerodynamically advantageous conditions of primary feather separation.

Admittedly, stalling in many flying animals does not carry the same cata-strophic consequences as for fixed-wing aircraft, as we shall see in Section 3.2. This is essentially because, in slow flight, they can give their wing a horizontal component of flapping velocity, and so increase its speed relative to the air during the part of the cycle which is important for weight support. Nevertheless, the stalling speed defined by equation (1), with $C_L = C_{L\max}$, remains an important characteristic airspeed in animal flight, if only because flight modes do have to change at or around that airspeed.

Another important property of C_L (to be discussed in detail in Section 2.3) is that for a given wing a certain "optimum" value C_{Lmd} (typically of the order of 0·5) yields *maximum lift-drag ratio*; accordingly, given the weight *mg* to be supported by lift, it produces minimum drag (aerodynamic resistance). With C_L equal to this value, equation (1) defines a *cruising speed*. This is the speed at which the animal can (in still air) traverse a given distance at lowest energy "cost", as estimated from the minimum work needed to overcome aerodynamic drag.

Intermediate between cruising speed and stalling speed is the minimum-power speed, again defined by a particular value C_{Lmp} (typically around 0·8). This is the speed which enables a flying animal to remain airborne for a given length of time at lowest energy cost. It may represent an optimum for animals using flight for food seeking, or for migrating insects which rely primarily on natural winds to achieve dispersal.

It will be noted that any one of the stalling speed or the cruising speed or the minimum-power speed, as defined by Eqn (1) with a particular value ($C_{L\,max}$ or C_{Lmd} or C_{Lmp}) for C_L, tends to vary as the square-root of the wing-loading (mg/S). Accordingly, the broad trend is for such a characteristic airspeed to vary as $m^{1/6} = m^{0.167}$.

This classical conclusion, from arguments based on geometrical and dynamical similarity, is again well supported by the latest work of Greenewalt,[3] who takes into account as far as possible the departures from geometrical and dynamic similarity in particular groups of birds. After prolonged analysis he infers minimum-power speeds varying in proportion to powers of the mass *m* (his Table 23) as follows: $m^{0.16}$ for Ducks (1), $m^{0.18}$ for Ducks(2), $m^{0.15}$ for Shorebirds and $m^{0.12}$ for the other groups (listed at the top left of my Fig. 2). Furthermore all these speed–size relationships for birds, when extrapolated down to a mass of 1 g characteristic of many common insects, give speeds around 4 m s^{-1} which indeed are characteristic of insects of that size.

We may conclude, then, that size and speed in flying animals are by no means unambiguously related, but that a general tendency for characteristic airspeeds to become progressively greater as the animal mass increases is fairly well represented by a proportionality to a power of *m* close to $m^{1/6}$. The scatter about that scaling law probably depends primarily on the scatter of the data in Fig. 2 for wing-loading (proportional by Eqn (1) to the square of a characteristic airspeed). This is illustrated by the fact that the numerous data for Diptera in Fig. 1 lie at the top of the band, in agreement with their position in Fig. 2.

2.2. Introductory Discussion of Power Requirements for Flapping Flight

On the basis of the extremely elementary level of aerodynamic discussion in Section 2.1, and before any attempts at detailed analysis of how flapping generates thrust (Sections 2·5 and 3.1), it is possible and helpful to give an introductory discussion of the power requirements for flapping flight. First, I set out the classical arguments for the existence of a maximum size in flapping flight.

These arguments are based, like those of Section 2.1, not only on geometrical similarity but also on a dynamical similarity which makes all aerodynamic forces proportional to the Bernoulli pressure $\frac{1}{2}\rho U^2$ as in Eqn (1). This latter similarity implies that a lift-drag *ratio L/D* at any particular characteristic value for C_L (such as that for minimum power) is scale-independent.

A lower limit to the power requirement for flapping flight is given by the minimum rate of working DU to overcome aerodynamic drag. This can be written as

$$\frac{mg}{(L/D)} U, \tag{2}$$

with both the lift-drag ratio L/D and the airspeed U given their minimum-power values. The *minimum specific power* (power divided by mass) therefore takes the form

$$\frac{g}{(L/D)_{mp}} U_{mp}; \tag{3}$$

namely, a scale-independent factor multiplied by the minimum-power speed. According to Section 2.1, then, the minimum specific power required for flapping flight must increase with size like a power of m close to $m^{1/6}$; the general form of this conclusion would not be changed if expression (3) were divided by various *efficiencies* (chemical and mechanical) provided they were scale-independent.

From this conclusion it is straightforward to infer the existence of a greatest possible mass for sustained flapping flight. Several arguments given elsewhere in this Symposium indicate the existence of an *upper limit* on the sustained power that can be generated per unit mass of muscle. Weis-Fogh's arguments for an upper limit of the order of 200 W per kg of muscle have been broadly accepted; any reservations have been concerned with whether the true maximum may progressively become even smaller for larger flying animals. Extensive data, summarised by Greenewalt (for birds, see Figs. 7 and 8 below), give 20% as a maximum for the proportion of an animal's total mass taken up by flight muscle (hummingbirds, specialized for hovering, are excepted from this; see Section 4). We infer, then, 40 W kg^{-1} as the greatest sustainable specific power in flapping forward flight, with possibly so~ reduction below that value as size increases.

In either of these cases, the previous conclusion that the minimum specific power to make flapping flight possible goes up (something like $m^{1/6}$) means that animals above a certain size can no longer sustain the output needed for flapping flight. Pennycuick[10] gives these arguments, and then uses data from four different orders of birds to suggest that this upper limit on mass may be about 12 kg.

I want to stress that the above line of argument is important even if details of the assumptions behind it are incorrect. In fact, the argument itself must lead us to *expect* departures from the assumptions.

Consider, for example, the assumption of geometrical similarity: the argument tells us that evolution towards larger but geometrically similar flying animals would have tended to leave them less and less power to spare over and above the minimum for sustained flapping flight. We might expect, then, that increase of size would have been accompanied by certain departures from geometrical similarity (within limits imposed by other constraints) such as would reduce the minimum specific power (3) by increasing the lift–drag ratio.

Aerodynamically, the way to increase lift–drag ratio as body size increases is to *increase the wing-span s* more rapidly than the square-root of wing area $S^{1/2}$ (as shown in Section 2.3 below); that is, to increase what aerodynamicists call the *aspect ratio s^2/S*. Accordingly, we might expect increases in size to have been accompanied by departures from geometrical similarity in such a way that s increases more rapidly than $S^{1/2}$, though only within the limits imposed by other constraints: for example, there might be a skeletal-mass penalty, imposed by much larger wing bending moments, which might leave a reduced proportion of mass available for flight muscle. At a certain level of aspect-ratio the possibilities for benefit would run out and the general trend, that the specific power available becomes inadequate for flapping flight as size increases, would reassert itself.

Greenewalt[3] finds in fact that s increases like $S^{0.57}$ for his Ducks, $S^{0.56}$ for Shorebirds and $S^{0.53}$ for his other bird groups, the departures from constant aspect-ratio being statistically significant. He also postulates that further small improvements in lift-drag ratio as size increases can arise from departures from *dynamical* similarity. Indeed, a slight additional tendency for the lift–drag ratio to increase with size should exist as the Reynolds number R increases, in the range $10^4 < R < 10^6$ applicable to bird flight; however, Greenewalt probably exaggerates this because he uses aerodynamic data for turbulent skin-friction applicable to the especially smooth surfaces used in aeronautical engineering. The wings and bodies of most flying animals must be regarded as "aerodynamically rough", so that the frictional drag will be rather closely proportional to $\frac{1}{2}\rho U^2$ and only slightly sensitive to Reynolds number. (Admittedly, the two effects here considered, aspect-ratio and Reynolds number, tend to reinforce one another in somewhat mitigating the problems of larger animals by reducing the minimum specific power requirement (3). By contrast, they tend to oppose one another in their effect on those characteristic values of C_L which influence the size-speed relationships discussed in Section 2.1; see for example Eqn (6) below where the effects on the minimum-drag speed U_{md} of increasing the wing semispan b and decreasing the frictional drag coefficient C_{Df} tend to cancel.)

We can apply similar arguments to the size-dependence of the specific power available in flight musculature. Although at a given temperature the specific-power expenditure for resting metabolism is found to fall with increasing size, Weis-Fogh argues that the maximum additional specific-power expenditure available for active flight need not so fall. Admittedly, in the design of a large mass of musculature so that it can realize its full potential for sustained power output, there are far greater problems (especially those of fuel and oxygen supply) than in the design of a small mass. The need to overcome these problems may not have fully arisen in the evolution of, say, terrestrial animals, because as we have seen the specific power actually required for terrestrial locomotion decreases with size in a manner broadly similar to that required for resting metabolism. By contrast, in the evolution of flight musculature, the strongest possible incentive to overcome all the problems of maintaining maximum specific power (in conditions of sustained output) as size increases must have come from the fact that the specific power needed to fly actually *increases* with size.

It is true that the necessary developments may have been possible only within a range of sizes up to some limit fixed by other considerations; for example, by the flapping frequency at which muscle operating at its strain-rate for maximum power output reaches its maximum allowable strain. Nevertheless, the above discussion may perhaps suggest the potential usefulness to comparative physiologists of even a very elementary analysis of the scaling of animal flight.

2.3. Soaring Flight

I set out in Section 2.2 (and developed a little) the classical line of argument that the gap, between the minimum power requirements for flapping flight and the greatest sustainable output from flight musculature, narrows with increase of size. In further support of this argument, I shall show how it seems to give a clear interpretation of the observational data regarding the incidence of two specialised modes of bird flight: soaring flight, found in larger birds, and bounding flight, found in smaller birds.

Even though larger birds can remain airborne by flapping only at a power cost near to their greatest sustainable output, such a cost is of course fully appropriate to "emergency" conditions of various kinds. Since, however, larger birds in general benefit from remaining airborne for extended periods, at convenient heights for surveying predatory opportunities, it is not surprising that they have evolved some less costly methods for achieving this. In these, the bird does not flap its wings, but sustains itself in forward flight through the air by extracting energy from natural winds: in a word, by soaring.

Because the aerodynamics of outstretched wings in a fixed position relative to the body is simpler than that of flapping flight, an analysis of soaring flight is valuable not only for its own sake but also as a first introduction to certain more widely important aerodynamic concepts.

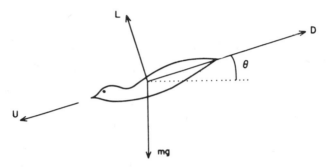

FIG. 3. Diagram of the movement of a soaring bird relative to the air, and of the forces acting on the bird (for explanation see text).

I shall give the theory of soaring in that frame of reference in which the air around the bird is at rest. This means that we fix our attention on the motion of the bird relative to the surrounding air. In steady conditions that is a glide downwards with speed U at a certain angle θ to the horizontal (Fig. 3). In such diagonally downward motion we continue to use "lift" L to mean the force on the wing perpendicular to its path through the air and "drag" D to mean the force opposing the bird's motion through the air. These forces are in equilibrium with the bird's weight mg if

$$L = mg \cos \theta, \qquad D = mg \sin \theta, \qquad (4)$$

equations which specify the angle of glide θ as the inverse tangent of the ratio of drag to lift.

For any understanding of flight it is vital to recognise that the drag D consists of two components,

$$D = \tfrac{1}{2}\rho U^2 S C_{Df} + \frac{KL^2}{\tfrac{1}{2}\rho U^2 b^2}, \qquad (5)$$

(where $b = \tfrac{1}{2}s$ is the wing semi-span) which are physically generated in quite different ways and have different scaling properties. The first term $\frac{1}{2}\rho U^2 S C_{Df}$ may be called the frictional drag (on wings and body combined): it is the rate of transfer of momentum from the surface of the bird to the mass of air very close to that surface (in the so-called "boundary layer") which is dragged forward as a result of the bird's motion through the air. The dependence on U^2 results from momentum proportional to U being exchanged at a *rate* proportional to U: such dependence is especially close (corresponding to a constant value of the frictional drag coefficient C_{Df}) when the mixing processes governing the rate of momentum exchange result primarily from airflow disturbances generated by surface roughness elements, which then define the thickness of the turbulent boundary layer.* For exceptionally smooth surfaces, however, there is a slow decrease with Reynolds number in the boundary-layer thickness, and hence in the value of C_{Df}.

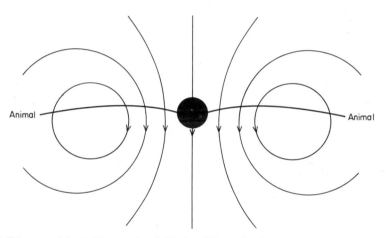

FIG. 4. Diagram of the trailing vortices behind a gliding animal.

I need to do everything possible, I believe, to avoid misunderstandings about the second term in Eqn (5), known as the induced drag. This represents a first example of what I called (Section 1) the nonlinear interaction between weight support and air resistance; for it is an additional drag, proportional to L^2, resulting from the generation of lift L.

The aerodynamic force L perpendicular to the wing's motion through the air is the rate of transfer of perpendicular momentum from the wing to the air in an extensive region (whose diameter is approximately equal to the wing-span s) around its path through the air. This region (quite different

* This is because the rate of transfer of momentum to the boundary layer must balance the rate at which boundary-layer momentum is left behind as the bird moves forward at airspeed U.

from any thin frictional boundary layer) is illustrated in Fig. 4 from a viewpoint behind the flying animal. The perpendicular motions left behind the animal have the form of a trailing vortex pair.

These long trailing vortices (which, in the case of aircraft, sometimes become visible as a result of condensation) always increase in length at a rate equal to the airspeed U and in perpendicular momentum at a rate equal to the lift force L. Their momentum per unit length is therefore L/U. The mass of air involved in the motion scales as the square of the wing semi-span b: it is proportional to ρb^2 per unit length, where ρ is the air density.

Now, induced drag represents the extra force which the wing has to exert on the air to generate the *kinetic energy* of these trailing vortex motions. In other words, induced drag is the kinetic energy per unit length in the trailing-vortex system. However, kinetic energy is always proportional to momentum squared divided by mass: here, to $(L/U)^2$ divided by a term proportional to ρb^2. It is not surprising, then, that induced drag tends to take the general form of the last term in Eqn (5). In gliding flight, the constant K tends to take values close to $0{\cdot}10$, somewhat greater than a theoretical value $(1/4\pi)$ corresponding to an optimum distribution of vorticity with *least* energy for a given momentum and semi-span. (It should be noted, furthermore, that rather greater departures from such an optimum distribution, leading to somewhat enhanced values of K, are to be expected in flapping flight.)

For analysing gliding at a fixed angle θ, we may take the lift L as fixed since it must balance the weight component $mg \cos \theta$. Then for giving wing semi-span b, the total drag (Eqn 5) increases when the airspeed U becomes either large or small, and takes a minimum at some intermediate minimum-drag speed U_{md}. For example, on the crude approximation which ignores any variations in C_{Df} or K, U_{md} takes the value

$$U_{md} = \left(\frac{4K}{\rho^2 C_{Df}} \frac{L^2}{Sb^2} \right)^{1/4}, \tag{6}$$

which makes both terms in (5) equal. (This value, incidentally, should be close to the minimum-drag speed defined in Section 2.1 for flapping-flight conditions, except that as just mentioned a somewhat enhanced value of K may be appropriate to those conditions; note also that the corresponding equation for the minimum-power speed U_{mp} makes it about $0{\cdot}76\, U_{md}$ because of an extra factor $1/3$ appearing inside the bracket in (6) when we minimise the product of the drag (5) with the airspeed U.)

Birds could not glide stably at speeds much below their minimum-drag speed (6), since a further decrease in speed would produce an increase in drag, leading to still more deceleration until stalling occurred; similarly, an increase in speed would produce a decrease in drag, leading to acceleration towards the minimum-drag speed. With variations in C_{Df} and K neglected, the condition $U \leqslant U_{md}$ for stable gliding is that the induced drag does not exceed the frictional drag. It can be written as a condition

$$b > \left(\frac{KS}{C_{Df}}\right)^{1/2} \left(\frac{L}{\frac{1}{2}\rho U^2 S}\right) \tag{7}$$

on the wing semi-span b. Subject to the limitation that the last bracket in (7) cannot exceed $C_{L\,max}$, the condition states that at lower and lower velocities the bird must increasingly spread its wings.

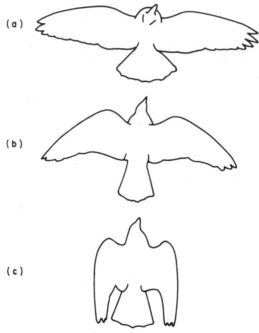

FIG. 5. Pigeon gliding in a wind-tunnel at speeds of (a) $8\cdot6$ m s^{-1}, (b) $12\cdot4$ m s^{-1} and (c) $22\cdot1$ m s^{-1} (Pennycuick[9]).

This is borne out, for example, by Pennycuick's[9] determinations of wing planform in a pigeon gliding at three different airspeeds (Fig. 5). These were obtained in a wind-tunnel with the airstream inclined upwards, so that the bird, with its motion inclined downwards, could be stationary in the laboratory frame of reference.

That experiment exactly parallels one type of soaring flight, where a bird finds an airstream of speed U inclined upwards at an angle θ (usually on the windward sides of cliffs, hills or tall vegetation) and is able to remain poised motionless within it, observing opportunities for prey. Relative to the air, it is gliding downwards as in Fig. 3. The requirement that the frictional drag $\frac{1}{2}\rho U^2 SC_{D_f}$ exceeds the induced drag and so is over half the total drag $D = mg \sin \theta$ fixes the airspeed U between

$$\left(\frac{mg \sin \theta}{\rho SC_{D_f}}\right)^{1/2} \quad \text{and} \quad \left(\frac{2mg \sin \theta}{\rho SC_{D_f}}\right)^{1/2}. \tag{8}$$

From the scaling standpoint, the interesting thing is that these critical airspeeds exhibit the same proportionality to the square-root of wing-loading as do the characteristic airspeeds studied in Section 2.1. This is important because the bird can remain airborne only if the wind speed U, in addition to satisfying (8), is in excess both of its stalling speed and also of its minimum-drag speed with the wing-span fully spread. Because all the different criteria scale on the square-root of wing-loading, this type of soaring (on an inclined airstream) is scale-independent: the range of inclinations of the airstream for which soaring is possible is independent of size.

Among birds, then, the greater tendency to soar as size increases cannot be ascribed to any change in the aerodynamic feasibility of this type of soaring. Rather, it must be interpreted, *either* in terms of greater incentives to the evolution of soaring in larger birds, that can remain airborne by flapping only at a power cost near to their greatest sustainable output, *or*, possibly, in terms of the advantages to smaller birds (Section 2.4) of the alternative strategy of "bounding flight". Again, the fact that some insects (such as the Red Admiral butterfly *Vanessa atalanta*) are also observed to soar is consistent with the aerodynamics being scale-independent.

In hot climates large birds soar alternatively by use of "thermals": regions of convective updraught with velocity, say, v. In such an updraught a bird can sustain continuous *horizontal* motion (often making horizontal *circles*) without doing work. At airspeed U the sine of its angle of glide θ relative to the air is v/U. Therefore, condition (8) implies that circling is possible at an airspeed U between

$$\left(\frac{mgv}{\rho SC_{D_f}}\right)^{1/3} \text{and} \left(\frac{2mgv}{\rho SC_{D_f}}\right)^{1/3}, \tag{9}$$

provided again that U exceeds the stalling speed and the full-span minimum-drag speed.

Note that although these last two characteristic speeds still scale as the square-root of wing-loading, the criterion (9) for given updraught velocity v scales only as the *cube-root* of wing-loading. Therefore, increasing wing-loading (shown in Section 2.1 to depend mainly on increase of size) makes it progressively more difficult for a circling velocity (9) to be consistent with a usable lift coefficient. In the morning as the ground warms up and the up-draught velocities v increase, the necessary conditions for circling become compatible first for the birds with lower wing-loading and later for the birds of higher wing loading. This is well illustrated in the meticulous observations of Hankin,[4] who regularly saw kites soaring earlier in the day than vultures.

At sea, thermals are not found but sheared winds commonly blow. The albatross *Diomedea* uses "dynamic soaring" to remain airborne without wing flapping for long periods by extracting energy from a sheared wind. Vultures, again, when a strong wind blows away their thermals, are observed to use this same method of remaining airborne. Lighthill[7] gives a full discussion and analysis of dynamic soaring.

From the scaling point of view, the interesting conclusion from his analysis is that in a given set of dynamic-soaring motions the average airspeed U varies as the *first* power of wing-loading. In fact, the shear $d\bar{u}/dz$ (rate of increase of wind velocity with altitude) augments the kinetic energy of the bird's motions relative to the air at a rate proportional to $mU^2\,d\bar{u}/dz$, while aerodynamic resistance decreases that kinetic energy at a rate proportional to $\rho U^3 S$; these rates balance at a velocity proportional to $(m/\rho S)\,d\bar{u}/dz$, which for a given shear is proportional to the wing-loading mg/S. Here, then, is a soaring mode which (for given mean wind shear) becomes aerodynamically more feasible (specifically, it becomes compatible with lower lift coefficients) as wing-loading increases: perhaps, the one type of animal flight where size confers a positive advantage!

2.4. Bounding Flight

I want now to give a brief analysis of flight modes in those smaller birds for which a far bigger margin may be available between the minimum power needed for flapping flight and the maximum sustainable output from their flight musculature. This means that smaller birds may be able to sustain flapping flight at speeds well in excess of not only their minimum-power speed but also their minimum-drag speed U_{md}.

I want to ask the question: what scope is there for energy saving by birds flying at speeds well in excess of U_{md}? Those are speeds where the induced drag

$$D_i = \frac{KL^2}{\frac{1}{2}\rho U^2 b^2} \tag{10}$$

(the energy cost of weight support for unit distance travelled) is relatively small, but where the frictional drag

$$D_f = \tfrac{1}{2}\rho U^2 S C_{Df} \tag{11}$$

is very large. Note that this frictional drag includes a contribution from the *body drag* D_b together with an extremely big contribution associated with loss of momentum to the boundary layer around the outstretched wings, often called the wing profile drag D_p.

This identification of D_p as a major contribution to energy cost suggests a strategy of *intermittent flight*, in which periods when the wings are folded are alternated with periods of flapping. This is the type of flight which ornithologists often call *bounding flight*, or "leaping flight", in which the bird traverses the relatively flat upper part of a projectile's parabolic trajectory during its periods with wings folded, and regains kinetic energy and an upward component of momentum during its periods of flapping.

Mr S. B. Furber (unpublished) suggested the following simple approximate analysis of bounding flight: if f is the fraction of time with wings outstretched, then an enhanced lift

$$L = (mg)/f \tag{12}$$

is needed during those periods if the long-term average lift force is to balance the bird's weight mg. This enhances the induced drag (10) although from an initially small value. More important is the fact that profile drag (as well as the induced drag) is acting for only a fraction f of the total time of flight.

Accordingly, the average drag (which determines the average energy cost per unit distance travelled) is

$$D_b + f\left[D_p + \frac{K(mg/f)^2}{\frac{1}{2}\rho U^2 b^2}\right] = D_b + fD_p + \frac{1}{f}D_i. \tag{13}$$

The value $f = 1$ corresponds to normal flapping flight. However, in any circumstances with $D_p > D_i$, it is possible to adopt a bounding-flight strategy, with a value of $f < 1$, at a net saving in energy. In fact, when $D_p > D_i$, the minimum of (13) is

$$D_b + 2(D_p D_i)^{1/2}, \text{ attained when } f = (D_i/D_p)^{1/2}. \tag{14}$$

For example, a small bird seeking to travel at an average airspeed 50% above its minimum-drag speed might make a large saving of energy by using the "bounding" mode of flight. This is because, if (10) and (11) are equal at the minimum-drag speed, then $D_f = 5D_i$ at $U = 1\cdot5\,U_{md}$, with perhaps $D_b = D_i$ and $D_p = 4D_i$. Then bounding flight, with wings outstretched for half of the time and folded for half of the time, would reduce the average total drag from $D_b + D_p + D_i = 6D_i$ to $D_b + \frac{1}{2}D_p + 2D_i = 5\,D_i$, a saving of 17%.

The relevance of the above analysis to scaling lies in the fact that such average speeds, well above the minimum-drag speed, are sustainable only by relatively small birds, which are therefore the ones that can save energy by adopting the bounding mode of flight at such enhanced speeds. These considerations, along with those of Section 2.3, make it more readily intelligible that, on the whole, soaring flight is observed in relatively large birds and bounding flight in relatively small birds. We may remark also that certain birds of intermediate size are observed to adopt a mode of flight intermediate between these two, involving periods of flapping alternatively with periods of *gliding*. The whole subject of intermittent flight is returned to later in this Symposium in a discussion contribution by Mr J. M. V. Rayner, with considerable references to observational data.

2.5. A Preliminary Crude Analysis of Thrust Generation

Hitherto I have tried to include almost everything that I could say about the scaling of aerial locomotion without going beyond elementary aerodynamic notions regarding lift and drag, and in particular without broaching in any way the question of how wings when flapped generate the thrust needed to counteract drag (while, of course, continuing to provide lift). Now, however, I can make no further progress without tackling this aerodynamically much more complicated question. Nevertheless, to avoid confusing my audience by going simultaneously into too many of the different complications involved, I shall postpone a more general theory to Section 3.1 and confine myself here to a preliminary crude analysis which highlights a few points that are important for fast horizontal flight. This treatment, analysing wing-beats of small amplitude in a vertical stroke plane, turns out to be least inappropriate under those flight conditions, at speeds well above the stalling speed, to which Section 2 is devoted.

Note that the assumption of small amplitude reduces to a minimum the nonlinearity of the interaction between thrust generation and weight support. Lighthill[7] gave a diagrammatic representation of *linear* interactions between thrust generation and weight support for wing-beats of small amplitude in a vertical stroke plane (Fig. 6), with the aim of making such crude considerations as clear as possible within an elementary survey. The linearity of the interaction is emphasized by writing it as an equation, in which a thrust-generating mode and a lift-generating mode are "added up" to give a mode typical of flapping flight.

A cetacean tail generates thrust by up-and-down movements like those illustrated on the left of Fig. 6. There is a twist at each extreme of the oscillation so that the tail's movement, whether it is down or up, has always a backward-facing component. This is a standard method of generating thrust without lift. A similarly standard method of generating lift without thrust, by a wing at a constant positive angle of pitch (as in a gliding bird), is shown in the middle column. The linear combination of the two modes (obtained by adding up their angles of pitch at each phase of the wing-beat) is shown on the right.

This mode for simultaneous generation of thrust and lift involves (i) a downstroke with the wing nearly parallel to the direction of mean motion (pronated) and (ii) an upstroke with the wing supinated to give a large positive angle of pitch, precisely as observed in a wide range of animals in fast forward flight.

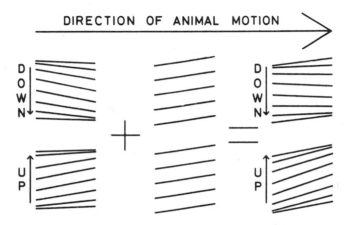

FIG. 6. Linear combination of thrust-generating and lift-generating modes (Lighthill[7]).

Note that in the pure-thrust mode the air forces always oppose the motion, both in the downstroke and in the upstroke. On the other hand, in the pure-lift mode the air forces always act upwards. Therefore, the different air forces add up (because their directions of action are the same) in the downstroke, which accordingly is very heavily loaded. They subtract, however (because their directions of action are opposite), in the upstroke, which therefore is relatively lightly loaded.

Such crude considerations already throw some light on the distribution of flight musculature. In birds, for example, a single muscle, the pectoralis major, is entirely responsible for this heavily loaded downstroke. Its mass is plotted against bird mass in Fig. 7, where the points cluster closely about a line representing a mean 15% ratio of pectoralis-major mass to total mass: a very high percentage. Fig. 8 gives the comparable results for the supracoracoideus muscle that powers the upstroke: the mean ratio is here 1·5%, only a *tenth* of that for the downstroke, though admittedly the scatter is greater; see Lighthill[7] for a fuller discussion.

A quantitative theory corresponding directly to the diagrammatic representation in Fig. 6 can be attempted after we first recognise that some kind of description of the cycle of movement like the right-hand column in Fig. 6 has to be applied to each separate wing section, from shoulder to wing tip.

The amplitudes of flapping motion are obviously greater for wing sections nearer to the tip; also, the observed twisting movements are greater for those. After the different movements of different sections have been recognised and analysed, some sort of averaging is required to estimate total aerodynamic forces.

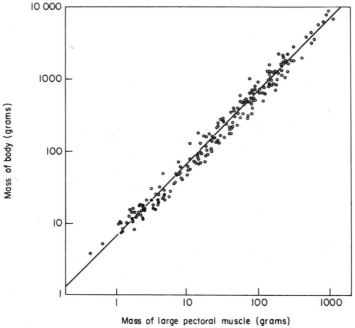

FIG. 7. Mass of pectoralis-major muscle (abscissa) plotted against total mass (ordinate) for birds (Greenewalt[2]).

To carry out this programme quantitatively, we need to define various angles relevant to the motion of any particular wing section, for an animal flying horizontally forward at airspeed U (Fig. 9). First, θ is defined as the angle of twist of the wing section, measured *positive in supination*, from a zero value in a "zero-lift configuration"; that is, one in which no wing section would experience any lift if the wings were held steady in that configuration.

Next, we define a certain angle ψ for the case when the flapping wing section has downward velocity W (this means that the velocity component W is positive during the downstroke and negative during the upstroke). The combination of the section's downward motion with velocity W and the whole animal's horizontal motion with velocity U is a movement of the section along

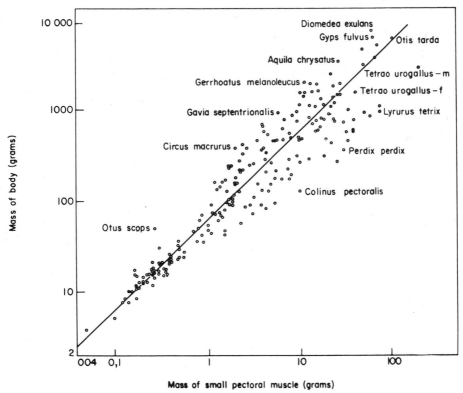

FIG. 8. Mass of the muscle powering the upstroke in birds (abscissa) plotted against total mass (Greenewalt[2]).

a certain path at an angle ψ below the horizontal. Here $\tan \psi = W/U$, but in a small-amplitude theory we can ignore trigonometry: for a small angle ψ, if measured in radians, ψ and $\tan \psi$ are practically equal and we can write

$$\psi = W/U. \tag{15}$$

Note that θ was defined so that, if the wing section were simply moving horizontally with velocity U, its angle of attack (measured from the zero-lift configuration) would be θ. However, because the wing section moves along a path through the air at an angle ψ below the horizontal, its angle of attack to the air is

$$\alpha = \theta + \psi. \tag{16}$$

This is what primarily affects the *lift L on the section*, which (as in Section 2.3) we define to mean the component of aerodynamic force perpendicular to the section's path through the air (Fig. 9).

In the present crude theory we assume a sectional lift coefficient which simply increases in direct proportion to α, although only up to a maximum usable value $\alpha = \alpha_{max}$ (beyond which stalling would occur). Thus, we take

$$C_L = a\alpha \quad \text{for} \quad \alpha < \alpha_{max}, \tag{17}$$

implying a lift force $\frac{1}{2}\rho U^2 a\alpha$ per unit area of wing around that section. A value of a around 5 is appropriate if α is measured in radians.

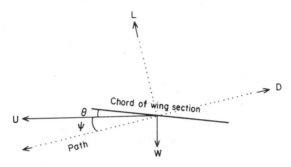

FIG. 9. Diagram of the path through the air followed by the chord of a wing section of an animal in horizontal flapping flight at airspeed U. Here, W is the section's downward velocity of flapping, and θ is its angle of twist relative to a zero-lift configuration. The directions of the lift and drag forces acting on the wing section are shown.

Such a crude formula is used primarily because even with very good observational techniques it is hardly possible to measure the angles (especially θ) accurately enough to justify an aerodynamically more precise expression. Such an expression would be, for example, $a_0(\alpha - \alpha_i)$ where a_0 is around 6 (just a little below the theoretical value 2π with the boundary layers neglected) and α_i is the reduction in effective angle of attack due to the "induced downwash" w_i. This is the downward flow of air at the wing section induced by the trailing vorticity shown in Fig. 4. It changes the downward velocity of the wing relative to the air to $W - w_i$, which reduces the section's angle of attack to the air by $\alpha_i = w_i/U$ (see (15) and (16)). Although typical reductions are by around one-sixth, as assumed in (17), the use of a constant fractional reduction represents only a crude and over-simplified approximation; so too does our use of an overall formula for induced drag (Eqn (19) below) instead of the more precise distribution of sectional induced-drag coefficients $C_{Di} = C_L \alpha_i$ indicated by the concept of induced downwash.

With all the assumptions listed above, we can now write down force-balance equations for the steadily flying animal. Its weight mg must balance the average lift force. Thus,

$$mg = \tfrac{1}{2}\rho U^2 Sa\bar{\alpha} \tag{18}$$

where the bar over α is used to signify an average over the whole area S of the wing and over the wing-beat cycle. Also the average drag $\bar{D}$, which by (5) can be written

$$\bar{D} = \tfrac{1}{2}\rho U^2 SC_{Df} + \frac{K}{\tfrac{1}{2}\rho U^2 b^2}\overline{(\tfrac{1}{2}\rho U^2 Sa\alpha)^2}, \tag{19}$$

must be overcome by the average thrust; that is, by the average *forward component of sectional lift force*,

$$\tfrac{1}{2}\rho U^2 Sa \cdot \overline{\alpha(W/U)}. \tag{20}$$

In order to analyze this thrust–drag balance further I do not hesitate to use some basic statistics of a kind well known to biologists. In the drag formula (19), the mean square angle of attack $\overline{\alpha^2}$ is greater that the square of the mean by an amount equal to the mean square deviation from the mean:

$$\overline{\alpha^2} = (\bar{\alpha})^2 + \overline{(\alpha - \bar{\alpha})^2}. \tag{21}$$

Again, in the thrust formula (20), I can replace $\overline{\alpha W}$ by $\overline{(\alpha - \bar{\alpha})W}$, because the mean value of the flapping velocity W is zero (there is no net vertical displacement of the wing as it flaps up and down). Then the balance between (19) and (20) can be rewritten, with some regrouping of terms and use of (18) and (21), as an equation

$$\tfrac{1}{2}\rho U^2 SC_{Df} + \frac{Km^2 g^2}{\tfrac{1}{2}\rho U^2 b^2} = \tfrac{1}{2}\rho U^2 Sa\left[\overline{(\alpha \quad \bar{\alpha})(W/U)} - \frac{KaS}{b^2}\overline{(\alpha - \bar{\alpha})^2}\right], \tag{22}$$

which represents the main conclusion of this analysis.

The two terms on the left-hand side represent the frictional drag and the "true" cost of weight support: this is what the induced drag would be if the lift remained constant and equal to mg instead of fluctuating due to flapping. The additional induced drag due to flapping is put with opposite sign on the right-hand side of the equation. It represents the penalty, or "wasted effort", associated with flapping: any thrust which flapping generates is effectively reduced by this consequential extra drag.

We can verify that the first term on the right of (22) represents the *work* done by the flapping wing against the opposing lift forces, per unit distance travelled; namely, the rate of working $\tfrac{1}{2}\rho U^2 Sa\overline{\alpha W}$ divided by the forward velocity U. However, part of this work of flapping is "wasted effort", used to overcome excess drag due to flapping itself, and only the remainder is used to overcome the drag that would be present even in a glide. The reduction is considerable: even with the value $K = 0.10$ (suggested in Section 2.3 for induced drag in gliding conditions) the ratio (KaS/b^2) would be at least 0.2, and the lift variations in flapping are likely to have a distribution much further from the optimum distribution for minimum induced drag than those in gliding, so that a considerably larger value of K should be used on the right-hand side of (21).

Cutting down the variations $\alpha - \bar{\alpha}$ in angle of attack to reduce the wasted effort would be self-defeating because it would seriously reduce the first term on the right of (22); that is, the thrust itself. Statistical theory tells us, however, that this term, involving the mean product

$$\overline{(\alpha - \bar{\alpha})W}, \tag{23}$$

can be largest for given mean square fluctuations of the variables if $\alpha - \bar{\alpha}$ and W are as well correlated as possible; that is, *if excess lift is well correlated with downward velocity*. This is observed in the heavily loaded downstrokes and lightly loaded upstrokes.

I will note also that in flight at airspeeds well above the stalling speed Eqn (18) allows $\bar{\alpha}$ to be much less than α_{max}. Then relatively large variations of $\alpha - \bar{\alpha}$ are possible, and an adequate thrust, proportional to the mean product (23), can be achieved without excessively large amplitudes of flapping velocity W. In crude terms, the small-amplitude theory is least inappropriate in such fast forward flight.

In slower flight, on the other hand, $\bar{\alpha}$ must be much larger. This constrains the permissible range of variation $\alpha - \bar{\alpha}$ *at the upper end*, where α is limited by α_{max}, so it is not surprising that larger flapping amplitudes tend to be observed. Furthermore, the one-sided nature of the constraint on the variations $\alpha - \bar{\alpha}$ puts a premium on the generation of mean square variations of $\alpha - \bar{\alpha}$ sufficient for good mean thrust through an unsymmetrical wing-beat (Fig. 10) with $\alpha - \bar{\alpha}$ (and the well correlated W) positive for relatively longer periods and negative for relatively shorter ones. This means an extended heavily loaded downstroke and a brief lightly loaded upstroke, as are commonly observed (see, for example, Fig. 11).

3. SLOW FORWARD FLIGHT

3.1. A More Refined Analysis of Thrust Generation

I use the phrase "slow forward flight" to mean flight at speeds comparable with the stalling speed. From the standpoint of the very restricted aerodynamic considerations of Section 2.1 it may seem paradoxical that so many animals are able to remain airborne at or below their stalling speed. In the mean-

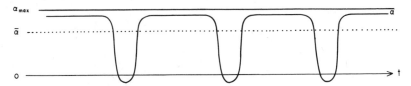

FIG. 10. Illustrating a distortion in the variations of angle of attack α about its mean $\bar{\alpha}$, which may increase thrust in slower flapping flight through an increase in the mean product (Eqn (23)); this requires an increased mean square variation of $\alpha - \bar{\alpha}$, which for large $\bar{\alpha}$ can be achieved only with an unsymmetrical wing-beat as shown.

time the crude analysis of Section 2.5 suggests that, as the stalling speed is approached, flapping amplitudes may need to become large—in which case, however, that analysis itself soon becomes invalid. Actually, slow forward flight is observed to involve both large flapping amplitudes and a stroke plane that deviates considerably from the vertical one assumed in Section 2.5 (Fig. 12).

In this section, I shall sketch a method of analysis of this kind of flapping flight; then, in Section 3.2, I shall use it to suggest why flight at or around the stalling speed by movements like those shown in Fig. 12 is neither anomalous aerodynamically nor excessive in its metabolic demands. Part of the end-product of Section 3 is to contrast those demands with the far greater metabolic requirements of hovering flight, to be analysed in Section 4.

Section 3, however, is the one which readers interested primarily in comparative zoology should omit. I claim in Sections 2 and 4 that aerodynamic analysis of the two extreme conditions, fast flight and hovering flight, has reached the stage where it can significantly help in interpreting scaling data. I cannot claim that this situation has been reached as yet in the intermediate range of airspeeds. Nevertheless, in areas of the scaling analysis of animal locomotion like this where most of the work still remains to be done, it is important to suggest to workers in the field how a beginning can be made.

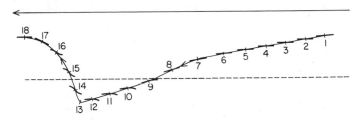

FIG. 11. Movement relative to the air of the mid-section of the forewing of the locust *Schistocerca gregaria*. The arrow shows the direction of mean motion, and the numbers refer to equally spaced instants of time. Note that the downstroke occupies two-thirds of the total wing-beat period. (Weis-Fogh and Jensen[13]).

o

FIG. 12. Eight successive positions (A to H) in the slow forward flight of a pigeon (Brown[1]).

In this spirit, I feel encouraged to include a piece of aerodynamic analysis which, with the object of interpreting flight modes like that of Fig. 12, has to be considerably more refined than those of Sections 2 and 4, although it still ignores much that is important in the observed motions. I include it partly to emphasise that such an analysis need by no means be long drawn out (in the lecture it was all on one transparency!) if the three-dimensional geometry is handled with economy.

The theory applies to an animal flying horizontally forward at airspeed U by flapping a single effective pair of wings in a stroke plane making an angle Θ with the vertical (Fig. 13). I use coordinates with x horizontally forwards, z vertically downwards and y "into the paper" (in aeronautical terms, to starboard). Note the separate diagram showing the wings' configuration in the stroke plane, both at an angle ϕ (which may of course be positive or negative) below the position in which both are stretched out horizontally.

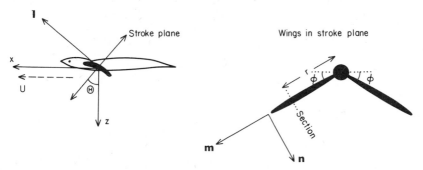

FIG. 13. Diagram of slow forward flight with a diagonal stroke plane (for explanation of symbols, see text).

In that stroke plane, I want to define two *directions* of importance for flight; in each case, by specifying a vector of unit magnitude pointing in the direction concerned. They are the *spanwise* direction

$$\mathbf{m} = (\sin \phi \sin \Theta, \cos \phi, \sin \phi \cos \Theta), \tag{24}$$

which is the direction of a median line stretching out along the span of the starboard wing, and the *tip-velocity* direction

$$\mathbf{n} = (\cos \phi \sin \Theta, -\sin \phi, \cos \phi \cos \Theta), \tag{25}$$

which would be the direction of motion of the tip of the same wing if (as assumed in this analysis) wing bending at joints were insignificant. In both (24) and (25) the unit vector is specified by giving its components in the x, y and z directions, which are easily calculated. I use also the direction

$$\mathbf{l} = (\cos \Theta, 0, -\sin \Theta) \tag{26}$$

perpendicular to the stroke plane, and note the fact (implied by (24) (25) and (26)) that one way of writing the animal's forward velocity $(U, 0, 0)$ is as

$$(U \cos \Theta)\mathbf{l} + (U \sin \phi \sin \Theta)\mathbf{m} + (U \cos \phi \sin \Theta)\mathbf{n}. \tag{27}$$

Consider now a wing section at a distance r along the span of the starboard wing, measured from a suitable central point within the animal's body (Fig. 13). Its flapping velocity is $r\dot{\phi}\mathbf{n}$ relative to the body. Since the *body* moves through the air at a velocity (27), the wing section's velocity relative to the air is

$$(U \cos \Theta)\mathbf{l} + (U \sin \phi \sin \Theta)\mathbf{m} + (U \cos \phi \sin \Theta + r\dot{\phi})\mathbf{n}. \tag{28}$$

In this expression the spanwise component (the term in **m**) can be regarded as aerodynamically insignificant (especially at low forward speeds when induced drag dominates over frictional drag). Only the **l** and **n** components, perpendicular to the span, generate significant lift or drag forces. Those two components by themselves form a vector whose magnitude is

$$V = [(U \cos \Theta)^2 + (U \cos \phi \sin \Theta + r\dot{\phi})^2]^{1/2}: \qquad (29)$$

the section's airspeed perpendicular to the span.

With the aim of expressing the aerodynamic forces on the wing section in a convenient form, I use a sectional lift coefficient C_L as in the earlier analyses. For drag, I use C_{Dw} as the wing section's total drag coefficient, including both induced drag and profile drag. Since the drag and lift forces are by definition directly opposed to, and perpendicular to, the section's velocity relative to the air (28) (with the **m** term omitted for reasons discussed above) we can write down the total aerodynamic force per unit area acting on the wing section as

$$-\tfrac{1}{2}\rho C_{Dw} V[(U \cos \Theta)\mathbf{l} + (U \cos \phi \sin \Theta + r\dot{\phi})\mathbf{n}]$$
$$-\tfrac{1}{2}\rho C_L V[(U \cos \Theta)\mathbf{n} - (U \cos \phi \sin \Theta + r\dot{\phi})\mathbf{l}]; \qquad (30)$$

where the drag and lift terms are arranged, respectively, to be in the above-mentioned directions and to have magnitudes $\tfrac{1}{2}\rho C_{Dw} V^2$ and $\tfrac{1}{2}\rho C_L V^2$.

Since the remainder of the analysis will be straightforward balancing of average horizontal forces and of average vertical forces, of a type familiar from Section 2, it is worth pausing here to note the main conceptual difference in this theory from that of Section 2·5. The present theory is put forward primarily as a foundation for future discussions of scaling. It is not proposed as a method for analysing meticulous observations of angles of wing twist, like those of Weis-Fogh and Jensen;[13] indeed, no definition of angles of wing twist, or of angles of attack derived from them, appears in the theory. The distribution of lift coefficient C_L along the wing-span is taken as the fundamental variable, rather than as one derived from a distribution of angle of attack α (as exemplified in Section 2.5).

Aerodynamically, I justify this because it is the distribution of C_L (not that of α) which determines the very important induced-drag contribution to C_{Dw}. I may remark that approximate expressions of this relationship (which I refrain from writing down) are relatively feasible to handle in the present theory. Quantitatively, they depart from results for straight wing-pairs (the case of small ϕ) much less than might be expected when ϕ is large; this is primarily because of a cancelling between (i) a certain increase in the induced flow over the starboard wing due to the increased proximity of the port wing and (ii) a reduction in the relative importance of the *downwash* component in any such induced flow, resulting from the geometry of vortex lines shed from the port wing in a large-amplitude heavily loaded downstroke. Note that treating C_L as fundamental requires the trailing vorticity to be considered only in the single context of inferring C_{Di}, and not also in determining the relationship between C_L and α. Note also that a limitation such as $C_L < C_{L\max}$ gives the most fundamental expression to the condition that local stalling must be avoided.

Biologically, I justify the present approach on the grounds that a relevant aim of such an analysis may be to work out the ultimate limitations which aerodynamics places on a particular mode of flight, as regards airspeed or power consumption or skeletal strength or different combinations of these. It may be appropriate, in estimating such limitations, to assume that whatever distribution of C_L would give "optimal" results might in principle have been achieved in the course of evolution, without inquiring too closely into the precise distributions of wing twist that would have been required. I hope that this philosophical digression may have left readers anxious to return now to the practical implications of the expression (30) for the sectional aerodynamic force per unit area!

First, let us average the x-component of (30) (that is, the horizontal forward component, which evidently is the same for the port as for the starboard wing) over the whole wing area and over the whole wing-beat cycle as in Section 2.5. This average must be equal to $\frac{1}{2}\rho U^2 C_{Db}$: the drag of the animal's body divided by the total wing area S. Using Eqns (25) and (26) for the x-components of the vectors $\mathbf{n}$ and $\mathbf{l}$, we can readily express that balance as

$$C_{Db}U^2 + \overline{C_{Dw}V[U(\cos^2\Theta + \cos^2\phi \sin^2\Theta) + r\dot{\phi}\cos\phi\sin\Theta]}$$
$$= \overline{C_L Vr\dot{\phi}}\cos\Theta, \qquad (31)$$

where all drag terms have been grouped together on the left-hand side, and the $(\frac{1}{2}\rho)$ factor as well as the wing-area factor S has been taken out of all the terms.

Equation (31) identifies the mean product $\overline{C_L Vr\dot{\phi}}$ as the true thrust (divided by $\frac{1}{2}\rho S$); this corresponds to $\overline{a\alpha UW}$ in (20), although with some enhancement because U is replaced by the true sectional airspeed V. Its effect on the balance (31), however, is reduced by the $\cos\Theta$ factor, which is required because the thrust is applied in the direction $\mathbf{l}$ (perpendicular to the stroke plane). Thus, only one component of it acts to overcome drag; the other, as we shall see, helps significantly to support the weight in low-airspeed conditions.

Indeed, when we average the z-component of (30) (that is, the vertically downward component of aerodynamic force per unit area, again the same for both wings) and add on the wing-loading mg/S (the gravitational force per unit area, which also acts vertically downwards), the total must be zero. Once more dividing through by $(\frac{1}{2}\rho)$, we deduce that

$$\frac{mg}{\frac{1}{2}\rho S} = \overline{C_L V r \dot{\phi} \sin \Theta} + \overline{C_L V U \cos \phi} + \overline{C_{Dw} V (r\dot{\phi} \cos \phi - U \sin^2 \phi \sin \Theta) \cos \Theta}.$$

(32)

Here, the first term represents the vertical component of thrust just discussed. The second term is the true lift, enhanced by the replacement of one forward-speed factor U by the true sectional airspeed V, but reduced by a factor $\cos \phi$ because the wings in the stroke plane are at an angle ϕ to the horizontal (Fig. 13). The third term is of some conceptual interest, because it emphasizes that one of the components providing weight support in large-amplitude flapping may be drag! The fact that when $\dot{\phi} > 0$ (in the heavily loaded downstroke) the induced-drag component of C_{Dw} is enormously greater than when $\dot{\phi} < 0$ (in the lightly loaded upstroke) can make the mean product $\overline{C_{Dw} V r \dot{\phi} \cos \phi}$ significantly positive, corresponding to upward values exceeding downward in the vertical components of resistance to the wing's movement through the air. Normally, however, the lift terms are a good deal more important because C_L is much bigger than C_{Dw}.

It is instructive to calculate the mean rate of working of the wings by averaging the product of a wing section's velocity relative to the body, $r\dot{\phi}$ in the **n** direction, with the component of aerodynamic force per unit area (30) in the opposite $(-\mathbf{n})$ direction, and then simplifying by use of Eqns (31) and (29). The mean rate of working, calculated in this way, is

$$\frac{1}{2}\rho S (\overline{C_{Dw} V^3} + C_{Db} U^3).$$

(33)

Alternatively, bold physical reasoning might have been used to deduce (33) in a different frame of reference (using velocities relative to the undisturbed air) by arguing that the average power input must balance the average power loss, due to wing sections moving with airspeed V against the air resistance opposing their motions, and to the body moving with airspeed U against its own drag.

3.2. Implications for Slow Forward Flight

Having put on record in Section 3.1 an indication of a broad type of analysis that may become increasingly popular in a variety of future applications, I limit myself here to an indication of its relevance to a case of slow forward flight similar to that illustrated in Fig. 12. This section simply involves a brief numerical estimation of various terms in Eqns (32) and (33) in a significant special case.

I take the stroke plane at an angle $\Theta = 45°$ to the vertical: a condition halfway between the vertical and horizontal orientations characteristic of fast forward flight and hovering flight respectively. Also, I indicate that a large amplitude of flapping is under consideration by (i) taking the tip velocity $b\dot\phi$ during the downstroke as equal to the animal's forward speed U, and (ii) simplifying the averaging process by giving $\cos\phi$ everywhere a mean value 0.9 (note that $\cos\phi$ is 1 for $\phi = 0$, 0.9 for $\phi = 26°$, 0.8 for $\phi = 37°$ and 0.7 for $\phi = 46°$).

With these rather conservative assumptions, it is interesting to note the significant enhancement of the airspeed V of the wing sections during the downstroke over the animal's airspeed U. At the tip ($r = b$) we have $r\dot\phi = U$ and then (29) gives $V = 1.78U$; at the mid-span position ($r = \frac{1}{2}b$) we have $r\dot\phi = 0.5U$ and $V = 1.34U$; at the root ($r = 0$), on the other hand, $r\dot\phi = 0$ and $V = 0.95U$.

Consider now the contribution of the *downstroke* to weight support. Being conservative again by ignoring the positive contribution of sectional drag (the C_{D_w} term) on the right-hand side of (32), we estimate only the contribution from the other two terms to an effective weight-support coefficient

$$C_L^{\text{eff}} = \frac{mg}{\frac{1}{2}\rho U^2 S}. \tag{34}$$

Those contributions are estimated as

$$\bar C_L[(\bar V/U)\cos\phi + (\overline{Vr\dot\phi}/U^2)\sin\Theta] \tag{35}$$

with the mean values taken during the downstroke only. Now, with the above distribution of $r\dot\phi$ and of V along the span, $\bar V = 1.35U$ and $\overline{Vr\dot\phi} = 0.75U^2$, so that (35) gives

$$C_L^{\text{eff}} = \bar C_L(1.22 + 0.53) = 1.75\bar C_L. \tag{36}$$

The increase of 75% in the effective weight-support coefficient above an average sectional lift coefficient arises from two sources: 22% arises from the wing's enhanced airspeed, but 53% arises from the vertically upward component of thrust.

I have emphasised this enhancement of the effective weight-support coefficient (by a factor of $1.75 = 7/4$) with the partial aim of resolving the paradox, mentioned at the beginning of Section 3.1, concerning animals' abilities to sustain flight at airspeeds around their stalling speeds. An animal whose aerodynamics was adequately represented by the above theory could, for example, fly at the stalling speed by means of a downstroke with $C_L = C_{L\max}$ occupying only 4/7 of the cycle and a completely unloaded upstroke with $C_L = 0$.

Furthermore, since the mean cube $\overline{V^3}$ takes the value $2{\cdot}7U^3$ during the downstroke, which occupies 4/7 of the wing-beat, the main term

$$\tfrac{1}{2}\rho S C_{D_i} \overline{V^3} \tag{37}$$

in the average rate of working (33) would be enhanced to 4/7 of $2{\cdot}7$ times the induced power loss in gliding, $\tfrac{1}{2}\rho S C_{D_i} U^3$. This enhancement of power cost, by a factor just over $1{\cdot}5$, is a reasonably modest penalty for airspeed maintenance by flapping in slow forward flight. We should bear this in mind when considering the much heavier enhancement of power cost demanded of animals hovering in still air (Section 4).

4. HOVERING IN STILL AIR

4.1. Induced Power Requirement: Its Enhanced Value in Hovering Flight, and its Modified Scaling

Weis-Fogh[12] gave a major review of sustained hovering flight in still air; at this Symposium, furthermore, his paper "Dimensional Analysis of Hovering Flight" immediately follows my own introductory lecture. I propose, therefore, to make an exclusively aerodynamical contribution to a preliminary discussion of this subject, mentioning certain unique aerodynamic features common to all hovering motions in still air, and then arguing their importance in the context of scaling.

Weis-Fogh has given the name "normal hovering" to a very common mode of sustained hovering flight, characteristic of hummingbirds and found extremely widely among insects, although he also showed how insects which are exceptional in this respect are of great interest, both generally and from the standpoint of scaling. Normal hovering motions are an adaptation of the motions used for forward flight (see for example Fig. 14, a sequence of pictures of the sphinx moth *Manduca sexta* viewed from above). In this adaptation the body has become nearly erect (so that the stroke plane is nearly horizontal) and the twisting movements at each extreme of the stroke are very greatly enhanced.

In principle, normal hovering flight could be analysed by regarding it as a limiting case of slow forward flight, so that the theory of Section 3.1 might be used with zero forward velocity ($U = 0$) and horizontal stroke plane ($\Theta = 90°$), giving $V = |r\dot{\phi}|$. This limiting process does in fact yield Weis-Fogh's own analysis of normal hovering. I prefer, however, to make this section completely independent of section 3 and I derive the analysis of normal hovering quite simply *ab initio* in Section 4.2 below.

Furthermore, I want to draw attention to one extremely important sense in which hovering flight is *not* a limiting case of slow forward flight. The im-

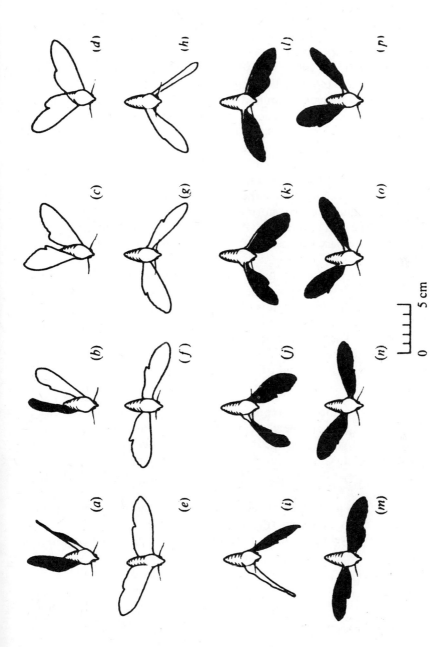

FIG. 14. Normal hovering of the sphinx moth *Manduca sexta* as traced from a slow-motion film, with the undersides of the wings shaded black (Weis-Fogh[12]).

portant induced-drag contribution C_{Di} to a wing section's drag coefficient C_{Dw} is much greater in hovering flight than in slow forward flight.

We shall see that this is true of any mode of hovering flight in still air: the induced drag resisting the wing's motion through the air is very substantially enhanced over its values for forward flight. Furthermore, it scales differently, and in a manner particularly important for larger animals, whose hovering capabilities may be limited primarily by the induced-power requirement (the power output needed to overcome induced drag). The fundamental difference which produces the different scaling is the fact that *airflow across the stroke plane* of a hovering animal has to be generated entirely by the induced downwash, whereas an animal in slow forward flight pulls its inclined stroke plane through the ambient air.

The arguments which determine the scaling and the minimum value of the induced-power requirement for normal hovering are just as compelling and fundamental as the similar ones for induced drag in gliding flight (Section 2.3) and, like those, are based on simple fundamental properties of momentum and kinetic energy.

The weight *mg* of a hovering animal can be supported only by the creation of a downward airflow below it which carries away downward momentum at a rate *mg*. The important point is that the *same* downward components of air velocity (say, *u*) determine *both* the momentum *and* the speed at which it is being removed. The air mass per unit height in the downward jet is ρA_j if A_j is the cross-sectional area of the jet. Hence,

$$mg = \rho A_j \overline{u^2};\tag{38}$$

the weight *mg* is balanced by the mean transport at downward velocity *u* of an air mass, ρA_j per unit height, carrying momentum *u* per unit mass.

The induced power P_i is required to generate the kinetic energy which this downward flow of air also carries away. Thus, it takes the value

$$P_i = \rho A_j \left(\overline{\tfrac{1}{2}u^3}\right);\tag{39}$$

this is the mean transport at downward velocity *u* of an air mass, ρA_j per unit height, which carries kinetic energy $\tfrac{1}{2}u^2$ per unit mass.

If the downward velocity *u* were uniform, Eqns (38) and (39) would imply that

$$P_i = \tfrac{1}{2}mg \left(\frac{mg}{\rho A_j}\right)^{\frac{1}{2}}.\tag{40}$$

Quite generally, however, the relationship

$$\overline{u^3} \geqslant (\overline{u^2})^{3/2}\tag{41}$$

holds for any variable positive quantity u; the two sides of (41) are equal when u is constant, but variability of u increases the left-hand side more than the right-hand side. This suggests that the induced power P_i has a minimum value given by (40), which is achieved by an animal making hovering motions that create a jet of almost uniform velocity u below it.

Equation (40) for minimum induced power is already sufficient for a scaling analysis, since the area A_j of downward airflow must evidently be proportional to b^2, the square of the wing semi-span. In the minimum-power case, generating a jet of uniform velocity u, a value of A_j *close* to b^2 is suggested by various arguments (essentially it should be around half the area swept out by the horizontally beating wings because about half of the acceleration of air to velocity u is achieved above the wings and half below (McCormick[8]). This gives minimum induced power per unit mass of animal as

$$\frac{P_i}{m} = \tfrac{1}{2}g \left(\frac{mg}{\rho b^2}\right)^{\frac{1}{2}}. \tag{42}$$

Three features of this result are important. First, it is completely independent of the speed, frequency or type of wing motions. Secondly, it varies in direct proportion to $(m/b^2)^{\frac{1}{2}}$. Thirdly, for geometrically similar animals it scales as $b^{\frac{1}{2}}$ or (equivalently) as $m^{1/6}$.

This last conclusion greatly restricts the possibilities for sustained hovering flight in larger animals. One of the largest animals known to be capable of sustained hovering is the big hummingbird *Patagona gigas* with mass $m = 20$ g and wing semi-span $b = 15$ cm. These data give, for the estimate (42) of minimum induced power, a value of 13 W kg^{-1}, which is uncomfortably near to the expected value of the greatest sustainable output from flight musculature (Section 2.2).

On the basis of that result alone, of course, it could be argued that the size of animals capable of sustained hovering could be elevated well beyond 20 g by appropriate departures from geometrical similarity. However, it must be recognised that *Patagona* at 20 g has already gone quite far in this respect with its 30 cm wing-span. Indeed, Greenewalt[3] shows that for the hummingbirds generally the semi-span b increases with m so much faster than it should according to the $m^{1/3}$ "law" of geometrical similarity that the factor $(m/b^2)^{\frac{1}{2}}$ in (42) remains practically constant. There must be limits (for example, structural limits) to how much farther such departures could go. Again, the hummingbirds have outstripped all other flying animals in their ratio (around 30%) of flight-muscle mass to total mass.

Strong evidence that the size of animals capable of sustained hovering is indeed limited by their maximum sustainable power output is provided, paradoxically, by the fact that animals of enormously larger size can make the same motions for brief periods (for example, during take-off or landing).

This proves that there is nothing to stop the wings of large animals having enough structural strength for those motions, *or* developing enough lift coefficient. On the other hand, any factor (say, 3) by which the anaerobic power output might exceed the maximum sustainable output would by (42) allow an increase in the *linear dimensions* (strictly, in m/b^2) by the same factor *squared* (say, 9); this represents a very large difference in size, exactly as observed.

Another interesting comparison is with the estimate, made at the end of Section 3.2, of the power per unit mass needed to overcome induced drag in slow forward flight at the stalling speed; that value is reduced below the value (42) by a factor $1.5\,K(8C_{L\,\max}S/b^2)^{\frac{1}{2}}$, where 1·5 is the enhancement due to flapping. For typical values of K, $C_{L\,\max}$ and the aspect-ratio $4b^2/S$, this reduction factor is well under 0·5, in agreement with the observation that sustained slow forward flight is found in animals much larger than the largest capable of sustained hovering.

4.2. The Scaling of Normal Hovering

In Section 4.1, I argued first that the induced power requirement for sustained hovering sets an upper limit (say, $l = l_{\max}$) on any particular linear dimension l (for example, the semi-span b) within a series of geometrically similar animals. Then I argued that the effects of departures from geometrical similarity, limited as they must be by various structural constraints, would change the conclusion not qualitatively but only quantitatively, by somewhat increasing the value of $l_{\max}$.

Although the existence of such a constraint is particularly important since it applies to every mode of hovering, it would be a most misleading introduction to the scaling of hovering flight which implied that no other important constraints exist. In this concluding section, I draw attention to several of these, in the special case of "normal" hovering for which large amounts of observational data exist. Here, I follow in most respects the fundamental mechanical and aerodynamic analysis pioneered by Weis-Fogh (Ref. 12; see also the paper immediately following this). However, I regard each particular constraint as confining a hovering animal not to the *general neighbourhood* of one particular line in the frequency-size diagram but rather as confining it to *one side* of a particular line, just as the constraint already discussed confines l to one side,

$$l \leqslant l_{\max}, \tag{43}$$

of the line $l = l_{\max}$. I hope to indicate that the general spread of points in the frequency-size diagram representing observations of normal hovering can be made reasonably intelligible in the light of a number of such constraints.

I again crudely analyse the constraints on the basis of geometrical similarity, but assume that departures from it change them quantitatively rather than qualitatively.

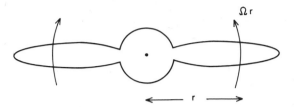

FIG. 15. Diagram of normal hovering flight (for explanation, see text).

Figure 15 shows diagrammatically an animal in normal hovering viewed from above; the phase of its motion corresponds to tracing (*m*) of Fig. 14. As in Section 3, I use *r* for the distance of wing section from a suitable central point within the animal's body. Now, however, I use Ω for the instaneous angular speed of rotation of the wing about a vertical axis through that central point (this would be $|\dot\phi|$ in the notation of Section 3); the airspeed V of the wing section is therefore Ωr.

I shall go rapidly through the list of constraints. Writing the lift per unit area on a wing section as

$$\tfrac{1}{2}\rho V^2 C_L = \tfrac{1}{2}\rho(\Omega r)^2 C_L, \tag{44}$$

we see that the total lift can be written $\tfrac{1}{2}\rho\Omega^2 S_2 \bar C_L$, where S_2 is the *second moment* of wing area about the central axis and $\bar C_L$ is a weighted average lift coefficient. This leads to Weis-Fogh's form of the condition that mean lift balances weight:

$$\tfrac{1}{2}\rho\overline{\Omega^2} S_2 \bar C_L = mg. \tag{45}$$

Here we have a new $\bar C_L$ involving a further averaging over the wing beat, while $\overline{\Omega^2}$ is the mean square angular speed, which scales as the *square* of the wing-beat frequency *n*. Under conditions of geometrical similarity S_2/m scales as the linear dimension *l*. Therefore, *because C_L is bounded above,*

n lies above a value proportional to $l^{-\frac{1}{2}}$. $\qquad$ (46)

Aerodynamics by itself imposes one other important constraint. Conventional wing aerodynamics, which an animal making the motions of normal hovering relies on for the generation of lift (as analysed above), becomes seriously impaired below a Reynolds number *of the order of* 100 (see Thom and Swart[11]). Therefore, *because* a characteristic Reynolds number,

$\overline{\Omega} l^2/v$ (with $\overline{\Omega}$ proportional to the frequency n), is bounded below,

$$n \text{ lies } above \text{ another value proportional to } l^{-2}. \tag{47}$$

I want to note now that power has to be expended by a hovering animal not only to overcome induced drag (Section 4.1) but also to overcome wing profile drag, which scales quite differently. A wing section's rate of working against profile drag can be written as

$$\tfrac{1}{2}\rho V^3 C_{Dp} = \tfrac{1}{2}\rho(\Omega r)^3 C_{Dp} \tag{48}$$

per unit area. Therefore, *because* the associated power per unit mass of animal

$$\frac{\tfrac{1}{2}\rho \overline{\Omega}^3 S_3 \overline{C}_{Dp}}{m}$$

(where S_3, the *third moment* of wing area, scales as ml^2) is bounded above,

$$n \text{ lies } below \text{ another value proportional to } l^{-2/3}. \tag{50}$$

Each of these conditions (43), (46), (47) and (50) (there will be two more!) can be represented diagrammatically on a log–log plot like Fig. 16, which I shall use to suggest that values for frequency n and length-scale l in normal hovering may be confined within some region roughly like the hexagonal area shown. Note how the conflict between the last two conditions (47) and (50) at the top left-hand corner of the hexagon leads to an absolute *lower* limit on the length-scale l for normal hovering. This part of the diagram is consistent with the views that: (i) evolutionary opportunities for hovering insects that could remain functionally effective at very small sizes gave rise to the development of "myogenic" or "fibrillar" musculature which made possible the beating of wings at extraordinarily high frequencies; but that: (ii) further size reduction was not possible until the Reynolds-number limit

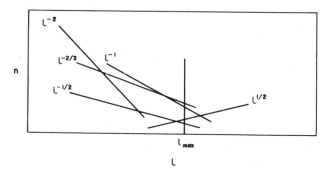

Fig. 16. Log–log plot of wing-beat frequency n versus a length–scale l; a set of constraints like (43), (46), (47), (50), (53) and (55) suggests that points, representing sustained normal hovering in still air, may be confined within a region such as the hexagon here shown.

could be reduced, below that for normal hovering, through the adoption of special hovering modes such as those involving Weis-Fogh's clap-and-fling mechanism.

Finally, there are important constraints provided by the strength of the animal's skeletal materials. The profile drag of a wing section, per unit area, generates at the wing root a bending moment

$$r(\tfrac{1}{2}\rho V^2 C_{Dp}) = \tfrac{1}{2}\rho\Omega^2 r^3 C_{Dp}, \tag{51}$$

so that the total wing-root bending moment due to profile drag is

$$\tfrac{1}{2}\rho\overline{\Omega^2}S_3\overline{C_{Dp}}. \tag{52}$$

For geometrically similar animals this bending moment must be divided by something proportional to l^3 to give the associated skeletal stress, which therefore is proportional to $n^2 l^2$. Hence, *because* skeletal stress is bounded above.

n lies below another value proportional to l^{-1}.   (53)

A further, possibly more important, constraint based on the strength of skeletal materials is provided by the need for wing-root bending moments to generate also the large angular accelerations $\dot{\Omega}$ required at the extremes of the wing oscillations. Those inertial bending moments can be written in terms of the wing's moment of inertia I about the central vertical axis as $I\dot{\Omega}$. The associated skeletal stress, however, is again proportional to $n^2 l^2$, so that, as Weis-Fogh has pointed out, the constraints on the bending moment required to overcome both inertia and profile drag take the *same* form (53) and we need not here enquire which dominates.

The bending moment which has to be provided to overcome induced drag scales differently: it can be written

$$P_i\overline{\Omega^{-1}} \tag{54}$$

where $\overline{\Omega^{-1}}$ (proportional to n^{-1}) is an average reciprocal angular speed of rotation weighted with respect to that rate of working against induced drag whose mean value is the induced power P_i, given by equation (42). Hence, *because* the skeletal stress due to the induced-power bending moment (54) is bounded above,

n lies above another value proportional to $l^{\frac{1}{2}}$.   (55)

Figure 16 is an attempt to show how a set of constraints like (43), (46), (47), (50), (53) and (55) can force the points in a frequency-size diagram for normal hovering into a flattened region (which has a hexagonal form according to the present crude analyses), broadly consistent with the known facts that the band of points representing observations is roughly parallel to a line $nl =$ constant but stops abruptly both at the upper left-hand and the lower right-hand ends. For further details on my concluding topic, see the next Lecture!

REFERENCES

1. Brown, R. H. J. The flight of birds. *J. exp. Biol.* **30**, 90–103 (1948).
2. Greenewalt, C. H. Dimensional relationships for flying animals. *Smithsonian Misc. Coll.* **144**, no. 2, 46 pp. (1962).
3. Greenewalt, C. H. The flight of birds. *Trans. Amer. Philos. Soc.* **65**, pt. 4, 67 pp. (1975).
4. Hankin, E. H. "Animal Flight". London: Iliffe & Sons (1913).
5. Karman, T. von "Aerodynamics", Ithaca, N.Y.: Cornell Univ. Press (1954).
6. Lewis, T. and Taylor, L. R. "Introduction to Experimental Ecology". London: Academic Press (1967).
7. Lighthill, M. J. Aerodynamic Aspects of Animal Flight. *Bull. Inst. Math. Applics.* **10**, 369–393 (1974) and also in the Proceedings of the 1974 Symposium "Swimming and Flying in Nature" (ed. by Wu, T. Y., Brokaw, C. J. and Brennen, C.). New York, Plenum Press (1975).
8. McCormick, B. W. "Aerodynamics of V/STOL Flight". London: Academic Press (1967).
9. Pennycuick, C. J. A wind-tunnel study of gliding flight in the pigeon *Columba livia. J. exp. Biol.* **49**, 509–526 (1968).
10. Pennycuick, C. J. "Animal Flight". London, Edward Arnold (1972).
11. Thom, A. and Swart, P. The forces on an aerofoil at very low speeds, *J. Roy. Aero. Soc.* **44**, 761–770 (1940).
12. Weis-Fogh, T. Quick estimates of flight fitness in hovering animals, including novel mechanisms for lift production. *J. exp. Biol.* **56**, 79–104 (1973).
13. Weis-Fogh, T. and Jensen, M. Biology and physics of locust flight, I–IV. *Phil. Trans. Roy. Soc. B*, **239**, 415–584 (1956).

24. Dimensional Analysis of Hovering Flight.

TORKEL WEIS-FOGH†

Department of Zoology, University of Cambridge, England.

ABSTRACT

Active flapping flight, and in particular hovering, is now sufficiently well understood to permit realistic estimates both of the effects of absolute size and of the relationship between the main parameters, wing length l and wingstroke frequency n. However, analyses of only a few of the dynamic functions involved are bound to result in misleading results, as has happened in the past. On the other hand, when all the major functions are taken into account, and in our case this means 16 simultaneous relationships, even a highly simplified approach can lead to a general understanding of why flying animals are "designed" as they are and how this "design" has made evolutionary progress possible with a minimum of genetic change.

Three model laws will be discussed in some detail and related to our actual knowledge of the limits of their applicability. As to length and frequency they are represented by: (1) $n \propto l^{-\frac{1}{2}}$ (the aerodynamic rule); (2) $n \propto l^{-2/3}$ (the intraspecific rule); and the most general case (3) $n \propto l^{-1}$ (the interspecific rule).

INTRODUCTION

If an animal is going to remain in full control of its flight performance, effective gliding and soaring in natural wind requires a relatively large mass of not less than 0·1 kg and a wing span of 0·5 m or more. One of the largest soaring birds is the Wandering Albatross (*Diomedea exulans*) which has a mass of 10 kg and spans 3·5 m; it flies at a Reynolds number (Re) of about 2×10^5 (Tucker and Parrott[19]). Some of the extinct pterosaurs were larger and *Pteranodon* is estimated to have reached 18 kg and spanned 8 m (Bramwell[3]). These animals depend almost entirely on refined control movements

† Deceased.

405

and they extract energy for flight from air currents and velocity gradients in the atmosphere rather than by active flapping of their wings. At the other end of the size range, we have the tiny winged and unwinged arthropods which constitute the major part of the passive "aerial plankton" and which can "float" in turbulent air because of their large surface-to-weight ratio. Their mass is usually 1 mg or less. At both ends of this size scale, the design of a passively flying animal must be governed mainly by aerodynamical, geometrical and static relationships. These extremes are *excluded* from the present discussion in which I shall concentrate on active flapping flight powered by rapidly contracting wing muscles. This means that the main emphasis will be on dynamic relationships and our task is to analyse how the dynamic and structural parameters of an oscillating system depend on the size of the animal. Some are scale-dependent and some are independent of absolute size but may vary between taxonomic groups. The latter applies to skeletal and supporting materials like bone in vertebrates and cuticle in insects. In any case, it is important that we take into account *all major functions* rather than concentrating on one or two because our task is to understand the *integrated organism*. For this we need to know a large number of quantitative relationships and in a form amenable to dimensional analysis. In this respect flight is a particularly favourable biological activity for studies of scaling because active flapping flight is extremely common, it embraces a large range of forms and sizes, it is demanding in energy and it imposes an absolute, size-dependent requirement, that of lifting the body weight. Moreover, recent studies provide us with a coherent set of relationships needed even for a first-order approximation like the present (see Weis-Fogh[24-26]).

SIZE AND NUMBER OF ACTIVELY FLAPPING ANIMALS

If mere numbers of species and individuals can be accepted as a measure of evolutionary success, the development of active flapping flight must be considered a decisive advantage for terrestrial life. It must have evolved independently at four different occasions, in what became winged insects (Pterygota), pterosaurs (Reptilia, Pterosauria), birds (Aves), and bats (Mammalia, Chiroptera). According to Mayr, Linsley and Usinger,[13] more than one million animal species have been described of which only 4% are vertebrates and at least 750 000 are winged insects, mainly belonging to Diptera, Hymenoptera, Lepidoptera and Coleoptera, i.e. to insects which habitually practise *hovering* and *slow forward* flight rather than *fast forward* flight. More than half of the 43 000 known vertebrate species live in water and scale effects connected with these groups have already been discussed at this

Conference. The remaining 20 000 odd species are terrestrial or amphibious; of these 8600 are birds and about 1000 are bats. Most of the flying vertebrates are of modest size (like a sparrow) and are able to fly slowly, even to hover on the spot for brief periods of time, as well as to fly rapidly forward at speeds approaching 100 body lengths s^{-1}. The reason why hovering and slow flight is expensive in power requirement will be discussed in the following. At this point it is sufficient to point out that the induced drag of the wings increases with decreasing flying speed U, and is proportional to U^{-2} (cf. Lighthill's article, Eqn (10)). It is therefore justified to simplify the analysis by concentrating on hovering because, if an animal can hover, it can usually also perform fast forward flight and at less cost. Moreover, the number of species which depend on a high forward speed is relatively small and consists mainly of larger bats and birds, the Great Bustard *Otis tarda* of 10 kg mass being at the upper end of the scale. The design of the majority must then be governed by the flow patterns, the power requirements, and the stresses and strains needed for slow forward flight and hovering on the spot.

The masses of actively flying animals range from 1 μg in some tiny insects to 10 kg in large birds, or $1:10^{10}$. Taking the great differences in basic design into account, the flight contours are remarkably similar so that we can use the wing length R as a representative length l in dimensional considerations (the span is $2R$ plus the width of the body). R varies from 0·1 mm to 1 m, or $1:10^4$. However, if we confine ourselves to hovering animals using known aerodynamic mechanisms (Weis-Fogh[25]), the mass ranges from 25 μg in the parasitic wasp *Encarsia* to 300 g in the pigeon ($1:10^7$) and the wing lengths from 0·6 mm to 0·3 m, or $1:500$, against $1:200$ in the case of strict geometrical similarity. These animals are therefore only roughly similar but this will do for a first approximation. Over the same range of mass and length, Reynolds number (Re) varies from 10 to 10^5 and therefore includes regions where the properties of aerofoils alter significantly.

MUSCULAR POWER, A SIZE-INDEPENDENT PARAMETER

Even today, it is sometimes stated that continuously working striated muscle can deliver only 15–20 W of mechanical power per kg muscle, the specific power output P_m^*, as is found in man. I have previously discussed this fallacy (Weis-Fogh[23]). If this were the case, hardly any animal would be able to fly with wings as we known them. Nevertheless, some properties of striated muscle are remarkably constant in vertebrates and invertebrates: contracting muscles can develop a maximum stress of 200 to 400 kN m^{-2}, and the maximum power output occurs when they shorten against one third of this stress, i.e. against 70 to 130 kNm^{-2} (see Alexander[1]). The property which varies most and which then determines the power output is the shortening speed. It is largest at the beginning of a twitch and gradually declines to zero. When a striated muscle shortens against the stress for maximum power output, the shortening speed is about one fifth of its so-called intrinsic value, the intrinsic speed being the maximum speed when there is no external load. How large is the intrinsic speed and what determines its upper limit? There is no theoretical answer to this question at present but only a few empirical observations on isolated muscles. At 36°C, one of the fastest muscles of a mouse (finger extensor) shortens at 26 lengths s^{-1} (Close[6]) while locust wing muscles reach 13 lengths s^{-1} at 30°C (Buchthal, Weis-Fogh and Rosenfalck[4]); it normally operates at 35°C. It is characteristic of locust wing muscle that the relaxation starts early and is completed rather suddenly so that the work is done while the shortening speed is near to the maximum. As a reasonable estimate based on only a few comparative data, wing muscles in small vertebrates and insects seem to be characterized by intrinsic speeds of 10–20 lengths s^{-1}. Since the shortening speed declines with time after the onset of a twitch, one tenth of these values, or 1–2 lengths s^{-1}, is reasonable for estimating the power output. Furthermore, we must allow equal time for shortening during one half-stroke and the following relaxation during the next period, indicating that the average continuous power output is 70–260 W per kg of muscle. Whether the upper limit is determined by molecular parameters in the contractile proteins or by the rate with which the mitochondria can supply ATP is not known. Both factors may be limiting because, in the typical aerobic wing muscle of mature locusts, the mitochondria occupy no less than 30 % of the volume (Bücher[5]); in the fibrillar muscles of giant waterbugs the percentage is close to 50 (Ashurst[2]) so, as to specific power, a further increase in mitochondrial mass is likely to result in diminishing returns. It is obvious that much more information about basic mechanical parameters in a variety of fast power-producing muscles is needed.

How do 70–260 W kg^{-1} compare with the mechanical power now known to be produced during actual flight? Table I strongly indicates that the specific power of striated muscle has indeed reached this level in all actively flying animals, and it also conclusively shows that the level is independent of absolute size and systematic group. The power output is one order of magnitude higher than in man. The conclusion is that P_m^* of flying animals usually range between 50 and 200 W kg^{-1} and that $P_m^* \propto l^0$.

AERODYNAMIC POWER AND HOVERING

The virtual lack of buoyancy in air means that active flight requires continuous work. Running and swimming animals can stop moving without

TABLE I. Mechanical power output of wing muscle during continuous flight (cont.) or maximum performance (max.), and the methods of estimating the work (metabolic rate, met.; aerodynamic work, aero.; total work, tot.), in Watts per kg muscle mass.

	mass (g)	type of flight (cont., max.)	how estimated (met., aero., tot.)	muscle power ($W\,kg^{-1}$)	references (number)
Bats: Chiroptera					
Pteropus gouldii	780	forward, cont.	met., aero.	140	18
Phyllostomus hastatus	93	" "	" "	240	18
Birds: Aves					
Columba livia	400	climbing, max.	aero.	220	1, 15
Larus atricilla	322	forward, cont.	met., aero.	70	18
Corvus ossifragus	275	" "	" "	80	18
Melopsittacus undulatus	35	near hover, cont.	" "	170	18
Amazilia fimbriata	"	forward, cont.	" "	95	18
" "	5	hovering, cont.	tot.	150	24
Insects: Insecta					
Schistocerca gregaria	2	forward, cont.	tot.	60–90	9
Drosophila virilis	2×10^{-3}	hovering, cont.	"	160	24
Large range of insects	1×10^{-3} to 13	" "	"	range: 120–360	25
	" "	" "	"	average: 140	25

disastrous consequences and can therefore adjust their efforts almost continuously between zero and maximum output. For an animal to remain airborne it must cause its weight mg to be balanced by continuously imparting a downward momentum to the otherwise still air so that, during true hovering on the spot, the induced vertical air speed w at the level of the wings is given by

$$w^2 = mg/(2\pi\rho R^2), \tag{1}$$

where ρ is the air density (1·23 kg m^{-3} at S.T.P.). Since $m \propto l^3$, this general requirement means that $w \propto R^{1/2} \propto l^{1/2}$. Per Newton weight lifted the power transferred from the animal to the air is the specific aerodynamic power P_a^*; numerically it equals w Nm s^{-1} = w W. The relationship therefore shows both the minimum specific power in absolute measure and how P_a^* should vary with size.

This simple relationship, so unusual in biology, is illustrated in Fig. 1. In accordance with previous calculations on hummingbirds (Weis-Fogh[24]), the aerodynamic efficiency is about 0·5 and the actual power output is therefore twice as large as deduced from Eqn (1) because of losses mainly

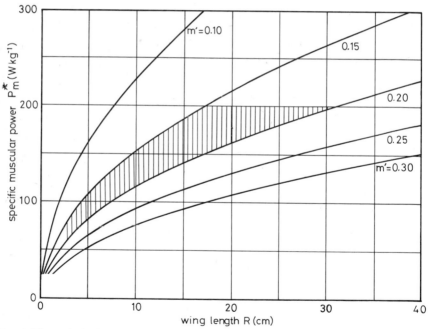

FIG. 1. Theoretical graphs of specific muscular power for hovering, P_m^*, against wing length R, for different values of m', the ratio of muscle mass to body mass. The shaded area represents the region in which the largest hovering animals are expected to occur (see text).

caused by tip vortices and the induced drag. All calculated figures have therefore been multiplied by 2. This factor also applies to fast forward flight of locusts and probably to small birds with large stroke angles (Weis-Fogh[26]). However, the efficiency decreases with decreasing size because the lift/drag ratio deteriorates at low Reynolds number, and the curves should *not* be used for wing lengths less than 2–3 cm. The vertical axis is the specific muscular power calculated by dividing the power needed for lifting the airborne mass by the relative muscle mass m', where m' equals wing-muscle mass divided by total mass. It should be emphasised that Eqn (1) is valid whatever mechanisms are used by the animal to hover, and the curves represent the maximum predicted sizes and not the average or minimum ones. Within the limits of 50 to 200 W kg^{-1} for P_m^*, the absolute size is highly dependent on m'. In many bats, m' is about 0·10 or slightly less (see Greenewalt[8]) and we should therefore not expect wing lengths in excess of about 10 cm; this is in fair agreement with the observation that the small *Plecotus auritus* $(R = 11·5$ cm) habitually hovers and that the larger species do not. Also, the bats fold their wings extensively during part of the stroke (Norberg[14]) so that inertial stresses are minimized. This is not the case in hummingbirds where the power needed to accelerate the wing mass is of the same magnitude as the aerodynamic power (Weis-Fogh[24]). They have an unusually high m' of about 0·3 so that the relevant curve to consider corresponds to $m' = 0·15$. It is then seen that the available power favours small forms with wing lengths somewhere between a few cm (min. 3·3 cm in *Calliphlox amethystina*, 2·8 g) and 15 cm (max. 13 cm in *Patagona gigas*, 20 g). (For a much fuller discussion of the power available from striated muscle in sustained activity, see the paper by Alexander and Weis-Fogh at the end of this volume.)

Taken as a whole, there appears to be good agreement between the specific muscular power discussed in the previous section and the maximum size of hovering birds and bats. This also applies to medium-sized and large insects. The elephant dung beetles *Heliocopris* $(R = 8$ cm, $m' = 0·1(?)$, mass $= 13$ g) come close to the upper limit, while the hawk moth *Sphinx ligustri* $(R = 5$ cm, $m' = 0·14$, mass $= 1·6$ g) is in the middle of the expected range, the largest Sphingid having a wing length of 9 cm (*Cocytius cluentis* Cr.; Seitz[16]). When all flying creatures are taken into account, the average m' is 0·17 (Greenewalt[8]) so that the hatched area in Fig. 1 represents the variation in size and power of the largest hovering animals. It is of particular interest because the very nature of flight, combined with the fact that P_m^* is of the same order of magnitude in all animals, makes it possible to use an absolute rather than a relative length scale, provided we realize that small insects which fly at low Reynolds number cannot be treated in such an elementary way. Moreover, the method omits reference to the wing stroke itself and refers only to the power needed.

TABLE II. Dimensional relationships to account for in hovering flight. The resulting quantities are the specific (*) forces (F), bending moments (Q) and power contributions (P) and their relation to the wing-stroke parameters n and ϕ as well as to other functional and structural parameters. Column (A) shows the complete expressions and Column (B) the simplified version used in Table III. For the aerodynamic expressions "conv." means that they are based on conventional aerodynamics and "fling" that the lift is assumed to be generated by the Weis-Fogh mechanism (Lighthill[11]).

Specific quantity	(A) Complete dimensional expression		(B) Expression after pruning and simplification, as explained in text
Aerodynamic vertical force: (average force)	conv.	$F_a^* \propto \bar{C}_L \times l^1 n^2 \phi^2$	$F_a^* \propto l^1 n^2$
	fling	$F_a^* \propto g(\alpha) \times l^1 n^2 \phi^1$	$F_a^* \propto l^1 n^2$
Bending moments of wings:			
aerodynamic	conv.	$Q_a^* \propto \bar{C}_D \times l^2 n^2 \phi^2$	$Q_a^* \propto l^2 n^2$
	fling	$Q_a^* \propto f(g(\alpha)) \times l^2 n^2 \phi^1$	$Q_a^* \propto l^2 n^2$
inertial		$Q_i^* \propto I \times l^{-3} n^2 \phi^1$	$Q_i^* \propto l^2 n^2$
elastic		$Q_e^* \propto E \times f(\gamma) \times l^0$	$Q_e^* \propto l^0$
damping		$Q_d^* \propto l^1 n^1$	$Q_d^* \propto l^1 n^1$
muscular		$Q_m^* \propto l^0$	$Q_m^* \propto l^0$
Power contributions:			
aerodynamic: induced	conv.	$P_a^* \propto l^{1/2}$	$P_a^* \propto l^{1/2}$
frictional	conv.	$P_f^* \propto \bar{C}_D \times l^2 n^3 \phi^3$	$P_f^* \propto l^2 n^3$
	fling	$P_f^* \propto f(g(\alpha)) \times l^2 n^3 \phi^2$	$P_f^* \propto l^2 n^3$
inertial		$P_i^* \propto I \times l^{-3} n^3 \phi^1$	$P_i^* \propto l^2 n^3$
elastic		$P_e^* \propto E \times f(\gamma) \times l^0 n^1$	$P_e^* \propto l^0 n^1$
damping		$P_d^* \propto l^1 n^2$	$P_d^* \propto l^1 n^2$
muscular		$P_m^* \propto l^0$	$P_m^* \propto l^0$
Reynolds number (Re)	conv.	$(Re) \propto l^2 n \phi$	$(Re) \propto l^2 n$

DIMENSIONAL ANALYSIS OF HOVERING FLIGHT

The majority of animals oscillate their wings in an almost horizontal plane when hovering and use them as aerofoils according to conventional aerodynamic principles, as I have discussed recently, but some insects (and probably some birds) are able to generate lift according to novel principles (Weis-Fogh[25]). Of the latter, the fling mechanism has been analyzed in detail by Lighthill.[11] However, for the present purpose these differences are of little importance because the two main parameters, length and frequency, vary in the same way with size in all cases. Being an oscillation, the wing movements involve five sets of relationships for force F, bending moment Q and power contribution P, which are listed in Table II. The five sets are marked by subscripts, a for aerodynamic (or induced), f for frictional (which can also be thought of as aerodynamic), i for inertial, e for elastic, d for damping, and m for muscular, as used in earlier treatments (Weis-Fogh and Jensen[27]). According to Weis-Fogh[25] the basic aerodynamic expression for sinusoidally oscillating wings is:

$$\text{vertical force } F_a = mg = \tfrac{1}{4}\rho\pi^2 n^2\phi^2\sigma cR^3\bar{C}_L, \tag{2}$$

where mg is the airborne weight, n the wing stroke frequency, ϕ the stroke angle, σ an integration factor, c the wing chord and $\bar{C}_L$ the average lift coefficient. In order to simplify the treatment, we shall use the specific force F_a^*, i.e. the force per unit weight lifted. Since $m \propto l^3$ and $cR^3 \propto l^4$,

$$F_a^* \propto \bar{C}_L \times l^1 n^2\phi^2. \tag{3}$$

The bending moments and the power contributions are treated in a similar way. In this context it is immaterial that the muscular power P_m^* in the previous section is the power per unit weight lifted and not per unit muscle mass since both P_m and m are proportional to l^3, and $P_m^* \propto l^0$. The bending moments caused by the acceleration of the wing mass depend on the moment of inertia I of all the wings about their fulcra, and the elastic bending moments depend on the elastic modulus E as well as on the specific variation of the elastic counter moment in relation to the angular position of the wing, described by the function $f(\gamma)$; $f(\gamma)$ is known only in a few insects (Weis-Fogh[24]). As to the damping moment in the working thorax Q_d, we shall assume simple viscous damping but the quantities involved are probably too small to matter significantly and the expressions for damping are included only for the sake of completeness. We then arrive at the basic relationships shown in column (A) of Table II. It should be noted that Reynolds number (Re) has been added at the bottom of the list because of its obvious importance in relation to size. Note too that the discussion below equation (1) indicates that P_a^* is always proportional to $l^{1/2}$.

In order to make progress, it is necessary to prune and simplify the 16 expressions as much as possible, and the result of this is seen in column (B) of Table II. At this point caution and scientific "tact" are needed if we are to avoid absurdities and false generalizations. Also, we are obliged to return to the complete expressions whenever specific cases are being studied in depth. However, as a first approach I have proceeded as follows. In most hovering species the stroke angle ϕ is about 120° and it does not exceed 180°, so that, compared with variations in length from 1 to 100, ϕ may be assumed to be independent of size. As to wing inertia, Greenewalt's[8] results indicate that I increases as $I \propto l^{5.33}$ and the rationale of this will be discussed elsewhere. Here we shall take the naive view and assume $I \propto l^5$. When dealing with a major systematic group, it is reasonable to assume that the material properties and therefore E are size-independent and that this also applies to the relationship between the positional angle γ of the wing and the elastic recoil, i.e. to $f(\gamma)$. It remains to discuss the coefficient of lift $g(\alpha)$ for the unsteady fling mechanism and its conjugated drag $f(g(\alpha))$. It has already been argued that if $n \propto l^{-1/2}$, $g(\alpha)$ is independent of size (Weis-Fogh[25]) so that the situation is similar to that for $\overline{C}_L$. Due to lack of other evidence we shall assume that this also applies to the fling drag. The main dimensional difference between the expressions for conventional and unsteady aerodynamic mechanisms is that lift depends on ϕ^2 in the first case and on ϕ^1 in the latter. However, this difference is eliminated when ϕ is assumed independent of size, as before. How do the simplified relationships in Table II match actual observations?

CONFRONTATION BETWEEN THEORY AND OBSERVATION

We shall make extensive use of the data compiled by Greenewalt[8] but at present only the rough outlines are clear. The main task is to analyze how frequency n varies with length l in freely flying animals. At first sight the scene is bewildering although Greenewalt's studies[7, 8] strongly indicate that $n \propto l^{-1}$ in the Animal Kingdom as a whole. As he points out, there are many deviations and I have recently analyzed the problem on the basis of his material and data accumulated later. It then appears that if, for instance, we choose well-defined insect families and groups like Apoidea, Vespoidea, Noctuidea, Sphingidae, Culicidae, Tipulidae and Lamellicornia, the frequency in each group varies almost exactly as $n \propto l^{-1}$. For the time being and as a provisional step, we may therefore call this rule, $n \propto l^{-1}$, the *general interspecific rule*. We shall discuss its significance shortly.

If, on the other hand, we investigate how n varies with l in a given species like *Schistocerca gregaria*, we find that $n \propto l^{-2/3}$ and this relationship is very precise (Weis-Fogh[20]). It is the same relationship found by Sotavalta[17] in a long series of experiments in which he increased or decreased the wing inertia I by increasing the wing mass or by mutilating the wings experimentally in a given specimen. We shall refer to this as the *intraspecific rule.*

Finally, and in a few cases, we find the relationship which one would expect if aerodynamic lift is the main concern, namely that $n \propto l^{-1/2}$. We shall refer to this as the *aerodynamic rule* for lift production and it is found for instance in the genus *Drosophila.**

The details will be illustrated and discussed elsewhere, but the consequences can be outlined here.

DISCUSSION OF RESULTS

When we apply these three rules to the simplified expressions in Table II, the results in Table III offer some clear and interesting answers.

TABLE III. The dimensional relationships between the most significant quantities and the size of hovering animals, when the wing-stroke frequency n varies with a representative length l as indicated.

Specific quantity	Aerodynamic rule: $n \propto l^{-1/2}$	Intraspecific rule: $n \propto l^{-2/3}$	Insterspecific rule: $n \propto l^{-1}$
Lifting force: F_a^*	$\{1^0\}$	$l^{-1/3}$	l^{-1}
Bending moments: Q^*			
aerodynamic	l^1	$l^{2/3}$	$\left.\begin{cases} l^0 \\ \end{cases}\right.$
inertial	l^1	$l^{2/3}$	l^0
elastic	l^0	l^0	l^0
damping	$l^{1/2}$	$l^{1/3}$	l^0
muscular	l^0	l^0	l^0
Power contributions: P^*			
aerodynamic {induced	$l^{1/2}$	$l^{1/2}$	$l^{1/2}$
{frictional	$l^{1/2}$	$\{l^0\}$	l^{-1}
inertial	$l^{1/2}$	$\{l^0\}$	$\{l^{-1}\}$
elastic	$l^{-1/2}$	$l^{-2/3}$	$\{l^{-1}\}$
damping	$\{l^0\}$	$l^{-1/3}$	l^{-1}
muscular	$\{l^0\}$	$\{l^0\}$	$\{l^0\}$
Reynolds number (Re)	$l^{3/2}$	$l^{4/3}$	l^1

* The phrase "aerodynamic rule" is here used as a shorthand for "rule based on an upper bound for the lift coefficient", or "lift-limited rule". It does not incorporate limitations based on induced or frictional power requirements, although they might also be thought of as aerodynamic.

A. *Aerodynamic rule, $n \propto l^{-1/2}$*

If the problem were merely to be airborne, one should have expected this rule to apply over a large range of form and size. However, it is obvious that one cannot design a set of similarly built animals of widely different sizes on this principle, partly because the specific bending moments for aerodynamic and inertial forces increase as l^1, partly because the inertial moment is balanced against the elastic moment only for one size, and also because $P_a^* \propto P_f^* \propto P_i^* \propto l^{1/2}$, i.e. as illustrated in Fig. 1. Animals larger than the optimum match are therefore at a disadvantage in almost all respects. In *Drosophila* a reasonably large range is possible because this genus has an unusually low wing-stroke frequency (because it probably uses the fling mechanism for lift generation) so that the inertial moment and power is insignificant, in contrast to most other insects (Weis-Fogh[24, 25]). Even so, a given design can only work effectively over a comparatively narrow size range.

B. *Intraspecific rule, $n \propto l^{-2/3}$*

The same applies in this case which is in accordance with the relatively modest variation in size found within members of any given species. The outstanding feature is that the frictional and inertial power contributions are size-independent, $P_a^* \propto P_i^* \propto l^0$, as is the power from the muscles (although the induced power increases as $l^{1/2}$). They have therefore been bracketed in the table. In a given experimental situation an animal tends to operate at a constant power output provided that the bending moments are kept within the elastic limits of the wings and the thorax. However, according to this rule a significant increase in size would be impossible without compensating increases in wall and beam thicknesses in order to counteract the stresses caused by aerodynamic and inertial forces. Even so, the elastic forces would not match the inertial forces unless the relative dimensions of the supporting structures were altered according to one rule for the elastic moments and another for the two other moments.

C. *General interspecific rule, $n \propto l^{-1}$*

The brackets in the last column of Table III indicate the rationale of this solution: all the major bending moments and stresses are independent of size. As to the aerodynamic force, F_a^* is proportional to l^{-1}, which means that if the frequency of the largest member of a "design group" permits active flight according to Fig. 1, all smaller members can be airborne. They can also be similarly built with respect to external contours, wall thicknesses, elastic properties and materials, damping, muscles, etc. This solution therefore makes it possible for a group to evolve up and down the size scale with a minimum of change other than a "command" to reach a given size below the maximum possible for that group. It is suggested that this simple solution to a complex engineering problem in biology has been of fundamental importance for the intense and highly successful speciation to which the enormous number and the large size range of flying species bear witness.

It should be noted that although $F_a^* \propto l^{-1}$, the aerodynamic advantage of being small only holds within a limited range of Reynolds number. For Re between 1000 and 100, it is still possible to create lift according to ordinary aerofoil theory, but $\bar{C}_L$ is decreasing and $\bar{C}_D$ increasing. Below this range some insects make use of the fling mechanism but it is also used in much larger forms (Weis-Fogh[25, 26]).

The main puzzle is then how we match the power requirements with the muscular output since $P_m^* \propto l^0$. First, the fact that inertial and elastic contributions (bracketed) are matched is very significant because they tend to counteract each other so that their summed contribution is zero in an ideal animal, a situation which has been approached by many insects (Weis-Fogh[22–25]). As to birds, it would be extremely interesting to study the possible elastic storage of energy in *active, twitching* wing muscle, for instance in hummingbirds; in the relaxed state there is no indication of storage in muscles or tendons. In insect wing muscle there is a strong elastic element, even in the non-active state (Weis-Fogh.[21] Buchthal *et al.*,[4] Machin and Pringle[12]) but the main energy is stored in and released from the elastic cuticle (Jensen and Weis-Fogh[10]).

How then can $P_f^* \propto l^{-1}$ when $P_m^* \propto l^0$? To answer this question we must go back to the factors pruned away in moving from column (A) to column (B) in Table II. It is then possible to suggest some answers. When a hovering animal increases its size, the aerodynamic load per unit wing area increases and $\bar{C}_L$ therefore increases. This means that the induced power increases, as indeed is shown by the dependence of P_a^* on $l^{1/2}$. To some extent and only over a limited range of Reynolds number this may be counteracted by a reduced frictional power P_f^*. The net result may well be that $P_f^* + P_a^*$ becomes almost independent of size over a considerable range of size and Re; this is actually found in more sophisticated calculations based on lift/drag diagrams of real wings and at the correct value of Re (Weis-Fogh[24]). Within the range of Re from 10 to 10,000, the change in aerodynamic power would not vary nearly as much with size as is indicated by the simple dimensional expressions: the conflicting changes tend to cancel each other out. However, much more work is obviously needed before we can be more precise.

The reason why the mechanical properties of the wing system, rather than the aerodynamic relationships, dominate its design is undoubtedly that the inertial and elastic bending moments are often several times larger than those caused by frictional drag in most hovering species, both in insects and in hummingbirds. Wing inertia must also be important in other small birds when they hover or perform slow forward flight. During faster flight, the wind speed and the air forces permit the kinetic energy of the wings to be utilized towards the end of each half-stroke so that the power loss is reduced, but the wings still have to resist the combined effect of aerodynamic and inertial forces. This is the price which must be paid by a locomotary system based upon oscillation of wings and limbs rather than continuous rotation with conservation of kinetic energy, as in the wheel or the propeller.

CONCLUSIONS

Active flapping flight, and in particular hovering, is now sufficiently well understood to permit realistic estimates both of the effects of absolute size and of the relationship between the main parameters, length and frequency. However, analyses of only a few factors are bound to lead to erroneous conclusions. On the other hand, when all the major functions are taken into account, and in our case we consider 16 simultaneous relationships, even a highly simplified approach can lead to a general understanding of why flying animals are "designed" as they are and how this "design" has made evolutionary progress possible on the basis of a minimum of "commands", i.e. a minimum of genetic change. With regard to insects, this applies only to the imaginal adult stage, the larval and nymphal stages not being winged. In the most advanced forms, the holometabolous pterygotes like Diptera, Hymenoptera, Lepidoptera and Coleoptera, this is reflected in their developmental biology: the diploid cells eventually forming most parts of the winged imago are tucked away in the larval body as islands of arrested cells. They develop into the adult only after the larval clles have broken down and the ancient "blueprint" of the winged form is released during the intense rebuilding process taking place within the pupa.

REFERENCES

1. Alexander, R. McNeill. Muscle performance in locomotion and other strenuous activities. *In* "Comparative Physiology. Locomotion, Respiration, Transport and Blood". (L. Bolis, K. Schmidt-Nielsen and S. H. P. Maddrell, eds). North Holland, Amsterdam, pp. 1–21 (1973).
2. Ashurst, D. E. The fibrillar flight muscles of giant waterbugs: an electron microscope study. *J. Cell. Sci.* **2**, 435–444 (1967).

3. Bramwell, C. D. Aerodynamics of *Pteranodon*. *Biol. J. Linn. Soc.* **3**, 313–328 (1971).
4. Buchthal, F., Weis-Fogh, T. and Rosenfalck, P. Twitch contractions of isolated flight muscle of locusts. *Acta physiol. scand.* **39**, 246–276 (1957).
5. Bücher, T. Formation of the specific structural and enzymic pattern of the insect flight muscle. *In* "Aspects of Insect Biochemistry". (Goodwin, T. W., ed.), Academic Press, London and New York (1965).
6. Close, R. The relation between intrinsic speed of shortening and duration of the active state of muscle. *J. Physiol.* **180**, 542–559 (1965).
7. Greenewalt, C. H. The wings of insects and birds as mechanical oscillators. *Proc. Amer. Phil. Soc.* **104**, 605–611 (1960).
8. Greenewalt, C. H. Dimensional relationships for flying animals. *Smithson. misc. Collns.* **144** (2), 1–46 (1962).
9. Jensen, M. Biology and physics of locust flight. III. The aerodynamics of locust flight. *Phil. Trans. R. Soc. Lond.* **B 239**, 511–552 (1956).
10. Jensen, M. and Weis-Fogh, T. Biology and physics of locust flight. V. Strength and elasticity of locust cuticle. *Phil. Trans. R. Soc. Lond.* **B 245**, 137–169 (1962).
11. Lighthill, M. J. On the Weis-Fogh mechanism of lift generation. *J. Fluid Mech.* **60**, 1–17 (1973).
12. Machin, K. E. and Pringle, J. W. S. The physiology of insect fibrillar muscle. III. The effect of sinusoidal changes of length on a beetle flight muscle. *Proc. R. Soc. Lond.* **B 152**, 311–330 (1960).
13. Mayr, E., Linsley, E. G. and Usinger, R. L. "Methods and Principles of Systematic Zoology". McGraw-Hill, New York (1951).
14. Norberg, U. M. Hovering flight of *Plecotis auritus* Linnaeus. *Proc. Ind. Int. Bat Res. Cong.* 62–66 (1970).
15. Pennycuick, C. J. Power requirements for horizontal flight in the pigeon *Columba livia*. *J. exp. Biol.* **49**, 527–555 (1968).
16. Seitz, A. "The Macrolepidoptera of the World". Vol. 6, p. 840. A. Kernen, Stuttgart (1940).
17. Sotavalta, O. The essential factor regulating the wing-stroke frequency of insects in wing mutilation and loading experiments at subatmospheric pressure. *Ann. Zool. Soc. "Vanamo"*, **15**(2), 1–67 (1952).
18. Tucker, V. A. Bird metabolism during flight: evaluation of a theory. *J. exp. Biol.* **58**, 689–709 (1973).
19. Tucker, V. A. and Parrott, G. C. Aerodynamics of gliding flight in a falcon and other birds. *J. exp. Biol.* **52**, 345–367 (1970).
20. Weis-Fogh, T. Biology and physics of locust flight. II. Flight performance of the desert locust (*Schistocerca gregaria*). *Phil. Trans. Roy. Soc. Lond.* **B 239**, 459–510 (1956).
21. Weis-Fogh, T. Tetanic force and shortening in locust flight muscle. *J. exp. Biol.* **33**, 668–684 (1956).
22. Weis-Fogh, T. Elasticity in arthropod locomotion: a neglected subject, illustrated by the wing system of insects. *15th Int. Cong. Zool.* (4, 29), 393–395 (1959).
23. Weis-Fogh, T. Power in flapping flight. *In* "The Cell and the Organism" (J. A. Ramsay and V. B. Wigglesworth, eds), 283–300. London: Cambridge University Press (1961).
24. Weis-Fogh, T. Energetics of hovering flight in hummingbirds and in Drosophila. *J. exp. Biol.* **56**, 79–104 (1972).
25. Weis-Fogh, T. Quick estimates of flight fitness in hovering animals, including novel mechanisms for lift production. *J. exp. Biol.* **59**, 169–230 (1973).

26. Weis-Fogh, T. Energetics and aerodynamics of flapping flight, a synthesis. *Symp. R. Ent. Soc.* no. 7 "Insect flight", 48–72 (1976).
27. Weis-Fogh, T. and Jensen, M. Biology and physics of locust flight. I. Basic principles in insect flight. A critical review. *Phil. Trans. R. Soc. Lond.* **B 239**, 415–458 (1956).

25. Some Scale Dependent Problems in Aerial Animal Locomotion.

NIKOLAI V. KOKSHAYSKY

Academy of Sciences of the U.S.S.R., A. N. Severtzov Institute of Evolutionary Animal Morphology and Ecology, Leninsky Prospekt 33, Moscow W-71, U.S.S.R.

ABSTRACT

The relationships between various biological parameters and the size of an organism have been presented by two-dimensional "clouds of points". The particular relationship between wing area and body mass is analysed here in detail. It is shown (from the examples of butterflies and gallinaceous birds) that the place which a particular group occupies in the "cloud" is determined by a long chain of interrelated features of a morphological, fluid-mechanical, physiological and ecological nature. The concept of transposition as one of the attributes of interspecific allometry is briefly discussed. As a cursory review of the system of avian wing-tip slots shows, size-required alterations in the form of animal flight organs are often masked by effects of a different nature. A wide set of variables must be considered in these cases. Differences between flying animals and flying machines are considered briefly with regard to maximum size limitation. The power available for flight scales up in flying animals at an unfavourable rate compared with aircraft, and no appreciable power loading reduction can be achieved in the former with scaling up. In animals representing an impressive size-range and capable of prolonged periods of sustained flight the power input scales nearly as the body mass, i.e. more intensely than the power available. This relation determines a low maximum size for flying animals. A many-sided approach to the study of aerial animal locomotion is advocated.

The size-dependence of many important fluid-mechanical and biological aspects of aerial animal locomotion have been understood for a long time. More than 60 years ago there appeared extensive reviews of the scaling of bird flight by August Putter.[22, 23] These reviews correctly reflected the most general trends, though they contained many inaccuracies and mistakes.

421

Nowadays it has become conventional practice to express various flight characteristics in terms of an animal's dimensions (body mass being used most often). To stress the importance of scaling effects in aerial locomotion, suffice it to say that it is difficult to understand properly many experimental facts, for example those concerning the power required for animal flight, without having due regard to the dependence of variables on Reynolds number (Kokshaysky[10]). Indeed, a theory which takes account of the influence of Reynolds number (Tucker[27]) fits the available data on flight metabolism better than one which does not (Pennycuick[20]).

There is a close connection between the scaling of biological form and function and various aspects of similitude: geometrical, mechanical, physiological. When one analyses the general size-induced trends in the correlation of form and function over vast assemblages of animals, one inevitably focuses on their common features, i.e. on their similarity. On the other hand if one diminishes the scale of the analysis and turns to individual groups and particular adaptations, one must focus then on the special features, on the departures from similarity laws. These approaches should not be opposed to one another, I believe, as any general size-linked relationship is made up from more or less dissimilar "units".

To consider the interrelation of these two approaches and to illustrate by some typical examples how scaling effects could manifest themselves in either situation are the principal aims of this paper. It is interesting also to examine how size-induced alterations in area-to-volume ratios show their influence in a wide range of naturally and artificially evolved systems.

SCALING OF WING AREA AND THE STRUCTURE OF SIZE-DETERMINED RELATIONSHIPS

Let us consider how the wing area of flying animals scales over a wide range of body mass (this range however is far from including the limiting cases). It would not be sensible to sacrifice many animals specially for this purpose, since extensive data have been accumulated by many authors in the past. A considerable portion of this material was compiled by Greenewalt.[8] Accordingly, Fig. 1 here is adapted from Greenewalt's Fig. 3. Some small inaccuracies in the latter (e.g. the number of points for butterflies on the graph exceeds the number of corresponding items in the tables; one can also note some discrepancies with his Fig. 4) are of no importance in the present case with its great degree of schematization.

In general, Fig. 1 shows that the line with slope 2/3 fits well the relationship between wing area and body mass of the flying animals considered. As in all other analogous cases the real relationships between wing area and body mass in the given aggregation of organisms are represented by a two-dimensional "cloud of points"; the principal line is only an idealisation of the general trend. The scatter is great here, and reflects the morphological and functional differences between separate representative types of flying animal, since the scatter due to variations in individual measurements is eliminated because of the small scale of the plot. So one can certainly take

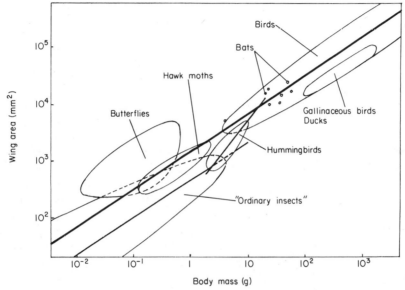

FIG. 1. Log–log plot of wing area versus body mass for insects, birds and bats based on Greene-walt's[8] Fig. 3. Only for bats are the separate points preserved, in all other cases merely general outlines of the fields occupied by the different groups are given. To gain a clearer picture the points for dragonflies have not been included in the insects' outline. The outline for gallinaceous birds was specified according to Greenewalt's Fig. 4. The principal line for flying animals as well as the line for insects in general have slopes of 2/3. The line for hummingbirds has a slope of 4/3 (values for wing areas in hummingbirds are not measured but calculated).

the difference in the positions occupied on the graph by the various groups, relative to the principal line, as real evidence of pronounced distinctions in the relative wing-size between members of the different groups. These distinctions can have far-reaching morphological, fluid-mechanical, physio-logical and ecological consequences of vital importance, as will be demon-strated below in the case of butterflies and of gallinaceous birds.

Because of their enormous wings butterflies are shifted well above the line for "ordinary insects". Compared, for example with hawk-moths, butterflies have a much smaller body (and therefore muscle) mass per unit wing area.

That alone would point to their low wing-beat frequencies and hence their ability to initiate flight at relatively low temperatures without warm-up, as the antagonistic flight muscles have time to complete their twitch during the relatively prolonged period of one stroke. The largely behavioural thermo-regulation, based on the utilization of solar energy (basking) and the conse-quent diurnal habits, are peculiar to butterflies. In contrast, hawk-moths, with their high stroke frequencies, are endotherms. Prior to flight they warm themselves up by shivering, during which the antagonistic flight muscles contract synchronously, i.e. against each other rather than on the wings (Heinrich[9]).

As butterflies' wings have very low aspect ratios (i.e. ratios of length to width), they can operate at large angles of incidence and produce considerable lift. Nachtigall's aerodynamic measurements on stationary butterflies[18] entirely corroborate this assumption. Moreover, he has shown that polar diagrams of butterfly wings are very sensitive to small variations of air speed. So slight alterations in wind velocity can produce appreciable fluctuations in lift and hence sharp changes of the insect's vertical position. Furthermore, it is unlikely that the vortex separation which must take place from butterfly wings in flapping flight is completely synchronized for both pairs of wings (because the lift coefficients are too high), and such functional asymmetry will also contribute to the insect's erratic flight.

The benefit of this familiar and apparently automatic pattern of erratic flight can be seen in the low vulnerability of butterflies to aerial predators, primarily birds. I cannot recall any bird in Middle Russia that regularly takes flying butterflies (the White Wagtail, *Motacilla alba*, often attempts to do so, but with limited success). The only bird in the fauna of the U.S.S.R. that to my knowledge depends on flying butterflies as an appreciable food source, and which I have observed in the Tajikistan, is the Paradise Fly-catcher (*Therpsiphone paradisi*) with its agile flight and (in the male) its very long tail. These observations were substantiated by Kovshar's detailed studies[13] carried out at the western outskirts of the Tien Shan mountain system, where *Argynnis pandora* proved to be the species of butterfly most often taken. Thus one may speculate that an exaggeratedly long tail, which is not infrequent among tropical birds, is not only, say, an attribute of display behaviour but simultaneously an aid to the exploitation of new sources of food.

As for gallinaceous birds, they are, as Fig. 1 shows, small-winged repre-sentatives of their Class. I wish to illustrate the flying capabilities of these birds by Popov's observations[21] on the Rock Partridge (*Alectoris graeca*), made during the relatively short seasonal movements peculiar to the species in Tajikistan. The partridges had to fly over a valley containing a river, bare ground and finally a native village. The rocky upgrade on the opposite side of the valley, where the partridges could continue on foot, did not begin until the village was passed. These autumn flights of the partridges proved to be quite regular; the natives were well aware of the event and hunted the migratory birds intensively, often only with sticks. At the approach of evening, the partridges ascended the bare hill rising above the river and set out from the top to cross the valley. The distance between the top of the hill and the safe landing spot behind the village was 2000 m, the hill's elevation above the landing place was 200 m. Only some of the birds starting the flight were able to accomplish it; others lost height faster, landed nearer the river, and were hunted on the ground. It is easy to see that if the partridges flew over the valley in gliding flight, the lift/drag ratio of the most successful birds would be as high as 10. However, they flew across actively, and therefore their performance was appreciably worse.

One can apply similar estimates to the flight capabilities of representatives of another family of gallinaceous birds. North American Ruffed Grouse (*Bonasa umbellus*) were released from a boat on a lake, and in general could reach land only if it was at least 350 m away; a distance of 460 m was once recorded, with a tail wind of about $10 \, \mathrm{m \, s^{-1}}$ (Palmer[19]). Micheev[17] has accumulated a vast amount of data, derived from his own observations and from the literature, on the Willow Grouse (*Lagopus lagopus*). The most northern populations of the species in the U.S.S.R. are migratory. Migratory flights turn out to be strictly dependent upon the winds, since the birds fly exclusively with a tail wind; a change of wind direction during the flight may have fatal consequences. The Willow Grouse of the Novosibirsky Islands do not dare to cross to the mainland until the strait has frozen up.

It is impossible to attribute the poor flying abilities of gallinaceous birds solely to their relatively small wing area: ducks and some other birds fall in the same region of the graph (Fig. 1). Apart from differences in the form of the wing, gallinaceous birds clearly have evolved an extremely specialized mode of flight, which permits them to achieve swift escape but severely limits their flight range. To this end their flight musculature is composed largely of white fibres and is capable of the immediate output of appreciable quantities of energy with a consequent oxygen debt. It must be stressed that the above dramatic examples, although very demonstrative of the flight performance of the birds, belong to situations (migrations) which are not typical of gallinaceous birds in general. It is worth mentioning that within the group the effect of scaling factors is quite appreciable: I have repeatedly observed that an undisturbed Hazel Hen (*Tetrastes bonasia*) is able to take off with much less noise and swiftness than a frightened one; the Quail (*Coturnix coturnix*), that is five times lighter than the Rock Partridge, accomplishes neary 300 km as it crosses the Black Sea on its seasonal migrations.

Let us now consider the relationship between wing area and body mass in more delicate detail. On Fig. 2 such a relationship is shown for three species of hawk. The choice of values to be compared is such that geometric similarity is closer, as the slope of the regression lines approaches unity. The effect of size-induced changes in the area-to-volume ratio on the relative dimensions of flight organs has in this case been expressed rather distinctly.

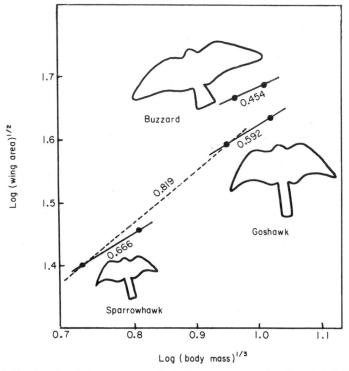

FIG. 2. Log–log plot of wing area versus body mass for three species of hawk. Solid lines represent intraspecific regressions, with indicated slopes. Points are the means for males and females. Broken line represents the interspecific regression. (After Meunier[16]).

It is seen that the intraspecific regression lines on Fig. 2 have unfavourably gentle slopes: as size increases the deficiency in wing area becomes more perceptible. If Goshawk (*Accipiter gentilis*) were scaled according to the Sparrowhawk's (*A. nisus*) regression line it could scarcely fly, having too high a wing loading. In the course of evolution, however, the situation has been moderated by a shift of the regression lines. As a result the slope of the interspecific regression line turns out to be more acceptable for the larger species. Karl Meunier, from whose paper[16]. Fig. 2 has been taken with only minor modifications, had designated such size-required regression shifts as transpositions. The concept of transposition can strictly be applied only to geometrically and functionally similar forms, eg. Sparrowhawk and Goshawk but not, for example, to Buzzard (*Buteo buteo*). Indeed, in the last case many more variables may be involved and more complex alterations are to be expected. Examples of analogous transpositions (for example in the graphs of wing loading versus cube root of the body mass) can be found in Gladkov's book.[5]

SCALING OF WING FORM

Size-required changes of relative dimensions of flight organs are, of course, not the only manifestation of scaling effects upon their morphology. More-over, such size-required allometry (an admirable review of its evolutionary aspects was given by Gould[6]) is hardly imaginable without accompanying changes in the shape of the relevant biological structures. There are in-numerable aspects of the scaling of wing shape in flying animals. I wish to consider only one, scaling effects on the emargination of a bird's wing tip, as it fairly well illustrates the complexity of the situation in general.

The system of slots in a bird wing certainly has no vital importance for the realization of flight itself (bats fly with unemarginated wings), but its develop-ment in different ways helps substantially to adjust the structure of the wing for particular types of flight in various birds. One can observe separation of the primary wing feathers in very disparate flight situations (Fig. 3), where it can evidently play different roles. It has been assumed that in birds which soar over land, the characteristically separated primaries have the effect of reducing the induced drag on the wing, and so neutralizing the unfavourable effect of low geometric aspect ratio in these birds (Cone[4]). This assumption is based on Cone's finding that branched wing-tips improve lift/induced drag ratios.[3] Separation of primaries can also often be seen during the downstroke of flapping birds and during the upstroke of slow flying or hovering birds. It is worth noting that the bending of primaries which can be seen on careful examination of filmed upstrokes, suggests, contrary to the adopted view and in support of Bilo,[1] that in ordinary flight the wing can produce some thrust during the upstroke. It seems also that some thrust may sometimes be produced during the upstroke of hovering passerines.

The twofold function of the system of separated primaries in flapping flight was first suggested by Graham:[7] it can withstand large angles of incidence without stall, like any multislotted wing, and it can enhance the effective twist of the wing at its tip by twisting the feathers separately. It can be seen from Fig. 4 that, during the downstroke, the maximum angles of incidence occur at the wing tip, and these angles are particularly enhanced when overall twist of the wing is limited, say on structural grounds. Thus it seems probable that the main function of separated primaries in flapping flight is an increase in the effective angle of incidence without stall, rather than a reduction in induced drag. Increase in circumferential wing velocity

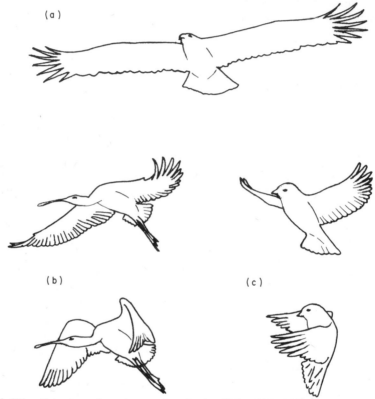

FIG. 3. Wing-tip patterns in various modes of avian flight. (A) Soaring Golden Eagle (*Aquila chrysaetos*); outlined from a photograph taken by the author. (B) Separate stages of downstroke (above) and upstroke (below) in a Spoonbill (*Platalea leucorodia*) in slow flight; from a 16-mm film by the author. (C) Early stages of downstroke (above) and upstroke (below) in hovering Pied Flycatcher (*Ficedula hypoleuca*); from separate successions of 1000 frames s^{-1} 35 mm film shot under the author's direction.

and in wing length will lead in principle to an increase of the angle of incidence at the tip, while shortening of the wing will lead to a decrease. But the real situation is more complicated. If in the course of evolution the wing becomes shorter, the flapping rate usually increases and the wing width also increases with the result that its twist decreases. Thus the final effect is not easy to assess, and kinematic measurements are needed in most cases.

To trace the effects of scale on the wing-tip slot system, one ought to choose an array of different sized species in which geometrical shape, functional characteristics and the ecological role of flight remain as similar as possible. It was shown (Kokshaysky[12]) in four open-habitat species of herons (mean body mass from 0·237 to 1·620 kg) that an increase in both relative and absolute wing length, although opposed by a decrease in the flapping rate,

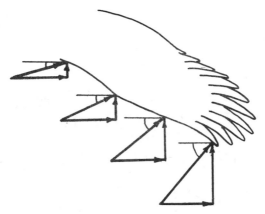

FIG. 4. Schematic distribution of angles of incidence along a bird wind during the downstroke. At any wing section the angle is measured between the sections' chord and the direction of the resultant air velocity, the latter being the vector sum of translatory and circumferential wing velocities.

produces an appreciable increase in values of wing-tip circumferential velocities and hence in angles of incidence. In accordance with these findings the number of slots at the distal part of the wing increases with size (from smaller to larger species the number of slots is: 0, 1, 3, 4). But the case of herons (apart from closed-habitat species) is a special one because of their uniform morphological and ecological specialization. A more typical situation is represented for example by the crows with their diverse specialization. Let us consider the dimensions of three familiar species of crow (the data given are means, mainly for Middle Russian populations):

		Mass (kg)	Wing-span (m)
Carrion Crow	(*Corvus corone*)	0·55	0·95
Jackdaw	(*Coloeus monedula*)	0·19	0·70
Jay	(*Garrulus glandarius*)	0·16	0·56

By analogy with the herons one might expect that the number of wing tip slots will decrease from Crow to Jay. But in fact this holds only from Crow to Jackdaw; the number of slots in Jay is the same as in Crow (Fig. 5). The

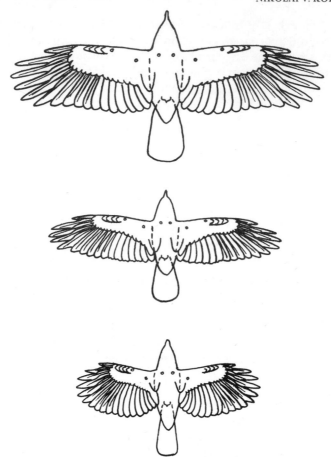

FIG. 5. General outlines of three representatives of crow family; from top to bottom: Carrion Crow, Jackdaw, Jay. Redrawn to scale from Stegmann.[25]

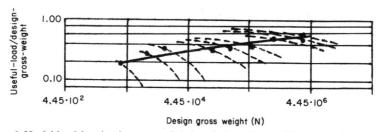

FIG. 6. Useful load fraction in commercial aircraft, from Wright Flyer to Boeing 747 and Lockheed C-5A "Galaxy". Actual values are shown by points for each aircraft (including those not specified). Hatched curves ("isotechnology" lines) represent square/cube law projections to both higher and lower design weights. The solid line is the median of the actual values. (After Cleveland [2]).

disorder is easy to explain: Crow and Jackdaw are open-habitat birds with more or less similar requirements on flight, while Jay is a forest bird.

The situations in which the effects of scaling on various aspects of wing form, or the form of other structures, are masked by other sharp distinctions caused by differences in specialization, are by no means rare. This naturally hampers the investigation of scaling effects on organic form and calls for versatile studies.

AIRCRAFT AND FLYING ANIMALS: POWER FOR FLIGHT AND SIZE LIMITATIONS

Any object, animate or inanimate, will if scaled up demonstrate a continuous decrease in area-to-volume ratio; in aircraft design this has been expressively named the "square/cube law". It implies that wing loading, and also stress, increase with size in similar structures. One may immediately infer from this that the tendency to gigantism is quite unprofitable for aircraft, as with continual scaling up the useful load fraction must eventually disappear. But that is not the case. Figure 6 is redrawn with slight modifications from Cleveland's paper[2] and shows that, in an array of nine real aircraft, from the Wright Flyer to the Lockheed C-5A "Galaxy", the useful-load/design-gross-weight ratio grows significantly with the aircraft's general size (and with the passage of time!). Is the square/cube law therefore invalid? Not at all. The decrease in value of the useful load fraction for enlarged versions of each aircraft considered, calculated by extrapolation from the square/cube law, is quite obvious. Historical progress, expressed as a continuous divergence of the individual points on Fig. 6, is determined by a wide range of factors (improvements in technology, systems etc.) and means continual departures from strict similarity. Suffice it to say that aircraft wing loading has increased a hundredfold between 1903 and 1970; this is much more than even the square/cube law requires. But it is of key importance that the concomitant power–loading reduction can be clearly demonstrated.

As to flying animals, all disadvantages connected with scaling up and dictated by the square/cube law are clearly visible. Long fast take-off runs to get off the water or level ground, dramatic shrinkage of the range of airspeeds which can be achieved, until only one speed is accessible, adaptation to different kinds of soaring—all this is typical for large birds. It seems that the upper limit of size for a bird capable of active flight is reached with a mass of nearly 12 kg, which is insignificant compared with the mass attained by terrestrial and, especially, aquatic animals. It is not easy in flying creatures to find something analogous to the continual departures from similarity which permit increasingly large aircraft to be developed. Transpositions of

wing area versus body mass of the type shown in Fig. 2 barely allow interspecific similarity to be maintained, and do not permit the deviations
required to mitigate size-induced difficulties. The main factor distinguishing
birds from aircraft lies in the deficiency of relative power, which grows as
the size increases.

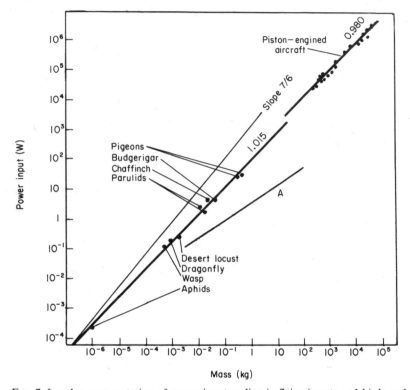

FIG. 7. Log–log representation of power input scaling in flying insects and birds and piston-
engined aircraft. Modified after Kokshaysky,[11] where (as well as in Ref. 10) the data are discussed in detail. Slopes of regression lines for animals and aircraft are marked on them. The line
for standard metabolic rate in nonpasserine birds with the slope of about 3/4 is also shown (A).

How does the flight power input (i.e. the total flight metabolic rate) scale
in animals? By now many measurements of active metabolic rate have been
made on birds and insects. But the conditions in which these measurements
are made are not always comparable with those of natural flight. So regression
lines fitted to such data (e.g. the one for insects presented by Sotavalta and
Laulajainen[24]) are hardly acceptable. It is of primary importance to
compare data only for animals that are in a similar state of flight activity (for

example, we should not compare those in prolonged sustained flight with those performing a forced take-off). If this requirement is satisfied it is probably permissible to calculate a common regression for such different organisms as insects and birds. An attempt of this sort, using data selected from the literature and pertaining to conditions of prolonged flight (Kokshaysky[10]), gave a slope of unity for the insect-bird regression line (Fig. 7): a result long anticipated by Weis-Fogh.[28] More recent measurements fit the regression reasonably well. For comparison the regression line of standard metabolic rate in nonpasserine birds (Lasiewski and Dawson[14]) has been marked on the graph. If the power available for flight scales up in a similar way (with a slope of the regression line of about 3/4), the energetic basis for size limitation in flying animals is obvious: it is a discrepancy between the rate at which the power required and the power available for flight scale up. The shift of the line for power required from a slope of 7/6 to one of nearly unity can be considered as a result of the influence of Reynolds number.

The total power input in aircraft, calculated on the basis of fuel consumption, varies with overall mass in a manner similar to that in animals. But to compare the effectiveness of flight in both cases one must also take into account the flying speeds. From elementary relations for flying animals or machines it is easy to obtain: $mg\ V/P = \eta L/D = K'$, where mg is weight, V = flight velocity, P = total power input, L = lift, D = drag, η = overall efficiency of the system.[10] It is not hard to see also that K' is a dimensionless number reciprocal to the often-used "caloric cost of transport" (Tucker[26]). As Fig. 8 shows, K' is a scalable parameter, both in aircraft and in flying animals, and its value becomes more favourable as size increases. So one might expect that if it were possible to increase the size of the animal's

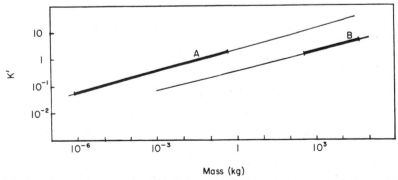

FIG. 8. Scaling of the parameter K' in flying animals (A) and piston-engined aircraft (B). Calculated from the data in Kokshaysky.[10] Thick parts of the lines correspond to the ranges of mass considered in deriving the lines.

flapping-wing system up to that of some modern aircraft, it would demonstrate superior efficiency. But that is hard to believe for many reasons. It is most likely that we would encounter some sort of regression line transposition, due, say, to the enormous growth of inertia forces as the flapping wing is scaled up.

CONCLUSIONS

Studies of the effects of scaling upon various biological functions will clearly develop further. Such studies presuppose a great degree of generalization, so the reverse side is worth emphasizing. Real relationships between various biological parameters and size (i.e. characteristic length, mass etc.) usually show some departure from the general regression line; the extent of this departure is ultimately a manifestation of the overall adaptability to diverse environmental conditions in the array of forms studied. Size-required alterations of biological form are often masked by effects of another nature, and therefore a many-sided approach is needed. It seems that this refers especially to various aspects of aerial animal locomotion, with its numerous unsolved problems and in which the profound study of some particular mode of flight (e.g. Weis-Fogh[29]) may serve as a basis for fluid-mechanical generalizations (Lighthill[15]) of fundamental importance.

REFERENCES

1. Bilo, D. Flugbiophysik von Kleinvögeln. II. Kinematik und Aerodynamik des Flügelaufschlages beim Haussperling (*Passer domesticus* L.) *Z. vergl. Physiol.* **76**, 426–437 (1972).
2. Cleveland, F. A. Size effects in conventional aircraft design. *J. Aircraft.* **7**, 483–512 (1970).
3. Cone, C. D., Jr. The theory of induced lift and minimum induced drag of non-planar lifting systems. *NASA techn. Rep.* R-139, 1–31 (1962).
4. Cone, C. D., Jr. The design of sailplanes for optimum thermal soaring performance. *NASA techn. Note* D-2052, 1–48 (1964).
5. Gladkov, N. A. "The Biological Foundations of Bird Flight". (Russ.). Publ. by Moscow Soc. of Naturalists, Moscow (1949).
6. Gould, S. J. Allometry and size in ontogeny and phylogeny. *Biol. Rev.* **41**, 587–640 (1966).
7. Graham, R. R. Safety devices in wings of birds. *J. Roy. Aeronaut. Soc.* **36**, 24–58 (1932).
8. Greenewalt, C. H. Dimensional relationships for flying animals. *Smithson. misc. Collns.* **144**, no. 2, 1–46 (1962).
9. Heinrich, B. Mechanisms of insect thermoregulation. *In* "Effects of Temperature on Ectothermic Organisms" (Wieser, W. ed.). Springer-Verlag; Berlin, Heidelberg, N.Y. (1973).

10. Kokshaysky, N. V. Flight energetics of insects and birds. *Zh. Obshch. Biol.* (Russ., Engl. summary) **31**, 527–549 (1970).
11. Kokshaysky, N. V. On energetic efficiency of flight in animals. *In* "The Problems of Bionics". (Russ.; Gaase-Rapoport, M. G. and Kokshaysky, N. V., eds). "Nauka" Publs., Moscow (1973).
12. Kokshaysky, N. V. Functional aspects of some details of bird wing configuration. *Syst. Zool.* **22**, 442–450 (1973).
13. Kovshar, A. F. "The Birds of Talassky Alatau". Trudy zapovednikov Kazahstana (Russ.) **1**, Alma-Ata (1966).
14. Lasiewski, R. C. and Dawson, W. R. A re-examination of the relation between standard metabolic rate and body weight in birds. *Condor*, **69**, 13–23 (1967).
15. Lighthill, M. J. On the Weis-Fogh mechanism of lift generation. *J. Fluid Mech.* **60**, 1–17 (1973).
16. Meunier, K. Die Grossenabhängigkeit der Körperform bei Vögeln. *Z. wiss. Zool.* **162**, 328–355 (1959).
17. Micheev, A. V. "The Willow Grouse". (Russ.). Moscow (1948).
18. Nachtigall, W. Aerodynamische Messungen am Tragflügelsystem segelnder Schmetterlinge. *Z. vergl. Physiol.* **54**, 210–231 (1967).
19. Palmer, W. L. Ruffed grouse flight capability over water. *J. Wildl. Manag.* **26**, 338–339 (1962).
20. Pennycuick, C. J. The mechanics of bird migration. *Ibis*, **111**, 525–556 (1969).
21. Popov, A. V. Migrations of Rock Partridge in Tajikistan. *In* "Trudy problemnych i tematicheskich sovestchanij ZIN" (Russ.) **9**, Acad. Sci. USSR, Leningrad–Moscow (1960).
22. Pütter, A. Die Leistungen der Vögel im Fluge. *Naturwiss.* **2**, 701–705, 725–729 (1914).
23. Pütter, A. Vogel und Flugzeug. (Ein Vergleich). *Naturwiss.* **2**, 861–865 (1914).
24. Sotavalta, O. and Laulajainen, E. On the sugar consumption of the drone fly (*Eristalis tenax* L.) in flight experiments. *Annls. Acad. Sci. fenn.* A (IV Biol.) **53**, 1–25 (1961).
25. Stegmann, B. K. Peculiarities of flying properties of corvine birds. *Zoologitsh. Zh.* (Russ.) **33**, 653–668 (1954).
26. Tucker, V. A. Energetic cost of locomotion in animals. *Comp. Biochem. Physiol.* **34**, 841–846 (1970).
27. Tucker, V. A. Bird metabolism during flight: Evaluation of a theory. *J. Exp. Biol.* **58**, 689–709 (1973).
28. Weis-Fogh, T. Power in flapping flight. *In* "The Cell and the Organism" (Ramsay, J. A. and Wigglesworth, V. B., eds). Cambridge University Press, London and New York (1961).
29. Weis-Fogh, T. Quick estimates of flight fitness in hovering animals, including novel mechanisms for lift production. *J. Exp. Biol.* **59**. 169–230 (1973).

26. The Intermittent Flight of Birds.

JEREMY RAYNER

Department of Applied Mathematics and Theoretical Physics, University of Cambridge, England.

Perhaps I should begin by stating in my own defence that this project was included in Part III of the Cambridge Mathematical Tripos last year, with the result that the time available was limited, and the physiological aspects of the findings were rather beyond the means of a mere mathematician!

We begin by making the assumption that a bird's main consideration on a long flight is to reduce its total fuel consumption, and that to achieve this it adopts a particular flight strategy; the "V" pattern of groups of migratory geese is a familiar example of such a strategy, and "intermittent" flight might also arise from this requirement. By "intermittent" flight, I refer to regular motion in a mean horizontal plane in which periods of wing flapping alternate with periods of idling. This can be further qualified:*

Undulating flight (Fig. 1(a)) in which the bird glides while unpowered and uses a few wing strokes to regain height; the mode is similar to that examined for fish by Weihs[4] and is typical of those birds larger than woodpeckers which glide well (gulls, crows, most birds of prey, etc.). In observation it is difficult to distinguish from static soaring, but we assume the air to be still.

Bounding flight (Fig. 1(b)) where the wings are folded during the unpowered phase, typical of many birds smaller than (and including) woodpeckers. This is modelled mathematically by assuming flapping on a circular arc followed by motion as a classical projectile neglecting lift and drag. While the bird might wish to save energy on a long flight (and this consideration encourages the adoption of gliding and soaring by migrating eagles, herons and cranes) it is not clear why this criterion alone should dictate strategy on a short flight, where the energy saved would be negligible. Furthermore, birds can achieve better performance (by some criteria) in

---

* The distinction between the two flight modes appears to have been first made by Vasil'ev in 1953[5].

437

continuous direct flight, for when startled the chaffinch will fly steadily at a high velocity until (presumably) it feels itself safe. However, assuming energy/fuel saving to be important, we attempt to estimate the percentage improvement as a function of the shape and speed of the path in each of the two modes.

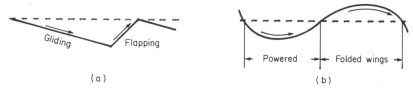

FIG. 1. Sketches of two possible modes of intermittent flight: (a) undulating flight, (b) bounding flight.

We shall need several intuitively-acceptable assumptions which are not detailed here, and in addition:

(1)   data: dimensions and sizes of birds (Greenewalt[1]);
(2)   data: aerodynamic constants etc., crudely estimated;
(3)   external power, P_r, required in level steady flight at speed u. After Pennycuick[3] a reasonable expression was derived for this:

$$P_r = D_1 u^3 + \frac{D_2 L^2}{u} + P_p,\tag{1}$$

where the three terms on the right hand side are the amounts of power required to overcome form drag, induced drag and profile drag respectively (L is the lift force — mg is steady level flight, and D_1, D_2 are known dimensional constants). The profile power P_p, expressing the rate of working against profile drag (essentially due to fluctuations in induced forces and their interaction with the bird's wake), has been shown by Pennycuick to have a constant value

$$P_p = \frac{4X}{3^{3/4}} (D_1 D_2^3 L^6)^{1/4}\tag{2}$$

where X is a constant whose value is probably about 1·2 but possibly as large as 2, the profile power ratio. It is hard to find a sound aerodynamic justification for the presence of a constant profile power, but it is included here as part of the only complete theory available. The three terms in (1) are plotted against airspeed in Fig. 2.

We define the *maximum range speed* u_1 as that which minimises fuel consumption over a given distance, and assume that this represents the optimum state of continuous steady flight (Fig. 3).

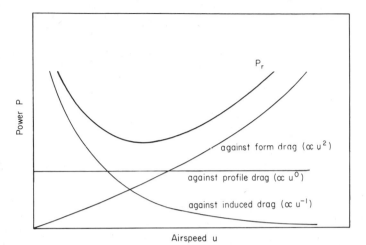

FIG. 2. Power required in level steady flight (after Pennycuick[3]).

If we reduce the various terms in the expression for P_r to allometric form with respect to the body mass we see that:

$$D_1 \propto m^{2/3}, \qquad D_2 \propto m^{-2/3}, \qquad L \propto m$$

and hence $P_p \propto m^{7/6}$. Thus at some characteristic speed $u \propto m^{1/6}$, we find the familiar result $P_r \propto m^{7/6}$. For later reference, note that

$$P_r(u_1) \approx 4\cdot78\,(D_1 D_2^3 L^6)^{1/4} \quad \text{when} \quad X = 1\cdot2. \tag{3}$$

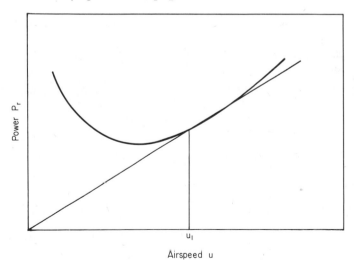

FIG. 3. Maximum range speed u_1 (minimum of P_r/u).

Also, steady flight at angle θ above the horizontal requires power

$$P_r(u, \theta) = D_1 u^3 + \frac{D_2 L^2}{u} + P_p + mgu \sin \theta \qquad (4)$$

where $L = mg \cos \theta$.

We calculate energy saving S from:

$$S = 1 - \frac{\text{Energy in bounding/undulating flight}}{\text{Energy in steady level flight at } u_1} \qquad (5)$$

and maximise S as a function of the flight path, obtaining the following results.

UNDULATING FLIGHT: 7 DIFFERENT BIRDS

$S = 52$–60% with a very flat maximum as path-parameters vary; even allowing for sizeable errors in data this is convincing. Including internal metabolic energy (data from Lasiewski and Dawson[2]) and assuming constant muscular efficiency of 40%:

$$S = \begin{cases} 47\text{--}54\% \text{ for non-passerines} \\ 14\text{--}29\% \text{ for passerines} \end{cases} \text{once again with a flat maximum}$$

Some of the larger birds might become anaerobic during the (relatively short) powered phase, recovering while gliding.

BOUNDING FLIGHT: THE CHAFFINCH*

The optimum flight mode has $S = 35\%$ with effective air-speed of 15 m s^{-1}, compared with $u_1 = 11 \cdot 6 \text{ m s}^{-1}$. Including metabolism $S = 25\%$; it is probably not legitimate to neglect lift and drag, as we can hardly expect both energy saving and increased speed!

To arrive this far we have made the tacit assumption that the bird can continuously vary its flight speed, and that the cost to the bird may be measured simply by the aerodynamic work done by the wings divided by a muscular efficiency, plus some (small) contribution from internal metabolic requirements. Further, this efficiency, η, would have to be relatively constant over a range of flight speeds. Previous papers have underlined the difficulties faced at this point in reconciling our knowledge of aerodynamic work and metabolic work and the problem of understanding the chemical efficiency of

---

* It is a feature of the model that the energy saving and the optimum path are independent of the animal concerned.

the muscles. Realising these problems, I shall attempt to show how this imbalance might in fact be the reason for intermittent flight.

A bird's pectoralis major muscle is composed solely of slow muscle, and we would therefore expect the efficiency η to depend heavily on contraction frequency, f, which is itself related to the air speed. It is readily observed that many small birds can only use one wing-beat frequency, and do not—and perhaps cannot—vary the pattern of their wing movements. We can thus deduce (although perhaps not rigorously) that $\eta(f)$ has a sharp peak at this value of $f = f_m$. The greatest strain imposed on the wing-system normally occurs at take-off, and therefore the aerodynamic power output at frequency f must be large enough to take account of the initial acceleration. (Those birds which cannot take off from a level surface may be considered as a special case.) Professor Weis-Fogh has shown how the maximum power available, P_a, from avian muscle is directly proportional to muscle mass m',

$$P_a = 200 \text{ W/Kg muscle mass}$$
$$= 200 \ m' \text{ W} \tag{6}$$

and we can reasonably assume that this is the aerodynamic power output at the optimum frequency f. Although similar frequencies might correspond to similar external rates of work, the actual cost to the bird would be much greater because of the low muscular efficiency.

By substituting the appropriate forms of D_1, D_2 and L into Eqn (3) we obtain an expression for the power required in sustained steady flight at the optimum velocity u_1,

$$P_r = 36 \ m^{5/3} \ b^{-3/2} \text{ W} \tag{7}$$

when $X = 1.2$, where m is the bird's mass in kg and b is the wing span in m. We have estimated the power for take-off as $P_a = 200m'$ W, and intuitively this will be rather larger than P_r; if the bird cannot work at rate P_r without an excessively low muscular efficiency we can see that a compromise may be to work at rate P_a for some fraction τ of the total flight time, given by

$$\tau = \frac{P_r}{P_a} \tag{8}$$
$$= 0.18 \ m^{5/3} \ b^{-3/2} \ m'^{-1}$$
$$\text{i.e. } \tau \propto m^{1/6}. \tag{9}$$

In view of the poor gliding capacities of small birds we would expect them to fold their wings during the unpowered portion of their flight, and to rely on the momentum gained during the powered phase. Larger species would glide if they were affected by a small value of τ, but are more likely to be able to vary their wing-geometry so as to avoid these complications, and undulating flight can only be tentatively justified by muscle-efficiency considerations. τ has been evaluated from Eqn (9) using data from Greenewalt,[1] with the results shown in Table I, from which we see, in particular, that $0.3 < \tau < 0.5$ for lapwing, tern, cuckoo, nightjar, swift, lark, swallow, martin, wagtail, and $0.5 < \tau < 0.7$ for starling, heron, curlew, sparrowhawk, gull, pigeon, nightingale. These conclusions are in striking agreement with observations of the birds concerned. Few birds which regularly use

TABLE I.

Bird		m (kg)	m' (kg)	b (m)	τ
Diomedea exulans	Wandering albatross	8·502	1·097	3·41	0·94
Ardea cinerea	Grey heron	1·408	0·233	1·73	0·62
Cygnus cygnus	Whooper swan	5·925	0·954	2·30	1·09
Anser anser	Greylag goose	3·065	0·630	1·63	0·91
Anas platyrhyncos	Mallard	1·105	0·248	0·90	1·04
Gyps fulvus	Griffon vulture	7·269	1·018	2·56	1·19
Aquila chrysaetos	Golden eagle	3·712	0·501	2·12	1·06
Buteo buteo	Buzzard	1·027	0·124	1·32	1·02
Accipiter nisus	Sparrowhawk	0·221	0·047	0·75	0·50
Falco tinnunculus	Kestrel	0·245	0·030	0·74	0·94
Crex crex	Corncrake	0·155	0·028	0·48	0·90
Vanellus vanellus	Lapwing	0·211	0·0481	0·75	0·45
Haematopus ostralegus	Oystercatcher	0·438	0·074	0·81	0·88
Numenius arquata	Curlew	0·768	0·163	1·04	0·69
Larus canus	Common gull	0·367	0·0506	1·08	0·62
Sterna hirundo	Common tern	0·118	0·021	0·83	0·28
Columba palumbus	Woodpigeon	0·495	0·137	0·75	0·65
Cuculus canorus	Cuckoo	0·104	0·021	0·58	0·46
Asio flammeus	Shorteared owl	0·390	0·049	1·08	0·71
Caprimulgus europaeus	Nightjar	0·092	0·018	0·57	0·45
Apus apus	Swift	0·0362	0·0076	0·42	0·36
Alauda arvensis	Skylark	0·028	0·0066	0·317	0·41
Hirundo rustica	Swallow	0·0184	0·00373	0·330	0·33
Delichon urbica	House martin	0·01435	0·00207	0·292	0·49
Motacilla cinerea	Grey wagtail	0·016	0·00399	0·252	0·37
Sturnus vulgaris	Starling	0·080	0·0165	0·391	0·65
Garrulus glandarius	Jay	0·160	0·02087	0·54	1·07
Corvus corone	Crow	0·470	0·06579	0·89	0·95
Troglodytes troglodytes	Wren	0·0101	0·00130	0·196	0·98
Sylvia atricapilla	Blackcap	0·01625	0·00198	0·238	0·85
Iuscinia megaryncha	Nightingale	0·0171	0·00264	0·255	0·62
Turdus merula	Blackbird	0·0915	0·01635	0·406	0·81
Parus major	Great tit	0·02145	0·003	0·233	0·92
Passer domesticus	House sparrow	0·030	0·00534	0·252	0·79
Fringilla coelebs	Chaffinch	0·02115	0·00535	0·285	0·37

"bounding" flight are omitted, and those in Table I which do not are all large enough to be unconcerned with limited efficiency.

An arbitrary "cut-off" point at $\tau = 0{\cdot}7$ has been chosen, as it is unlikely that a higher powered proportion of the total flight would be worthwhile. We could use this to derive a critical mass below which intermittent flight would be relevant and above which the bird would have to compromise with the best rate of working posible; however, there is not sufficient information in Table I to show the expected allometric equation $\tau \propto m^{1/6}$, and indeed some entries appear to contradict it. The small deviations on the usual $m' \propto m$ and $b \propto m^{1/3}$ laws might in this case be important, but further computation on a wider set of data will clarify the point.

We conclude, then, that bounding flight occurs as a result of an imbalance between available power output (which is limited by muscular efficiency) and the optimum power output for steady flight. By adopting such a mode the bird also benefits from a saving in energy, but at the expense of a reduction in flight speed. Larger birds might use intermittent flight for the same reason, but would have to glide to sustain height; as size increases the muscle geometry becomes more flexible and the problem of restricted energy is less significant. The question of intermittent flight depends as much on the geometry and structure of an individual species as on the laws of scaling. Much remains to be done to clarify several of the points raised here; in particular more information on the $\eta(f)$ relationship is needed.

ACKNOWLEDGEMENTS

The author would like to thank the many people who have helped and encouraged him, and in particular, Dr R. H. J. Brown, Professor Sir James Lighthill, F. R. S., Dr K. E. Machin, Professor C. J. Pennycuick and Professor Torkel Weis-Fogh.

REFERENCES

1. Greenewalt, C. H. Dimensional relationships for flying animals. *Smithson misc. collns.* **144**, 1–46 (1962).
2. Lasiewski, R. C. and Dawson, W. R. A re-examination of the relation between standard metabolic rate and body weight in birds. *Condor*, **59**, 13–23 (1967).
3. Pennycuick, C. J. The mechanics of bird migration. *Ibis*, **111**, 525–556 (1969).
4. Weihs, D. Mechanically efficient swimming techniques for fish with negative buoyancy. *J. Marine Res.* **31**, 194–209 (1973).
5. Vasil'ev, G. S. "Principles of Flight Models with Flapping Wings" (in Russian) (1953).

27. Model Tests on a Wing Section of an Aeschna Dragonfly.

B. G. NEWMAN
Department of Mechanical Engineering, McGill University, Montreal, Canada.

S. B. SAVAGE AND D. SCHOUELLA
Faculty of Engineering, McGill University, Montreal, Canada

ABSTRACT

The chord Reynolds number of a dragonfly wing flying at high speed is of the order of 10^4. It is well known that aerofoils at moderate incidence experience an increase of drag coefficient and a decrease of lift coefficient as the Reynolds number is reduced below a critical value of about 5×10^4 and that this is associated with complete separation of the laminar boundary layer.

The aerofoil section of the dragonfly is unusual, as indicated by Fig. 1a which is a sketch of a photograph of the section just inboard of the nodus on the wing of *Aeschna Eremita*.

Attention is drawn to the minute saw-teeth on each web of the T section which forms the leading edge (costa), and the spurs on both sides of the matrix which supports the rear membrane. The saw-teeth very likely act as turbulators to promote transition in the separated shear layer and subsequent reattachment of the boundary layer in a turbulent state. Smoke tunnel studies of the wing section support this hypothesis. Trapped vortices are observed in the V sections at low incidence, and at high incidence (about 10°) the flow is observed to reattach to the rear cambered membrane; the leading-edge separation bubble is then greater than half a chord in length. The purpose of the spurs which are roughly the height of the viscous sub-layer for the reattached flow, is less obvious.

The average flapping frequency in forward flight (about 10 m s^{-1}) is roughly 25 Hz (Nachtigall[12]) so that in one cycle the wing moves forward about 40 chord lengths. It is therefore postulated that the aerodynamics may be usefully studied, at least initially, on a static wing in steady flow.

Indoor free-flight tests have been made on two sizes of model glider with a similar wing section. Under stroboscopic illumination flights have been photographed using a long time exposure. With reflectors placed on the model and above the horizontal floor the flight altitude, flight-path angle and speed have been measured in steady flights for various trim conditions. From this data the lift coefficients and drag coefficients as a function of angle of attack and Reynolds number may be estimated. The results show the dragonfly wing section to be very efficient and comparable to a very-high-performance, low-Reynolds-number aerofoil, fitted with an artificial turbulator.

INTRODUCTION

In his book "Structure-Form-Movement" Hertel[7] describes the membranous wings of a large dragonfly (*Aeschna Cyanea*). He comments on the pleated structure near the leading edge which is composed of strong veins at top and bottom joined by thin membranes (Fig. 1(a)). He notes that the chord Reynolds number of *Aeschna* flight (Re = $U_\infty c/v$, where U_∞ is the flight speed, c the chord length and v is the kinematic viscosity of air), which is typically $0.5–1.5 \times 10^4$, conventional aerofoils become less efficient than curved plates because the laminar boundary layer separates completely. It is then very advantageous to promote transition from laminar to turbulent flow in the boundary layer by means of trip wires or other devices. The dragonfly wing probably does this by forcing separation near the leading edge. Transition occurs much sooner in a separated shear layer, which then reattaches well before the trailing edge as a turbulent boundary layer: in this form it is capable of advacing against a higher pressure rise (higher by a factor of about 10) before separating again. Thus maximum lift is increased and form drag is reduced. Hertel also comments on other features of the wing: the T shaped leading edge (costa) with three rows of serrations which may act as turbulators to promote transition still further in the separated shear layer; the hairlets at the base of each serration; the small spurs which are scattered on the polygonal matrix of veins that form the rear 2/3 of the wing; and also on the chordwise veins near the leading edge.

The present investigation was undertaken to obtain quantitative information on the aerodynamics of the dragonfly wing and to determine whether some of these particular wing features have any aerodynamic significance. The broad aim is to improve the design of aerofoils at Reynolds numbers of about 10^4, where there are applications to model aircraft, and to low Reynolds number turbines and compressors (Charwat[3]). A wing section

FIG. 1(a). Sketch of dragonfly wing section.

Fig. 1(b). Chosen dragonfly wing section.

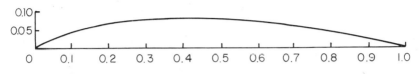

Fig. 1(c). McBride B-7 aerofoil section.

Fig. 2. McBridge B-7 and dragonfly smoke-tunnel models.

just inboard of the nodus (see Fig. 1b and Needham and Westfall[13] for a description of the various parts of the wing) of the fore wing of *Aeschna Interrupta* was chosen, since it was similar to the section of *Aeschna Cyanea* shown by Hertel. The wing was sectioned and scaled off by Paul Hayward from photographs which he took of a fairly old specimen (about 10 years old). A fresher specimen was not available at that time, but wings preserve well. This section was then used for flow visualisation on a similar tunnel model (Fig. 2) and for the wings of a model glider of aspect ratio 8 (Fig. 3).

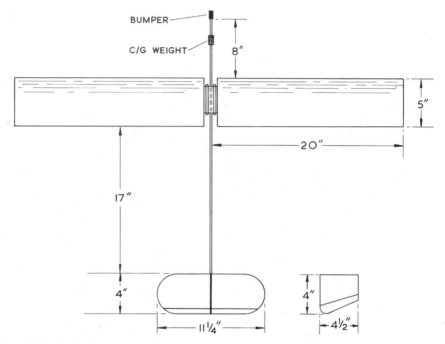

FIG. 3. Planform of large model glider (small model half-size but tail surfaces have rectangular planform). 1″ = 2·54 cm.

This glider was flown indoors using stroboscopic illumination and photography with long exposure to give flight speed, altitude and angle of attack. In addition, pictures of the detailed wing features of an Aeschnida and Libellulida dragonfly have been taken with a scanning electron miscroscope. Comparative pictures of a damselfly (a *Coenagrion*) have also been taken.

From an entomological viewpoint the above approach may seem naive, for Odonata are wonderfully complex. They have strong muscles which directly and independently flap the fore and hind wings, and weaker muscles which control the relative positions of the main structural spars (costa, subcosta, radius, media and cubitus) (Pringle[17]). Thus the angle of attack, twist and camber are controlled to some extent from the root. Unlike birds there are no muscles within the wing itself. However, the main ribs must be fed with blood to supply nutrient to the various sensors which detect structural stretch, or the movement or rate of movement of hairs and diaphragms. Moreover, the blood supply can probably be greatly increased in order to straighten the ribs after they have been accidentally bent, in much the same manner as the wing was originally deployed when the full-grown nymph (instar) blossomed into the dragonfly.* Wing shape, therefore, depends on muscle control and internal fluid pressure in the rib, as well as the aerodynamic load on the wing. In addition the inertia loads due to flapping change the wing geometry (Weis-Fogh and Jensen[22]). In this connection the build-up of chitin on the leading edge near the wing tip (pterostigma) is no doubt significant as a mass which changes the inertia loading (Norberg[15]). The carboniferous relatives of the dragonfly which had a span of almost 0·75 m were apparently gliders, and both the nodus and the stigma were absent (Corbet, Longfield and Moore[5]).

The wing shape of the living insect may therefore be different from that of a dead specimen. The remarkable photograph of an *Aeschna* landing, which is presented by Nachtigall,[12] indicates that the camber in flight may be somewhat less than that of the sectioned wing of the dead insect. Thus the geometry we chose may be too highly cambered, and moreover it is specific to a particular spanwise position.

The flapping frequency is about 25 Hz at typical fast forward speeds of $10 \, \text{m s}^{-1}$ (a maximum of $15 \, \text{m s}^{-1}$ is quoted in the literature). Thus with a chord of 0·1 m, the wing travels forward 40 chords in one complete flap of the fore wing. It is perhaps not unrealistic therefore to investigate the efficiency of a typical aerofoil section in *steady* flow. The same assumption was justified in detail for the locust by Weis-Fogh and Jensen.[22]

The low flapping rate of the wings of both dragonflies and damselflies makes them particularly suitable for a quasi-steady study of wing efficiency. The smaller insects generally have a much higher flapping rate (several hundred Hz, increasing to about 1000 Hz for midges: British Museum[1]), and this is presumably one way of increasing the Reynolds number. The finite time required to build up a wing stall is also probably used to advantage. The low-speed flappers such as butterflies and moths (Lepidoptera) hook their wings together to avoid mutual wing interference and to increase the wing chord; butterflies also appear to apply high wing loading (high lift

---

* One of us has bent the main ribs of both dragonflies and damselflies and found that they are restored within a second or so, and that the animal is capable of flight again in less than a minute.

coefficient) only for very short periods. Of all the slow flappers the dragon-flies (Anisoptera) are probably the fastest and most manoeuvrable (Chadwick[2]); they prey on the smaller insects which, if it were not for the decrease of wing efficiency with reduced Reynolds number, would be more manoeuverable and would probably be able to avoid capture. Thus the dragonfly apparently uses very ancient solutions to the problem of reducing the critical Reynolds number at values of about 10^4.

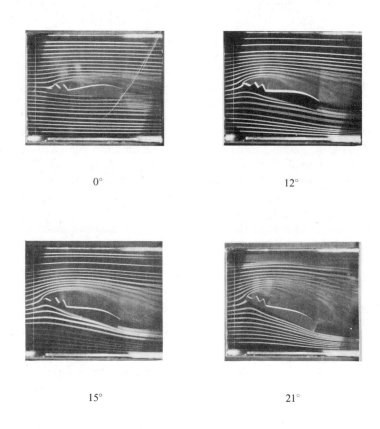

0°   12°

15°   21°

FIG. 4. Smoke tunnel pictures of the dragonfly aerofoil section at various angles of incidence

The present work relates to the theme of this conference in its investigation of optimum aerofoil shape at various Reynolds numbers around 10^4 and its subsidiary investigation of the maximum size of the spurs.

EXPERIMENTS AND RESULTS

Smoke Tunnel Observations

A two-dimensional model (12·7 cm chord, 8·9 cm span: Fig. 1) was built from thick balsa sheet and mounted in the McGill smoke tunnel (working section 8·9 cm × 17·8 cm): the flow was studied at various angles of attack at a Reynolds number of about 10^4. At 0° incidence of the chord line (see Fig. 4) a trapped vortex was observed in the first valley of the section. At 12° the flow was fully separated near the leading edge and reattached about 2/3 chord from the leading edge as a turbulent boundary layer. At 15° the flow was fully separated.

It was concluded that, *with the ends of the separation bubble sealed to maintain a low pressure within the bubble*, the wing section may indeed produce high lift at moderate incidence by promoting transition due to separation and reattachment of a large separation bubble.

Model Glider Tests

(i) *Experimental Procedures and Results*

As Reynolds number is decreased, most aerofoils at a fixed angle of attack experience a rather abrupt reduction of lift and increase in drag at a certain Reynolds number, which is usually somewhere between 3 and 8×10^4. This "critical" Reynolds number, which is analogous to the critical Reynolds number of about 3×10^5 for smooth spheres and cylinders in a low turbulence stream, depends upon aerofoil geometry, angle of attack, surface roughness near the leading edge and wind-tunnel turbulence level. Roughly speaking, raising the free stream turbulence level decreases the critical Reynolds number; the reasons are well understood. The higher turbulence level moves transition forwards, which delays separation. Since we thought that the unusual section of the Dragonfly wing with its pleats (which are found also on other insects, such as the locust), serrated leading edge (other types are also found) and spurs (not fully investigated) had evolved to have a low critical Reynolds number it was necessary to eliminate the effect of wind-tunnel turbulence. Accordingly free-flight tests were made indoors in the still air of the Sir Arthur Currie Gymnasium at McGill, thus effectively reducing the ambient turbulence to zero. Although the tests and data reduction are more tedious than wind-tunnel tests, one can obtain lift and drag coefficients as functions of angle of attack.

Two model gliders were constructed (Fig. 3), one having a span of 1 m, and designed to fly at a typical Reynolds number of 4×10^4, and the other half this size, more or less scaled as to weight and designed to fly at a Reynolds number of 1×10^4. The latter is the value typical of high-speed *Aeschna*. The first model also had interchangeable wings so that the "dragonfly" wing performance could be compared with that of a more standard curved-plate aerofoil, the McBride B-7, an old section aeromodellers nonetheless consider to have a high performance near its critical Reynolds number. A comparison of the two sections is shown in Figs 1 and 2.

The wing with the dragonfly aerofoil was tested both with and without a correctly-scaled leading-edge turbulator which was machined on a screw-cutting lathe to resemble the serrated costa of the dragonfly closely. Wing end plates could also be added, since it was felt that they would help to "cap" the leading-edge separation bubble in the smoke-tunnel photographs at large angle of incidence. Note too that, whereas the model wings had a constant chord and section, the dragonfly wing pleats are swept back towards the tips and there are spanwise variations of section and chord.

Only the dragonfly section was tested on the smaller model, the purpose being to investigate the effect of reduced Reynolds number on performance.

The models were trimmed to fly at constant speed in various straight glide paths, and thus at various angles of attack of the wing, by adjusting the horizontal tail surface and a moveable nose weight (on the larger models) to change the position of the centre of gravity while keeping the total weight constant. The models were launched from a height of about 5·5 m above the gymnasium floor. Although a special launch mechanism was built for the gliders it was found to be quicker, simpler and more accurate to have them hand-launched by one of us who is an experienced aeromodeller. The usual procedure was to trim the model close to the stall condition and then successively trim it for steeper and steeper glide path angles. Test data were obtained by panning the gliding model with stroboscopic illumination in the darkened gymnasium while having two still cameras (a 35 mm Nikon for accurate data analyses and a Polaroid Land camera for "immediate" checks of flight quality) with the open shutter focussed on the glide path. With high speed black and white (3000 ASA) film and small reflectors on both the nose and tail of the fuselage (also on the rear wall to provide a horizontal reference), the flight altitude, flight path angle and flight speed could be determined. A typical photograph is shown in Fig. 5. The model is faintly visible in the photograph; the longer reflector is mounted on the nose and the shorter one near the tail. The horizontal reference reflectors are clearly visible in the bottom left and right of the photograph. Special calibration tests determined that the maximum photographic distortion of distances was 1·5% and the angular distortion was nil. Each photograph was rated A, B, C or D, or discarded depending on how closely the flight approached equilibrium conditions (i.e. constant speed U_∞, flight path angle β, and angle of attack α). Excellent quality was denoted by A, while B and C correspond respectively

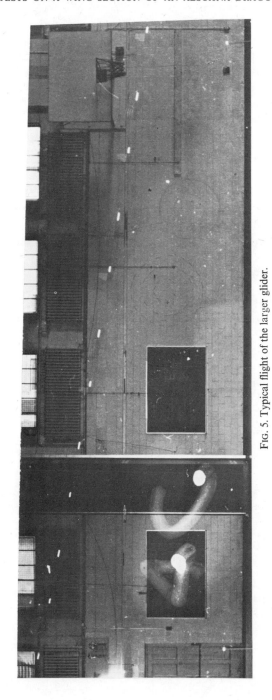

FIG. 5. Typical flight of the larger glider.

to most or half of the picture showing equilibrium flight and D corresponds to equilibrium being reached only at the edge of the field of view. The straightness of the flight also came into the assessment. In each case only the equilibrium position was used for data reduction purposes. This ranking, which is shown in some of the subsequent figures, is helpful in weighting various points when drawing curves through the data.

Each flight path yields particular values of α, β and U_∞ from which the lift and drag coefficients C_L, C_D for the complete glider can be obtained as follows

$$C_L = \frac{mg \cos \beta}{\frac{1}{2}\rho U_\infty^2 S}, \qquad C_D = C_L \tan \beta, \tag{1}$$

where mg is the weight of the model, S is wing surface area and ρ is the air density. These lift and drag coefficients include the effects of components other than the wing (i.e. the fuselage, horizontal and vertical tail etc.) and the effect of finite aspect ratio (i.e. the induced drag). Thus they are peculiar to the particular models tested and cannot in this form be compared with "standard" aerofoil data of other wing sections. Therefore the data have been further reduced to obtain the lift and drag, minus induced drag, for the wing alone. The lift on the horizontal tail was estimated from the calculated downwash at the tail and an assumed variation of lift with angle of incidence for the tail. This tail lift was subtracted from the total lift to obtain the wing lift coefficient C_{LW}; the corrections were rather small as shown in Figs 7 and 10(b). The drag coefficients of the various glider components aside from the wing were estimated following procedures used for the prediction of aircraft performance; typical values are shown in Table I. The drag coefficients of the

TABLE I. Estimated drag coefficients for large model (span = 1 m) at $U_\infty = 4\,\mathrm{m\,s^{-1}}$

Item	C_D
Horizontal tail	
Skin friction	0·00380
Induced	0·00014
Rudder skin friction	0·0016
Tail and reflector interference	0·0006
Wing guy wires	0·0050
Fuselage skin friction	0·00026
Nose weight and bumper	0·0032
Nose reflector and wing support	0·0013
Total	0·0159
End plates	0·0026

components were subtracted from the glider drag coefficient $C_{D\,(glider)}$ to yield the wing drag coefficient C_{DW}. The induced drag was calculated and subtracted to yield the drag coefficient of the wing alone, corrected to infinite aspect ratio, C_{D_0} (see Figs. 8(a), 8(b), 10(c) below).

(ii) *Discussion of Results*

Because of the manner of flight testing the data are not obtained at constant Reynolds number. It is simple to show that Re $\propto (mg \cos \beta / C_L)^{1/2}$. For the larger model (1 m span), where the trim was changed by adjusting the nose-weight keeping the total weight mg constant, the Reynolds number increased with decreasing C_L and increasing β (C_L increases faster than $\cos \beta$). The smaller model (0·5 m span) was so small that it was trimmed with small calibrated weights which were placed in a nose compartment. Thus the nose weight and with it the all-up weight was larger when the small model was trimmed at low C_L. Because of this the Reynolds number variation over a given range of C_L is rather more for the smaller model than for the larger models.

The results are displayed by means of three types of graph: plots of lift coefficient C_L versus angle of incidence α, plots of drag coefficient C_D versus α, and plots of C_L versus C_D (polar plots) derived from the first two. The former are straightforward to understand, and would show clearly, for example, the rise in drag and fall in lift which occur as α is increased above the stall angle (at any given Reynolds number). However, from the point of view of assessing a wing's aerodynamic performance, polar plots are extremely informative and are widely used by aerodynamicists; in particular, they allow one to see at a glance the maximum value of the lift–drag ratio, which, for constant efficiency of propulsion and constant rate of fuel consumption, determines the maximum range of a flying animal (see Lighthill's paper above), and the minimum gliding angle β (Eqn 1).

Large models (dragonfly and curved-plate sections). The Reynolds number for these tests ranged from about $3·5 \times 10^4$ to 6×10^4. A complete model polar plot (of C_L versus C_D) for the plain dragonfly wing section is shown in Fig. 6. The C_L versus α curves for this model, for the dragonfly wing with turbulator, dragonfly wing with both turbulator and endplates, and the McBride B-7 curved plate sections are shown in Fig. 7. The drag coefficients $C_{D\,(glider)}$, $C_{D\,(wing)}$ and C_{D_0} versus α for the plain dragonfly and the curved plate wings are shown in figs. 8(a) and (b).

Of particular interest are the "kinks" that occur in the C_L versus C_D, C_L versus α and C_D versus α curves for the plain dragonfly wing at a value of C_L of roughly 0·75 ,and $\alpha = 8°$. The drag coefficients also decrease with α

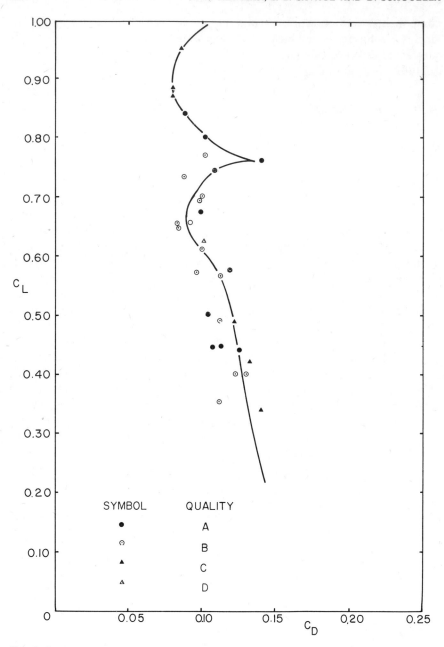

FIG. 6. C_L–C_D curve for plain dragonfly section—large model. Re $\approx$ 3·5 $\times$ 10⁴–6 $\times$ 10⁴.

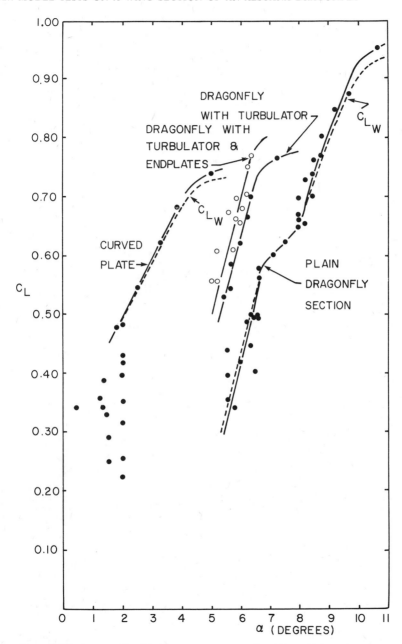

FIG. 7. C_L α curves for large models with and without modifications and also the McBride B-7 aerofoil. Re ≈ 3·5 × 10⁴–6 × 10⁴.

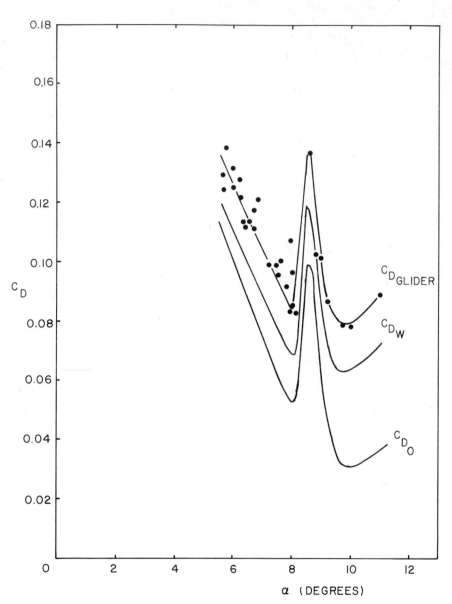

FIG. 8(a). C_D–α curves for large plain dragonfly section.

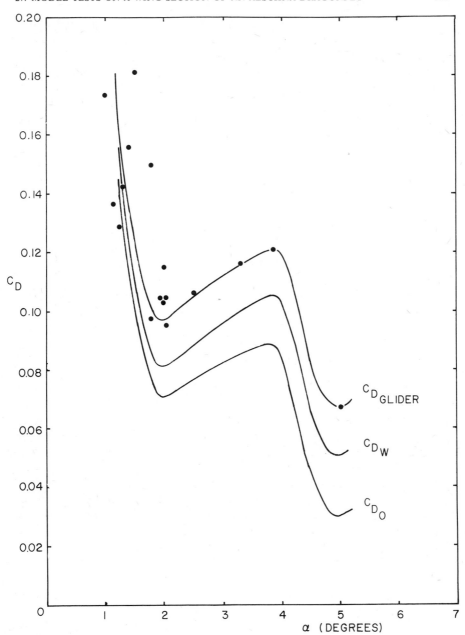

FIG. 8(b) $C_D-\alpha$ curves for the McBride B-7 aerofoil.

at smaller α. Both of these characteristics are sometimes noted for thin uncambered aerofoils at low Reynolds numbers (see for example Rabel[18]). There are relatively few aerofoil data for Re $< 10^5$ and many of these are unreliable because of wind-tunnel turbulence. Rabel has compiled most of this data.

It is seen that the kink in the C_L/C_D curve for the dragonfly model (Fig. 6) represents a significant variation in C_D at a more or less fixed value of C_L. In a steady glide the three forces, lift L, drag D and weight mg are in equilibrium. Moreover the lift is much greater than the drag and the glide angle β is small. The lift then approximately equals the weight and equations (1) become

$$C_L \simeq \frac{mg}{\frac{1}{2}\rho U_\infty^2 S}, \qquad C_D \simeq C_L \beta. \tag{2}$$

For any given configuration U_∞ is thus determined by C_L. Hence when C_D varies at a fixed C_L the model (and by inference the dragonfly) can fly at a constant speed and at various glide-path angles β. Such behaviour may be of great use to the dragonfly in pursuing and catching other insects in the longitudinal plane, a behaviour which is characteristic of the dragonfly when feeding.

The flow about the corrugated dragonfly wing section is a bit more involved than the flow about the plain wing, and a possible physical explanation of the aerofoil characteristics is as follows (see Fig. 9). At low α several separation bubbles are present, one (at least) in each pleat and a larger one on the rear undersurface. In region (1) the high lift-curve slope ($>2\pi$) may be associated with the change of effective camber as the bubble on the undersurface decreases in size with increasing α. C_D decreases as the wake narrows. In region (2) the bubbles are confined to the pleats, the lift-curve slope is reduced, probably due to boundary layer thickening over the aft portion of the wing. At (3) a large upper-surface bubble appears, increasing the lift and the drag, a feature which is aggravated by the decrease in flight Reynolds number with incidence. With further increase in α it is postulated that the critical Reynolds number is reached, transition in the separated shear layer occurs earlier, the bubble decreases in size and with it the drag. The flow finally separates completely at (4) and the wing stalls. These schematic diagrams should be compared with Figs. 7 and 8(a).

The effects of the turbulator are to increase the drag slightly and to increase the lift considerably at a given α before stall (Fig. 7). With the turbulator in place the "first" stall occurs at a higher C_L. It is likely that, as with the plain dragonfly wing, there is a "second" stall at higher C_L. The test procedure, in which the model was trimmed close to what appeared to be the stall and C_L successively decreased, may have caused us to miss the upper portion of the polar. In this region (near the first stall) C_D changes rapidly with α and it was difficult to launch the model for a straight glide path. Even a slight misalignment could cause one half of the wing to have considerably higher drag than the other, inducing the model to veer off strongly to the left

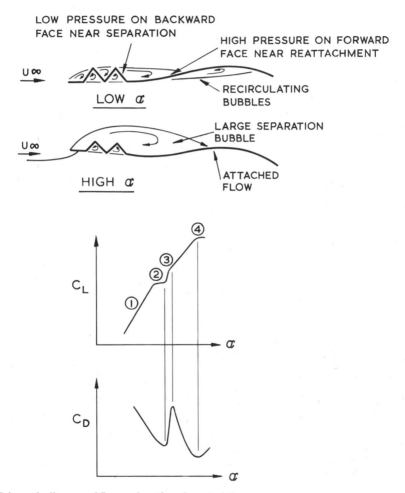

FIG. 9. Schematic diagram of flow and section characteristics.

or to the right. The cause of these difficulties was not realized during the flight tests and it was assumed that the model was turning because of loss of lift on one wing due to stalling.

Addition of the end plates to the dragonfly wing increased the drag because of the increase of area covered by the separation bubble. The lift at a given α was increased (Fig. 7), probably because the end plates closed or "capped" the leading-edge separation bubble and increased the camber effect.

The curved-plate wing has the same gross behaviour as the dragonfly section (Figs. 7 and 8(b)). The drag at low α is rather larger, probably because of a larger undersurface separation bubble. The stall may not have been reached; the curves at high C_L are uncertain because of the lack of data points. On the C_L versus α curves there is a great deal of scatter for $C_L < 0.5$ (Fig. 7). It is likely that this is associated with hysteresis for the lower-surface separation bubbles whose size may depend on the flight-path history previous to the equilibrium portion of the flight. The concavity of the surface must play a significant part in this hysteresis for it does not seem to be present for the upper surface separation bubbles.

It is seen from Fig. 7 that the correction to C_L to account for the tail lift is quite small and from Fig. 8(b) that the wing produces the major portion of the total drag. It may also be noted that the lift curve slope, $dC_L/d\alpha$ of the dragonfly wing is very large (11 per radian) compared with the theoretical value of 2π per radian. This is also attributed to the separation bubbles. With increasing α the lower bubble size decreases and the upper bubble increases in size and both effects increase the effective camber. The effective aerofoil shape thus changes with α, giving rise to a lift-curve slope greater than the theoretical value.

Small model with dragonfly section. The smaller model, whose linear dimensions were half those of the larger model, flew at Reynolds number roughly a quarter as large as the larger model. The test data, which were obtained only for the plain dragonfly section, were reduced in the same manner as for the larger models (see Figs 10(a), (b) and (c)). There appear to be kinks in the curves for C_L versus C_D and C_D and C_L versus α at around $C_L = 0.3$ ($\alpha = 4°$) and $C_L = 0.65$ ($\alpha = 7.5°$). From the curves it is not clear that the model was close to stalling in any of the glide tests and it is possible that higher lift coefficients could have been obtained before the wing stall. Curves of C_{LW} versus C_{D_0} for the plain dragonfly section on the large model (Re $\simeq 3.5 \times 10^4$ to 2.5×10^4) and the small model (Re $\simeq 1.2 \times 10^4$ to 2.5×10^4) are compared in Fig. 11. Surprisingly, the smaller model has a lower C_{D_0} for $C_{LW} < 0.56$. To give some idea of the performance of the dragonfly section compared with other aerofoils, Fig. 12 shows C_L versus C_{D_0} for probably the highest-performance, low-Reynolds-number aerofoil chosen from amongst the reliable data compiled by Rabel.[18] This aerofoil is the Göttingen 803 with a wire turbulator placed 8% of the chord ahead of the leading edge. Data are

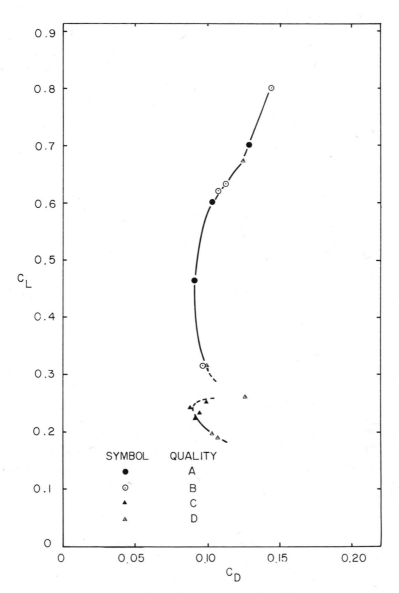

FIG. 10(a). C_L–C_D curve for plain dragonfly sections—small model.

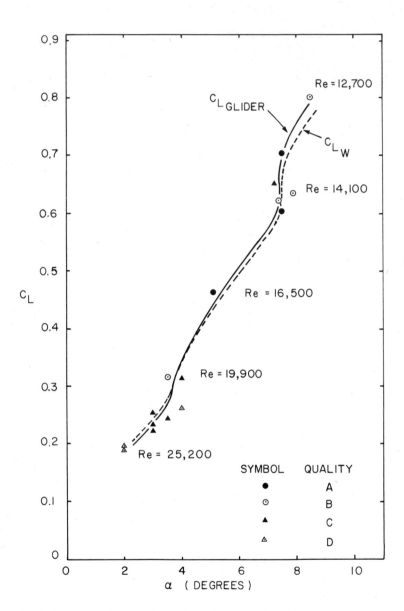

FIG. 10(b). C_L–α curves for plain dragonfly sections—small model.

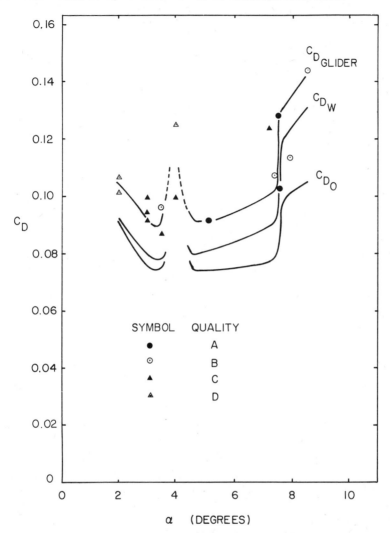

FIG. 10(c). C_D–α curves for plain dragonfly sections—small model.

shown for the aerofoil with and without the turbulator for several Reynolds numbers. The strong effect of Reynolds number and the beneficial effect of the turbulators are quite evident. After the abrupt performance drop at the critical Reynolds number, most aerofoils experience a more gradual, but still very marked, performance degradation with decreasing Reynolds number. Maximum lift to drag ratios (corrected to infinite aspect ratio) for standard,

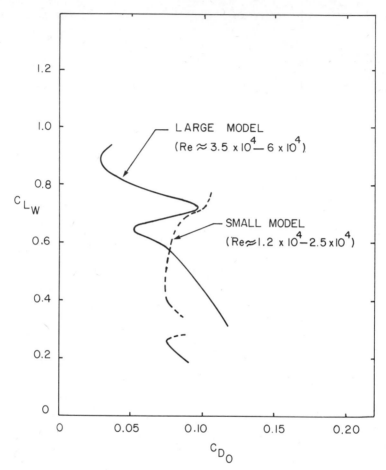

Fig. 11. C_L for the wing alone versus profile drag coefficient C_{D_0} of the wing alone, for both plain dragonfly models.

quite ordinary looking, aerofoils are typically as low as 5 (or even 2 for thick aerofoils) for Reynolds numbers around 2.5×10^4. By comparison the Göttingen 803 with turbulator and the dragonfly section are very efficient aerofoils. In fact, the small dragonfly section (without turbulator) at the higher C_{LW} corresponding to Re $= 1.2 \times 10^4$ compares favourably with the Göttingen 803 with turbulator at a higher Reynolds number of 2.5×10^4. It is very likely that the actual live dragonfly wing, containing features such as the serrated leading-edge, nodus with the change in aerofoil section, pleats swept back at wing tips, spurs or cones on wing surface etc., has a better performance than that shown in Fig. 10(a).

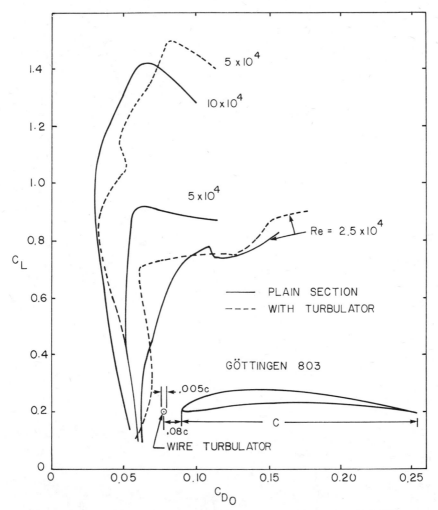

Fig. 12. C_L–C_{D_0} curve for Göttingen 803 aerofoil at low Reynolds numbers (Rabel[18]).

SCANNING ELECTRON MICROSCOPE PICTURES

Preliminary pictures were taken on a Cambridge Stereoxam 600 in the Soil Mechanics Laboratory at McGill University with the help of Dr. D. Sheeran. Uncoated specimens of a small *Libellula* were viewed at magnifications of up to 2000 and voltages of 7·5 kV. The specimens were damaged to some extent by the build-up of charge. However, the form of the spurs was

(a)

(b)

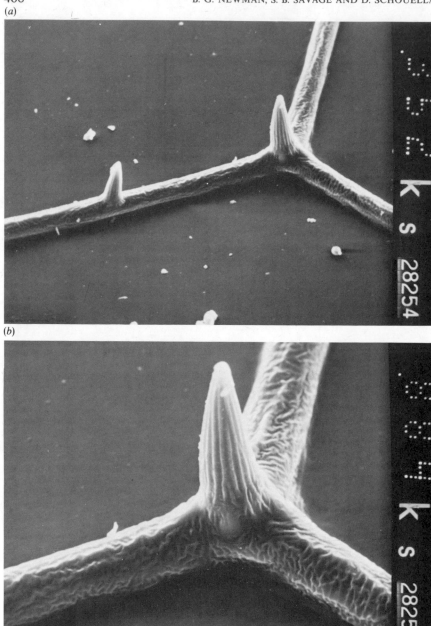

FIG. 13. Spurs on the upper surface of the fore wing of an *Aeschna*, at magnifications of (a) 352 and (b) 884.

clearly seen and they were essentially the same as the later ones shown here. The height of the spurs above the diaphragm was 55 μm.

Better quality pictures were taken on the larger instrument at the Pulp and Paper Research Institute of Canada by Gunther Seibel on specimens which were plated with gold–palladium. Two pictures of a spur at the joining of two veins about half way back along the upper side of the fore wing of *Aeschna Interrupta* are shown at magnifications of 352 × and 884 × in Figs 13(a) and (b). Similar pictures are shown of the underside of the hind wing of the same dragonfly in Figs 14(a) and (b) at magnifications of 144 × and 730 ×. The wing was viewed obliquely at 45°. The heights of the spurs were found to be 90 μm and 100 μm respectively. The alternate orientation of the spurs (about ± 10°) is interesting and was noted first on the *Libellula* specimen.

The corresponding spurs on the underside of the hind wing of a damselfly (a *Coenagrion* from California) are shown at magnifications of 145 × and 695 × in Figs 15(a) and (b). This is the side of the wing which is exposed when the wings are folded. The spur is 60 μm high in this case and has a distinctly different shape. We have not found spurs on the upper side of the wings of damselflies, although we have not studied Lestidae which do not fully fold their wings when resting.

The heights of the spurs, which appear to be independent of chordwise position, are nevertheless more or less proportional to the wing chord. Comparative visual estimates have been made using a magnifying glass on other species (Table II).

It is possible that the spurs are bumpers to protect the wing and are smaller for smaller species to save weight. However, they are also found inside the pleats which makes the hypothesis less plausible.

TABLE II. Spur height for various species

Species		Height of spur above membrane (k μm)	Local chord at nodus (c mm)	Semi-span of wing (mm)
Aeschna Interrupta and *Eremita*	Fore	90	10·7	47·8
	Hind	100	13·7	45·7
Libellula Quadrimaculata	Fore	≃80	7·9	35·6
	Hind		9·1	34·3
Gomphus	Fore	≃80	6·6	31·5
	Hind		8·6	29·7
Sympetrum	Fore	≃70	5·8	24·1
	Hind		7·4	22·9
Anax Junius	Fore	≃125	10·2	49·5
	Hind		13·7	48·3
Coenagrion	Hind	60	4·1	25·4

(a)

(b)

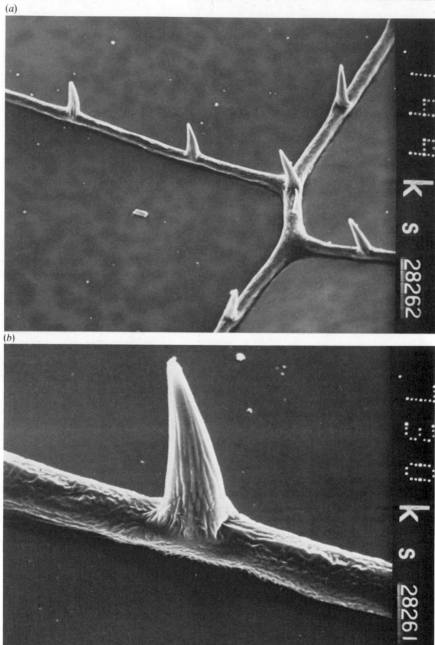

Fig. 14. Spurs on the lower surface of the hind wing of an *Aeschna*, at magnifications of (a) 144 and (b) 730.

(*a*)

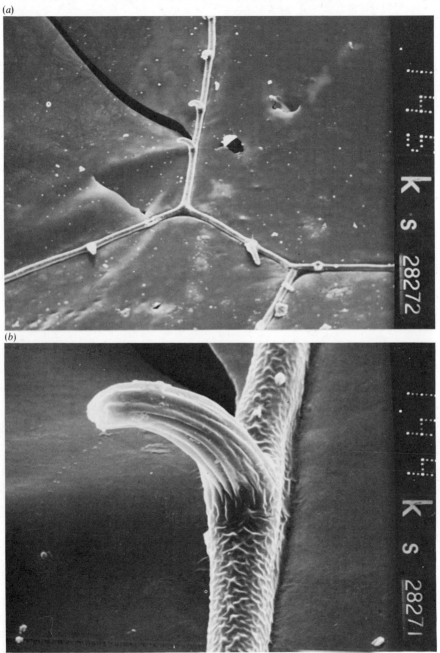

(*b*)

FIG. 15. Spurs on lower surface of the hind wing of a *Coenagrion*, (damselfly) at magnifications of (a) 145 and (b) 695.

Schlichting (Ref. 19, p. 617) quotes the following formula for the maximum roughness which renders a surface aerodynamically rough by protruding through the viscous sub-layer beneath the turbulent boundary layer. This is a sensitive region where the scales of Lepidoptera appear to affect the flow and reduce the drag (Nachtigall[12]). Schlichting's formula is

$$h_{max} = 100 \frac{v}{U_\infty}.$$

This admissable roughness is independent of downstream distance. Taking the kinematic viscosity $v = 15 \times 10^{-6} \, m^2 \, s^{-1}$ and the typical free stream velocity of $15 \, m \, s^{-1}$, $h_{max} = 100 \, \mu m$. It is no doubt significant that the height of the spurs for most of the cases listed is usually less than h_{max}.

Pictures of the serrated leading edge along the lower arm of the costa on the hind wing of *Aeschna Interrupta* are shown at three different magnifications in Figs 16(a), (b) and (c) ($60 \times$, $246 \times$, $1220 \times$). In the corner of each serration there is a hairlet about $3 \, \mu m$ in diameter and $35 \, \mu m$ long which protrudes forwards and slightly ahead of the serration. Smart has identified it as a sensory macrotrichea and Weis-Fogh thought that it was there mainly for the purpose of detecting impinging objects. However, this does not

(*a*)

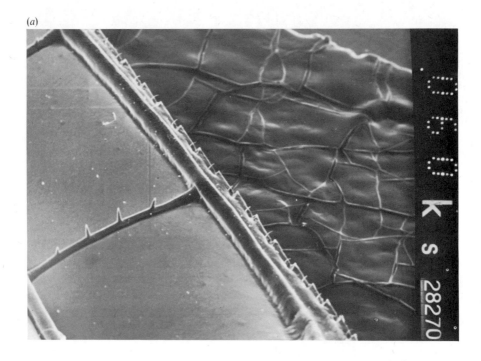

(b)

(c)

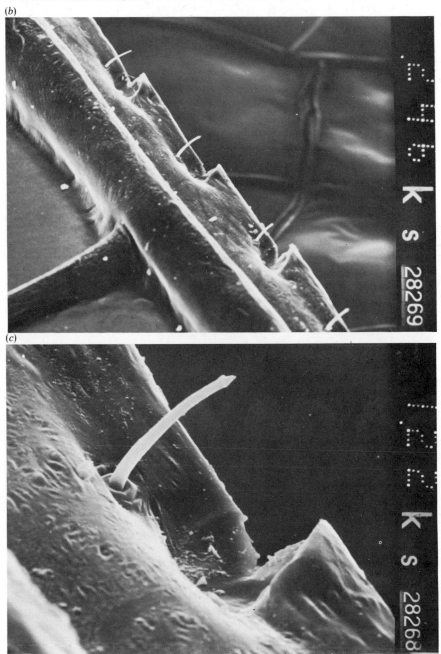

FIG 16. Leading edge serrations and hairlets on the hind wing of an *Aeschna* at magnifications of (a) 60, (b) 246 and (c) 1220.

preclude the possibility that the hairlets may also be available for sensing the airflow and that the serrations may act as turbulators to promote transition in the separated shear layer.

It might be noted that whereas the leading-edge serrations are sharp, very regular and "well-formed," the trailing edge serrations are much more irregular and rounded. This suggests that the leading-edge serrations serve to enhance transition to turbulence while the trailing-edge serrations are indentations which serve to protect the hairlets and their sockets from damage during inadvertent contact with other objects.

FIG. 17. Trailing edge serrations and hairlet on the hind wing of the same *Aeschna* at a magnification of 748.

It is possible that the sensory hairs at the base of the serrations in the costa have the additional role of sensing the direction of the air flow and perhaps keeping the stagnation point exactly at the leading-edge. Such a possibility would reduce the size of the leading-edge separation bubble and make it easier to understand how the bubble could be controlled on a finite wing. The entrainment underneath a large separated region may be supplied by inward flow from the wing tip and reattachment would not necessarily occur. With a small bubble confined to the pleats themselves, however, the sweeping back of these valleys, which is a typical feature of all dragonfly wings, would provide an effective aerodynamic seal. The change of the section at the nodus also acts as a small end plate. Moreover the almost rectangular planform of the hind wing tends to reduce the wing-tip loading, especially when the wing is flapped. Thus separated regions will tend to get smaller near the tips which helps to seal the bubbles inboard. This no doubt explains the good performance of our glider which had a rectangular wing.

There may also be an aerodynamic role for the trailing-edge hairlets (Fig. 17). These are well placed to detect whether the flow leaves the trailing edge smoothly (the Kutta–Joukowski condition). They could therefore monitor the rate of incidence or camber change, which is mainly controlled by muscles attached to the leading-edge ribs, to ensure that the lift is maintained and the drag does not become unduly high. The additional information may be important in controlling the flip mechanism which Weis-Fogh[24] has described.

CONCLUSION

General

The free flight glide tests proved to be a successful method of obtaining aerofoil data free from the distorting effects of free stream turbulence. The quality of the data obtained was comparable or better than that usually obtained in full-scale flight tests but it is more scattered and requires larger corrections than data usually obtained in wind tunnels.

The unusual pleated aerofoil, in addition to being an efficient structure, has a very high aerodynamic performance. The tests with a model having a constant-chord and constant section (modelled after the dragonfly just inboard of the nodus) compared well with a very high performance, low Reynolds number, conventional aerofoil fitted with an artificial turbulator. It is probable that the actual dragonfly wing, which contains several small features not present on the test models, has an even higher performance.

Devices which *may* Improve the Aerodynamic Efficiency

The following are possibilities:

(i) The sensory macrotrichea within the serrations along the three arms of the costa may serve to detect not only foreign objects but also the airflow direction, so that the position of the leading edge stagnation point may be controlled by the root muscle and the pressure of blood within the wing.

(ii) The leading edge serrations may act to produce trailing vortices which help destabilize the separated laminar shear layer and promote transition and reattachment. This is commonly done on model aircraft flying at somewhat larger Reynolds numbers, from about 10^4 to about 10^5.

(iii) The spurs, which appear to vary in height approximately as the first power of the wing chord from species to species, may be designed to act as anchors for the longitudinal vortices which form within the unstable shear layer downstream of separation. Such a possibility has however not yet been established experimentally and the suggestion is highly conjectural.

There is little doubt that the anatomical structure of the wing is largely determined by constraints of rapid deployment from the instar, and by structural considerations. The detailed geometry of the pleats, for example, may have no particular aerodynamic significance even though the pleats themselves may be important. They combine to make a strong main spar as the section changes in the spanwise direction. Moreover, the structure is broadly the same for all dragonflies, despite a roughly ten-fold change of Reynolds number from the largest to the smallest species, due to reduction of scale and flight speed. However, it is definitely possible that the pleats help to close the bubbles by impeding the back flow in the separated region (Wygnanski and Newman[25]), but their detailed geometric shape may be less important.

ACKNOWLEDGEMENTS

We wish to acknowledge helpful conversations with entomologists and paleontologists. Dr. John Smart of Cambridge University, Professor Vernon R. Vickery of McGill University, and the late Professor Torkel Weis-Fogh of Cambridge University. Help with scanning-electron microscopy was provided by Gunther Seibel and Dr D Sheeran, and is greatly appreciated.

REFERENCES

1. British Museum (Natural History) "Insect Flight". Leaflet No. 4, 1974.
2. Chadwick, L. E. The wing motion of the dragonfly. *Bulletin of the Brooklyn Entomological Society*, **35**, 109–112 (1940).
3. Charwat, A. F. Experiments on the variation of airfoil properties with Reynolds number. *J. Aero. Sci.*, **24**, 386–388 (1957).
4. Corbet, P. S. A Biology of Dragonflies. *In* "Aspects of Zoology". Witherly (1962).
5. Corbet, P. S., Longfield, C. and Moore, N. W. "Dragonflies". The New Naturalist Series., Collins (1960).
6. Frazer, F. C. A Reclassification of the Order Odonata. *Roy. Zool. Soc. of N.S.W.* (1957).
7. Hertel, H. Membranous Wings of Insects. *In* "Structure-Form-Movement". pp. 78–87. Reinhold (1966).

8. Hocking, B. The intrinsic range and speed of flight of insects. *Trans. Roy. Ent. Soc. London*, **104**, 225–345, (1955).
9. Imms, A. D. "General Textbook of Entomology". Methuen (1957) (revised by Richards and Davies).
10. Longfield, C. "Dragonflies of the British Isles". Worne and Co. (1949).
11. Miller, P. L. Regulation of Breathing in Insects. *In* "Advances in Insect Physiology". Vol. 3, pp. 279–344. Academic Press, London and New York (1966).
12. Nachtigall, W. "Insects in Flight". Allen and Unwin (1974).
13. Needham, J. G. and Westfall, M. J. "Dragonflies of North America (Anisoptera)". University of California Press (1955).
14. Neville, A. C. Aspects of flight mechanics in anisopterous dragonflies. *J. Exp. Biol.* **37**, 631–656 (1961).
15. Norberg, R. A. The pterostigma of insect wings, an inertial regulator of wing pitch. *J. Comp. Physiol.* (German), **81**, 9–22 (1972).
16. Oldroyd, H. "Insects and Their World". British Museum (Natural History), 3rd Ed. (1973).
17. Pringle, J. W. S. "Insect Flight". Cambridge University Press (1957).
18. Rabel, Von H. " Modellflug Profile". München (1965).
19. Schlichting, H. "Boundary-Layer Theory". 6th Ed. McGraw-Hill, New York (1968).
20. Teale, E. W. "Grassroot jungles". Dodd Mead, p. 34 (1958).
21. Tillyard, R. J. "Biology of dragonflies". Cambridge University Press (1917).
22. Weis-Fogh, T. and Jensen, M. Biology and physics of locust flight (in four parts). *Phil. Trans. Roy. Soc.* B **239**, 415–584 (1956).
23. Weis-Fogh, T. Diffusion in insect wing muscle, the most active tissue known. *J. Exp. Biol.* **41**, 229–256 (1964).
24. Weis-Fogh, T. Unusual mechanisms for the generation of lift in flying animals. *Scientific American*, **233**, 80–87 (1975).
25. Wygnanski, I. and Newman, B. G. The reattachment of an inclined two-dimensional jet to a flat surface in streaming flow. *C.A.S.I. Trans.* **1**, 3–8 (1968).

28. On the Aerodynamics of Separated Primaries in the Avian Wing.

HANS OEHME

Academy of Sciences of the German Democratic Republic, Station for Vertebrate Research (in the Berlin Zoopark).

ABSTRACT

The narrowed vane parts of the outer primaries, which in many bird species separate from each other when the wing is stretched, and which are referred to briefly as "separated primaries", were studied for their aerodynamic performance under conditions of steady state gliding by means of a simplified model calculation. It was thought that the separated primaries may cause reduction of induced drag, provided that the lift coefficient is equal to that in the inner unsplit wing. However, reduction of total drag and a resulting improvement in gliding performance is predicted to occur only with larger birds, if at all, since the Reynolds number of the flow past the separated primaries of smaller birds is low enough to cause an increase of profile drag. Reduction of drag by separated primaries will increase with their number in a given wing, with their area relative to the total wing area, with the lift coefficient of the wing, and with decreasing aspect ratio. In principle, these correlations also apply to non-accelerated powered flight, where a possible reduction of drag in the distal segment of the wing could increase the forward thrust of the downstroke. Studies into wing geometry and profiles of separated primaries in several bird species suggest that their main functional importance does not in fact lie in drag reduction, which is expected to occur only under certain conditions, with the absolute size of the animal concerned playing a substantive role. The suggestion that their principal function is to increase the total lift coefficient turns out to be much more plausible.

INTRODUCTION

The outer primaries in the completely stretched wing are usually separated in most bird species. In such primaries the distal part of the vane is more or less narrowed in comparison to the proximal part which remains covered

by the overlapping rear primaries. Adhesive or frictional surface characteristics (Sick[12]), typical of the proximal part where they keep the feathers
together to form one coherent wing surface, are absent from these distal
parts. It is tempting to assume that these "separated primaries" might have a
particular functional role. Their usual characterisation in terms of aerodynamics and aeromechanics is based on a proposition originally made by
Graham.[1] The split distal segment of the wing is thought to act as a high-lift
device (slotted wing system) and to reduce the induced drag of the wing.

Wing slots have several applications in aviation. The auxiliary wing of an
aircraft, employed for lower flight speeds and for take-off and landing on
short runways, is one of the known variants in this context. The component
wings are arranged very close to one another in such a slotted wing. The
alula will form such a system with the inner part of the distal wing segment,
as originally assumed by Graham[1] and more recently confirmed by experimental studies (Nachtigall and Kempf[4]). Yet, circumstances are different
in the separated primaries. They can be thought of as independent wings
since they are separated by half to one full chord length. Hence, a Handley–
Page–Lachmann effect will be less probable in the context of gliding or soaring
flight or of the downstroke in unaccelerated horizontal powered flight. The
extent to which the drag of the wing can be brought down will be the subject
of this study, conducted with reference to absolute wing size.

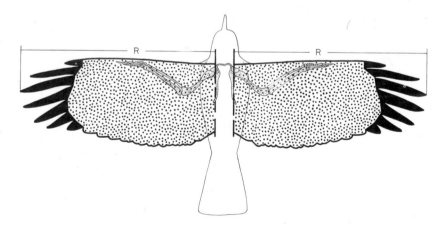

Fig. 1. Principles to determine characteristic dimensions of wing geometry; "span" $2R$, un-
separated wing area S_1 (dotted), area of separated primaries S_2 (black), total wing area $S =
S_1 + S_2$, fractional area of separated primaries $y = S_2/S$, aspect ratio $A = 4R^2/S$. Note that the
"wing span" in this context is less than the distance from wing tip to wing tip which is the wing
span used in aeronautics. The reason is the difficulty of measuring the total wing span of a
sacrificed bird with sufficient accuracy, while precise values of the wing length R can be easily
obtained.

MATERIAL

Serial measurement was applied to 13 fairly common species, with the view to deriving an average wing representative of each. The equipment used and the photographic techniques will not be expounded in this paper, as they have been described in the context of earlier investigations (Oehme,[6, 7] Oehme and Kitzler[8]). The dimensions needed for adequate treatment of the problem were obtained by the principles given in Fig. 1. Profile shapes

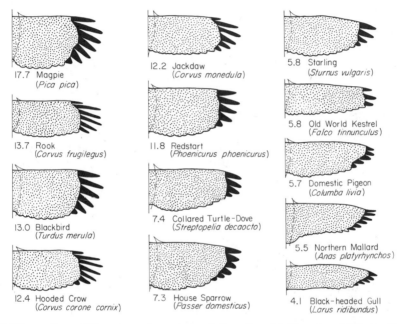

FIG. 2. Wing outlines of 13 species, redrawn to equal dimension of R. The value of 100y is added.

and structure of the separated primaries were determined as follows. The narrowed part of the vane was cut across centrally and perpendicularly to the rachis, with a sharp razor blade. Photomicrographs were taken from the cut surface in incident light. The cross-sections of the feathers remained unchanged by application of this technique, unlike wax or paraffin imbedding which regularly caused cross-sectional deviation. No attempt was made in the context of this study to find out if and to what extent a given profile camber of separated primaries may be changed in flight, as observed from profiles of the inner non-split wing (Nachtigall and Wieser[5], Oehme[7]).

Different Wing Shapes

A comparison of the wing shapes, given in Fig. 2, shows that the number of separated primaries and the fractional area of the split distal wing segment, $y = S_2/S$, vary within fairly wide limits. No regularity could be found, according to Fig. 3, for correlating y to data of aeromechanical relevance.

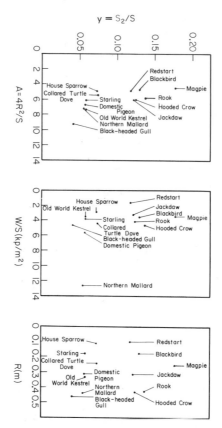

FIG. 3. Fractional area of separated primaries (y) plotted against aspect ratio (top), wing loading (middle), and wing length (bottom).

Consequently, no simple morphologico-functional interpretation will be possible of the wing structures under review. Such a situation seems to support the idea of studying the aerodynamic peculiarities of the "separated primaries" phenomenon.

DRAG OF A WING WITH OUTER PARTS SPLIT INTO TANDEM WINGS

Let us first elaborate briefly on what is called the tandem effect (Fig. 4). Assume an untwisted airfoil (area S, span $B = 2R$) with continuous profile and elliptic outline and, consequently, an elliptic distribution of lift. Then both lift and induced drag are determined by the air speed and the angle of incidence. Suppose that the airfoil is subdivided into two tandem wings, equal in size and positioned one behind the other, and with span equal to that of the

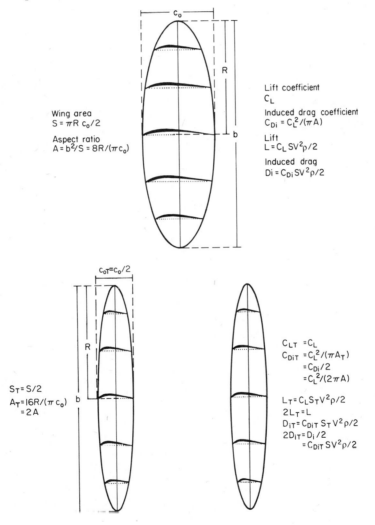

Wing area
$S = \pi R\, c_0/2$

Aspect ratio
$A = b^2/S = 8R/(\pi c_0)$

Lift coefficient
C_L

Induced drag coefficient
$C_{Di} = C_L^2/(\pi A)$

Lift
$L = C_L\, S V^2 \rho/2$

Induced drag
$Di = C_{Di}\, S V^2 \rho/2$

$S_T = S/2$
$A_T = 16R/(\pi c_0)$
$= 2A$

$C_{LT} = C_L$
$C_{DiT} = C_L^2/(\pi A_T)$
$\quad = C_{Di}/2$
$\quad = C_L^2/(2\pi A)$

$L_T = C_L\, S_T V^2 \rho/2$
$2L_T = L$
$D_{iT} = C_{DiT}\, S_T V^2 \rho/2$
$2D_{iT} = D_i/2$
$\quad = C_{DiT}\, S V^2 \rho/2$

FIG. 4. Induced drag in tandem wings (see text).

original wing. Angle of incidence and air speed will remain unchanged as well. If one assumes that there is no mutual interaction between the tandem wings the total lift produced by them will be equal to that of the original undivided wing, but the total induced drag of the system will be lowered by half. However, the tandem wings will actually affect each other. The front wing flies in the upwash of the rear wing, while the rear wing is affected by the downwash of the front wing. The effective angle of incidence and, consequently, lift and induced drag will go up in the front wing but decline in the

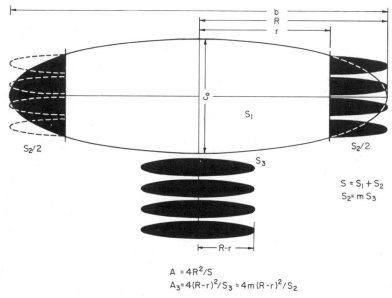

$$A = 4R^2/S$$
$$A_3 = 4(R-r)^2/S_3 = 4m(R-r)^2/S_2$$

FIG. 5. Geometric foundations for change of drag in response to splitting the wing ends into tandem wings (see text and appendix).

rear. Total lift and total induced drag, however, remain equal to their ideal values. Nothing will be changed, either, if the rear wing is set slightly above or below its original position. Only the distribution of lift between the component wings will vary, and at best will have the effect that the lift is equal in each of them. This principle can be applied to more than two tandem wings so that, theoretically, by subdividing a given wing into m equal wings of unchanged span and profile geometry the total induced drag of the wing system can be reduced to the mth part of the value which the wing would experience in its original undivided condition.

This phenomenon should also be effective on a wing in which the outer parts consist of small wings arranged one behind the other. This can be shown more conveniently by calculation on the basis of a simplified model rather than by using the actual bird's wing with all its peculiarities. Any geometric or aerodynamic twisting or sweep-back of the small wings will not be considered. Again, the original wing is assumed to be untwisted and of elliptic outline (Fig. 5). The outer part of the wing on either side will be cut off parallel to the plane of symmetry (c_0) at distance r. Its area $S_2/2$ is replaced by m semiellipses, equal in size and with major semiaxis $R - r$.

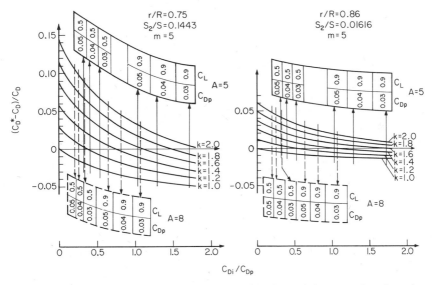

FIG. 6. Change of total drag in response to splitting the wing ends into $m = 5$ tandem wings; left hand for $r/R = 0.75$, $S_2/S = 0.1443$; right hand for $r/R = 0.86$, $S_2/S = 0.0616$; see text.

These semiellipses are assumed to be united to give elliptic wings arranged one behind the other and the coefficient of induced drag is calculated for such small wings. The coefficient of induced drag C_{Di} in the non-split part S_1 is assumed to be equal to that of the original wing. The coefficient of induced drag in the modified wing becomes

$$C_{Di}^* = C_{Di}(1 - S_2/S) + C_{Di_3}S_2/S$$

with C_{Di_3} being the coefficient of one of the small elliptic wings. The further assumption is made that the lift coefficient C_L is equal at all points of the wing span, both in the original condition and after modification, and that the coefficients of profile drag are independent of lift coefficient. The profile

R

drag coefficient is C_{Dp} for the original wing, constant over span $2R$ and over the inner part of the wing after modification. The profile drag coefficient of one of the small wings is C_{Dp_3}, constant over its span $2(R - r)$. The total drag coefficient is $C_D = C_{Di} + C_{Dp}$ in the original and

$$C_D^* = (C_{Di} + C_{Dp})(1 - S_2/S) + (C_{Di_3} + C_{Dp_3})S_2/S$$

in the modified wing. The calculation is given in the appendix.

The variation of total drag of the modified wing compared with that of the original wing is given by the expression $(C_D^* - C_D)/C_D$. In the graphs of Fig. 6 this ratio is plotted against the ratio of induced drag to profile drag in the original wing C_{Di}/C_{Dp}. The parameters are the aspect ratio (A), the lift coefficient (C_L), the profile drag coefficient of the original wing, equal to that of the non-split inner part of the wing (C_{Dp}), and the ratio of profile drag coefficient of a tandem wing to that of the original wing ($k = C_{Dp_3}/C_{Dp}$). The values of $x = r/R$ used in the calculations were 0·75 and 0·86 respectively.

Drag reduction in the modified wing increases with increasing number of tandem wings, with an increase in their fractional area, with increasing lift coefficient, and with decreasing aspect ratio of the whole wing. Growth of profile drag in the tandem wings relative to that in the mid-wing is accompanied by rapid decline in drag reduction and even by growth of drag in the modified wing, i.e. by a deterioration of gliding performance. In addition, it should be borne in mind that reduction of lift in the small wings may not only lower drag savings but may even cause drag rise, since the span-constant lift coefficient applicable to the original wing would be smaller.

The growth of profile drag in the tandem wings is a phenomenon which depends on the size of the entire wing. Every wing profile has its own characteristic critical value of the Reynolds number. The ratio of lift to profile drag is relatively poor below the critical value, on account of separation of the laminar boundary layer, which tends to occur on the suction side. The same ratio is improved above the critical value, since the boundary layer on the suction side will be turbulent (cf. Schmitz [10, 11]). The Reynolds number is given by $\mathrm{Re} = Vc/v$, where V is the air speed, c the chord length, and v the kinematic viscosity of the flowing medium. Since the chord lengths of the tandem wings are much smaller than those in the unsplit part of the wing, the latter may be above the critical value of Re and the former below, at any given speed. The considerable difference in the two values of Re is shown in Fig. 7, with particular reference to the value $\mathrm{Re} = 10^4$.

This is the situation in greater detail: the wing is assumed to have a profile which is insensitive to changes of Reynolds number over a wide range (a plane or cambered plate with sharp leading edge). Then, according to Schmitz,[11] development of a turbulent boundary layer on the suction side should be expected for $\mathrm{Re} > 10^4$. In this region variation of the Reynolds number will be accompanied by little variation of C_L/C_D. But for Re $< 10^4$ a growth of profile drag in such wing sections should be assumed to take place, along with a remarkable decline in lift (Fig. 8). If one further assumes that the bird's wing has such profiles—the sharp leading edge of the thicker inner segment of the wing and the surface roughness of the feather structures

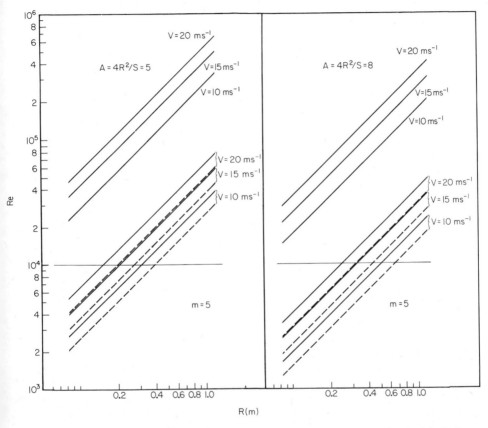

FIG. 7. Dependence on wing size of Reynolds number for mean chord lengths of original wing (top) and tandem wings of the split outer parts of the modified wing (bottom); full lines for $r/R = 0.75$, dotted lines for $r/R = 0.86$.

being interpreted as turbulence generators (cf. Nachtigall and Wieser,[5] Oehme[6, 7])—then calculation of Reynolds numbers for the chord length at $R/2$ and for an average chord length in the middle of all the separated primaries of the bird species under investigation, for one and the same speed, gives fairly good agreement with the theoretical model (Fig. 9). Hence, for smaller birds, up to thrush size, the lift–drag ratio will not be improved by means of separated primaries. On the contrary a deterioration of gliding performance is very likely to occur. The potential benefit of separated primaries is questionable even in larger species for low flight speeds.

This conclusion really applies only to the model, in other words, to an approximation of gliding or soaring flight of the bird in general. In fact, none of the species will reach the upper limit of $20~\mathrm{m\,s^{-1}}$, and some of them will do no steady-state gliding or soaring at all, or at least not with wings fully stretched and hence outer primaries separated. Yet, for the time being, let us

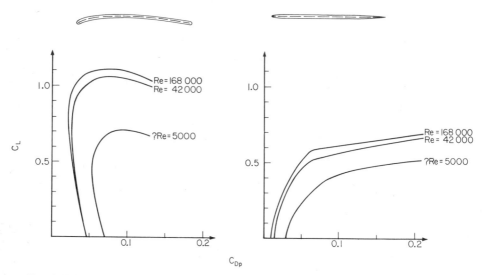

FIG. 8. Polar curves for thin wing sections with different Reynolds numbers, according to Schmitz;[11] a probable curve for Re $\approx$ 5000 is added.

still insist on this assumption in order to touch upon two more problems, profile design and the fine structure of separated primaries. The transverse sections of primaries are not homogeneous, but three major types may be differentiated (Fig. 10): the blackbird type (including redstart, house sparrow, magpie, jackdaw, rook, hooded crow) with thin, cambered plates (maximum camber between three and seven per cent of chord length); the pigeon type (domestic pigeon, collared turtle-dove) with profiles thin and almost symmetrical, somewhat comparable to flat plates; the mallard type (including starling, kestrel, black-headed gull) with a flat plate in the foremost large primary, but cambered plates in the others. A fairly large and more or less constant lift coefficient up to the wing tip, the prerequisite for a positive tandem effect, can be expected for the first type, and perhaps for the third as

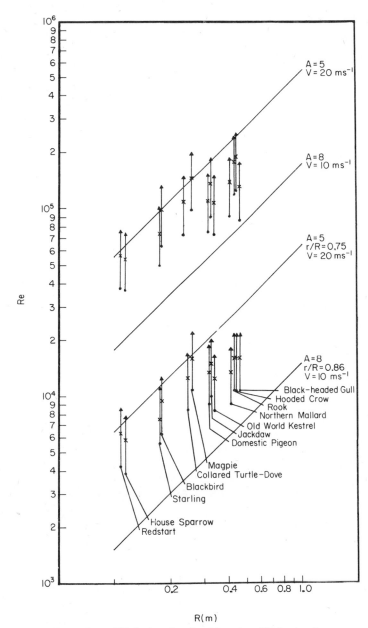

FIG. 9. Reynolds numbers of bird wings for chord length at $R/2$ (top) and average vane width at mid-point of separated primaries (bottom) plotted against wing length R for $V = 10\,\mathrm{m\,s}^{-1}$ (●), $V = 15\,\mathrm{m\,s}^{-1}$ (×), $V = 20\,\mathrm{m\,s}^{-1}$ (▲); also plotted for comparison are maximum and minimum values of model example, cf. Fig. 7.

well. But a positive tandem effect (improvement of gliding performance) should be expected only for birds of the size of jackdaw or larger, and it should not occur to any great extent unless the distal segment of the wing is deeply split. As to the species quoted above, it should be justified for magpie, jackdaw, rook, and hooded crow, and it probably applies to many larger birds, such as eagles, vultures, buzzards, harriers, storks, herons, and cranes, although the profiles of their primaries are unknown. The pigeon type seems to represent a different trend. Strong lift would cause high profile drag in the

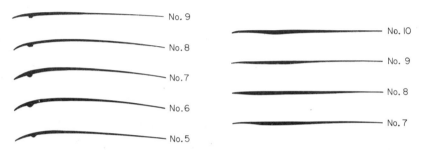

FIG. 10. Adjusted outlines of transverse sections at mid-point of separated primaries from *Turdus merula* (left hand) and *Columba livia* (right hand).

split part of the wing, and render impossible any reduction of total drag. If, on the other hand, the action of the separated primaries took place with smaller angles of incidence and hence lift coefficients below those in the inner part of the wing, the profile drag in the primaries would be lower. However, no reduction of total drag would be possible either, as we saw in the context of the model example. Separated primaries of such design probably have no significant tandem effect, especially as in their case the distal segment of the wing is obviously less split.

While at first glance a classification of structural peculiarities in the distal segment of the wing appears to be feasible, through their influence on drag variation, an additional uncertainty unfortunately arises. The transverse sections of the primaries so far described are idealised outlines. The real appearance of such a "profile" may be seen in Fig. 11. Surface roughness generated by the barbs is up to one per cent of the chord length on the suction side and up to four per cent on the pressure side. The question of whether such a structure still has the lift-drag characteristics of idealised profiles with similar contour is unanswered. It remains unelucidated even if smaller Reynolds numbers are considered, which would apply to such "wings" and for which turbulence generators, on principle, would be favourable. Therefore, further studies into the drag problem of the bird's wing are necessary to complement this theoretical concept or to propose some modification of it.

Finally we have to consider the efficacy of the tandem effect in powered flight. While in unaccelerated horizontal flight the flow acting upon the wing can be assumed to be steady (von Holst and Küchemann,[2] Oehme and Kitzler[8]) and the wing in its downstroke can be treated like a propeller with a high advance ratio (cf. Nachtigall,[3] Oehme and Kitzler[9]). The direction of the aerodynamic force generated will increasingly tilt forward, from wing

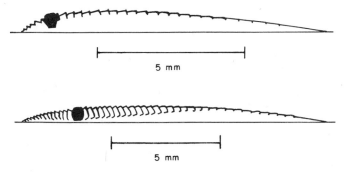

FIG. 11. Microstructure of transverse sections of separated primaries, primary No. 7 of *Turdus merula* (top) and primary No. 9 of *Anas platyrhynchos* (bottom).

base to wing tip, and the effective air-speed will also grow relative to flight speed. An elliptic lift distribution, such as that used in the model, can no longer be expected. Nevertheless, it is plausible that if drag in the outer segment of the wing goes down the downstroke will give more forward thrust. The threshold beyond which drag reduction may be expected will not, even in such a case, be surpassed by smaller birds; the greatest possible value in the effective air speed in the area of the separated primaries being $20 \, \text{m s}^{-1}$. The thrust achieved in the downstroke, however, can actually be improved in a larger bird, such as a crow, and in this case definition of separated primaries as "forward-thrust-feathers" (Stresemann[13]) seems to be justified. The principle that separated primaries depend on size for their drag-reducing action, as expounded by the simplified model of a rigid wing, is valid for normal powered flight as well.

CONCLUSIONS

By analysis of the flight performance of separated primaries, a set of new questions is established which cannot be answered in this paper, but the major issues relating to them should be mentioned. A completely closed wing tip, as recorded in swifts and hummingbirds, is not displayed by the stretched wing of most bird species. More or less strongly pronounced separation of the outer primaries is common. The aerodynamic implications elucidated in this paper do not support an interpretation of this morphological structure as a device to reduce wing drag. Such a drag-reducing function *may* occur, and if it does it depends on a minimum size of the bird and on the shape of its primaries. The structural peculiarity of "separated primaries" is more likely to be explained by other functional relations. Partial answers may be found (i) in the context of asking whether a closed wing tip is consistent with the elastic stress experienced by it in the downstroke during powered flight, (ii) in the bird's ability to use its wings in a great variety of ways (folding and unfolding, use as an inversed blown wing system in upstroke while hovering, deceleration, and acceleration), and (iii) in the restriction of the basic design in the bird's forelimb to certain structural parts. However, the most likely explanation is that suggested by Kokshaysky above, to the effect that (iv) separated primaries enable the distal part of the wing, and indeed the individual feathers, to be twisted far enough to increase the angle of incidence substantially without stalling, thereby increasing the total lift coefficient. They may also be used to *decrease* the effective angle of attack at the wing-tip when the main wing is almost stalling, in order to prevent the stall (Nachtigall, discussion).

APPENDIX

CALCULATION OF DRAG OF A WING WITH OUTER PARTS SPLIT INTO TANDEM WINGS

(a) Geometry

Wing with elliptic outline, span $b = 2R$, aspect ratio $A = b^2/S = 4R^2/S$ with wing area S. Major semiaxis of ellipse R, minor semiaxis $c_0/2$, chord length in the plane of symmetry c_0. $A = 8R/(\pi c_0)$ and $c_0 = 8R/(\pi A)$, because $S = \pi R c_0/2$ and $S = 4R^2/A$. Wing ends are cut off parallel to c_0 at distance r and replaced by m semiellipses, equal in size and with their major semiaxis being $R - r$. Area of the two cut wing ends S_2, remaining area of original wing S_1, area of one of the small ellipses composed of both halves S_3; $S_2 = mS_3$. Chord length at r is

$$c_r = c_0 \sqrt{(1 - x^2)} = 8R \sqrt{(1 - x^2)/(\pi A)}$$

with $x = r/R$.

Area of segment of ellipse

$$S_2/2 = 4R^2 \text{ arc cos } x/(\pi A) - 4R^2 x \sqrt{(1 - x^2)}/(\pi A).$$

Therefore,

$$S_2 = 8R^2[\text{arc cos } x - x\sqrt{(1 - x^2)}]/(\pi A)$$

and

$$y = S_2/S = 2[\text{arc cos } x - x\sqrt{(1 - x^2)}]/\pi.$$

Area of one small ellipse

$$S_3 = S_2/m = yS/m = 4yR^2/(mA),$$

its aspect ratio

$$A_3 = 4(R - r)^2/S_3 = 4R^2(1 - x)^2/S_3 = mA(1 - x)^2/y.$$

Mean chord length of original wing $\bar{c} = S/2R = 2R/A$, mean chord length of small elliptic wing $\bar{c}_3 = S_3/[2R(1 - x)] = \bar{c}y/[m(1 - x)]$. ·

(b) Drag Coefficients

Total lift equal in original wing (S) and modified wing ($S_1 + mS_3$). Elliptic lift distribution is assumed in the original (untwisted) wing. Therefore, C_L is constant from wing tip to wing tip. Further C_L has to be of equal value in all wing sections of S_1 and S_3. Induced drag coefficient in the inner, non-split part of the modified wing is assumed equal to that in the original wing, a condition not completely satisfied in reality.
Values of induced drag:

in the original wing

$$C_{Di} = C_L^2/(\pi A),$$

in the small elliptic wing

$$C_{Di3} = C_L^2/(\pi A_3) = C_{Di} y/[m(1 - x)^2],$$

in the wing after modification

$$C_{Di}^* = C_{Di}(1 - y) + yC_{Di3} = C_{Di}\{1 - y + y^2/[m(1 - x)^2]\}.$$

Change of induced drag

$$(C_{Di}^* - C_{Di})/C_{Di} = y\{y/[m(1 - x)^2]\}.$$

Coefficient of profile drag in the original wing, constant over span $2R$, and in all sections of inner part of wing S_1 is C_{Dp}, while coefficient of profile drag in a tandem wing S_3, constant over span $2(R - r)$, is C_{Dp3}.

Drag coefficient in the original wing $C_D = D_{Di} + C_{Dp}$, in the modified wing

$$C_D^* = C_{Di}^* + (1 - y) C_{Dp} + y C_{Dp3}.$$

Given $k = C_{Dp3}/C_{Dp}$, then

$$C_D^* = C_{Di}^* + (1 - y) C_{Dp} + ykC_{Dp}$$
$$= C_{Di}\{1 - y + y^2/[m(1 - x)^2]\} + C_{Dp}(1 - y + yk).$$

Change in total drag becomes

$$(C_D^* - C_D)/C_D = C_{Di}y\{y/[m(1 - x)^2] - 1\}/(C_{Di} + C_{Dp})$$
$$+ C_{Dp}y(k - 1)/(C_{Di} + C_{Dp}).$$

REFERENCES

1. Graham, R. R. Safety devices in wings of birds. *J. Roy. Aero. Soc.* **36**, 24–58 (1932).
2. Holst, E. von and Küchemann, D. Biologische und aerodynamische Probleme des Tierfluges. *Naturwissenschaften*, **29**, 348–362 (1941).
3. Nachtigall, W. Biophysik des Tierfluges. *Rheinisch-Westfälische Akad. Wiss., Vorträge*, N **236**, 73–152 (1973).
4. Nachtigall, W. and Kempf, B. Vergleichende Untersuchungen zur flugbiologischen Funktion des Daumenfittichs (*Alula spuria*) bei Vögeln. I. Der Daumenfittich als Hochauftriebserzeuger. *Z. vgl. Physiol.* **71**, 326–341 (1971).
5. Nachtigall, W. and Wieser, J. Profilmessungen am Taubenflügel. *Z. vgl. Physiol.* **52**, 333–346 (1966).
6. Oehme, H. Vergleichende Profiluntersuchungen an Vogelflügeln. *Beitr. Vogelk.* **16**, 301–312 (1970).
7. Oehme, H. Die Flügelprofile von Star und Türkentaube. *Forma et Functio*, **2**, 266–287 (1970).
8. Oehme, H. and Kitzler, U. Zur Geometrie des Vogelflügels (Untersuchungen zur Flugbiophysik und Flugphysiologie der Vögel II). *Zool. Jb. Phys.* **79**, 402–424 (1975).
9. Oehme, H. and Kitzler, U. Die Bestimmung der Muskelleistung beim Kraftflug der Vögel aus kinematischen und morphologischen Daten (Untersuchungen zur Flugbiophysik und Flugphysiologie der Vögel III). *Zool. Jb. Phys.* **79**, 425–458 (1975).
10. Schmitz, F. W. Zur Aerodynamik der kleinen Reynolds-Zahlen. Jb. 1953 d.WGL, 150 166 (1953).
11. Schmitz, F. W. "Aerodynamik des Flugmodells. Tragflügelmessungen bei kleinen Geschwindigkeiten" I u. II. Carl Lange, Duisburg (1960).
12. Sick, H. Morphologisch-funktionelle Untersuchungen über die Feinstruktur der Vogelfeder. *J. Orn.* **85**, 206–272 (1937).
13. Stresemann, E. "Aves" (Handb. d. Zoologie VII, 2). De Gruyter, Berlin und Leipzig (1927–1934).

Discussion Contribution

Separated Primary Feathers

R. H. J. BROWN

Dept. of Zoology, Cambridge University, England

All separated primary feathers show emargination and move in response to aerodynamic forces. In particular a first primary of an emarginate series will bend and twist in an air stream as the shaft is rotated. The effect is to produce a curve of lift against shaft angle which rises for some 10–15 degrees and then remains constant for a further 15–20 degrees. A group of such feathers behaves similarly provided the series is complete, starting at the first primary; only the first is stable on its own. Such an array is able to produce high values of lift coefficient, though with high drag (measured), so that the overall 'angle of presentation' to the air stream is uncritical; this arrangement has obvious advantages. The maximum development of emargination is seen in two very different types of bird. Firstly, those like pheasants or partridges which live in thick low vegetation: short wings are inevitable, and a steep climbing take-off necessary. The very short wing (aspect ratio 2–3) can be flapped along a horizontal plane while the tip feathers assume an optimum (?) lift configuration. Secondly, the large birds of prey: they are primarily soaring birds and much discussion has taken place on the functions of the emarginated tip in soaring. Why does such emargination not occur in marine soarers? The large land soarers have to take-off from flat ground in conditions of little or no wind by flapping a large gliding wing, most of which cannot, due to its low air speed, develop useful lift. The tip can develop high values of lift, under overload conditions, until the forward speed has risen to a value where the rest of the wing can develop useful lift. Whether emargination has a function in the reduction of induced drag in gliding flight, as has been suggested, remains to be demonstrated. If it has such an effect then it must be of use only in the lightly loaded wings of low aspect ratio, otherwise how can one explain its almost complete absence in marine soarers?

29. Scaling and Avian Flight.

VANCE A. TUCKER

Department of Zoology, Duke University, Durham, North Carolina 27706, U.S.A.

ABSTRACT

A bird's morphology, and the properties of the materials from which it is constructed, determine its flapping flight characteristics. The properties of biological materials are relatively unchanged, regardless of the size of the animal from which the material comes. Given these properties, one may explore the relations between changes in morphology and changes in the characteristics of flapping flight; in particular we examine the effect of changes in body mass and wing span, which describe the size of the bird. This paper analyzes the power required for flapping flight and the power available from the flight muscles for birds of different sizes. The difference between the power available and the power required determines the flight characteristics of speed, rate of climb and acceleration, as well as the ability to fly at all. These characteristics are computed for hypothetical birds with wing spans up to 15·5 m, a span that has recently been attributed to a pterosaur, an extinct flying reptile, on the basis of fossil remains.

INTRODUCTION

How large a bird can fly by flapping its wings? A certain amount of power is required to remain aloft in flapping flight in still air, depending on the characteristics of the wings and body and the flight speed. But the characteristics of the wings and body themselves depend upon the size of the muscles that provide the power for flight. The relations between body size, power required for flight and the power available in flapping birds and bats have become much better known in recent years (for example see Pennycuick,[6, 7] Tucker,[8, 9]). In this paper I will utilize known relations where possible, and assumed relations otherwise, to investigate the size limitations for flapping flight. My aim is to provide a structure of relations within which the assumptions can be modified as new information becomes available.

That the present assumptions might sometimes lead to exotic flying machines is of secondary importance.

The power required for level flapping flight through still air can be calculated after dividing and subdividing the power required into several categories that can be analyzed separately (Pennycuick,[6, 7] Tucker[8, 9, 10]). The first division separates the power required by the flight muscles from that required by the rest of the bird. Then, the muscle power (power output, P_0) is divided into three categories; (1) induced power ($P_{0, in}$), which supports the bird's weight; (2) parasite power ($P_{0, par}$) which overcomes the aerodynamic drag of the bird's body; and (3) profile power ($P_{0, pr}$), which is the muscle power in addition to 1 and 2 that is necessary for flapping the wings.

The power input required by the bird exclusive of its flight muscles is also divided into three categories: (1) the basal metabolic rate ($P_{i, B}$) which accounts for maintenance activities; (2) respiratory power ($P_{i, r}$) which moves air in and out of the lungs; and (3) heart power ($P_{i, h}$) which pumps blood. Thus, the total power (P_i) required for flight is the sum

$$P_i = (P_{0, in} + P_{0, par} + P_{0, pr})/\eta + P_{i, r} + P_{i, h} + P_{i, B} \tag{1}$$

where η is the efficiency of the flight muscles and is taken to have a value of 0·2.

Each of the terms in Eqn (1) has been analyzed for birds flying in air at different temperatures and altitudes—i.e., air characterized by different values of viscosity (μ) and density (ρ) (Tucker,[9]). The induced power is given by

$$P_{0, in} = 2(mg)^2/(0·7 \, \pi\rho b^2 V), \tag{2}$$

where m is the bird's mass, b the wing span, V the speed of flight, g the acceleration due to gravity, and ρ the air density. The parasite power is given by

$$P_{0, par} = 434 \, (\rho\mu)^{1/2} A_0 V^{5/2}, \tag{3}$$

where A_0 is the "equivalent flat plate" area of the bird's body, and μ is the air viscosity. By comparison with the equations given by Lighthill for induced and frictional power, it can be clearly seen that Eqns (2) and (3) have the right form to represent the physical effects described. Unfortunately no such expression is available for profile power, which should depend on wing area (since it includes the power needed to develop the extra thrust to overcome wing frictional drag), and on wing moment of inertia or length (since it includes any power needed to overcome wing inertia). The following formula satisfies both of these criteria, and agrees with a number of experiments (Tucker[9]), but is not based on the physics in the same way that Eqns (2) and (3) are. We take the profile power to be given by

$$P_{0, pr} = k_4 \, \mathrm{Re}^{-1/2}(P_{0, par} + P_{0, in}) \tag{4}$$

where $k_4 \, \mathrm{Re}^{-1/2} = 471 \, (\mu/(\rho m^{1/3} \, V))^{1/2}$, Re being the Reynolds number. The sum of the power required for respiration and the heart is given empirically by

$$P_{i,r} + P_{i,h} = 0.1 \, P_i, \tag{5}$$

and the power for basal metabolic rate can be measured. All quantities are expressed in the base units of the SI system or in combinations of them. The base units used here are m, kg, s. Power is in Watts. After appropriate substitutions, Eqn (1) becomes

$$P_i = 1.11 \left\{ [2(mg)^2/(0.7 \, \pi \rho b^2 V) + 434(\rho \mu)^{1/2} A_0 V^{5/2}] \right.$$
$$\left. \left[1 + 471 \left(\frac{\mu}{\rho m^{1/3} V} \right)^{1/2} \right] / \eta + P_{i,B} \right\}. \tag{6}$$

The quantities wing span (b), equivalent flat plate area (A_0) and basal metabolic rate can be scaled on body mass (m) (Tucker,[8]). The average relationships for birds as a whole are

$$b = 1.1 \, m^{1/3} \tag{7}$$

$$A_0 = 0.00334 \, m^{0.66} \tag{8}$$

and, for birds of the order Passeriformes,

$$P_{i,B} = 6.15 \, m^{0.724}. \tag{9}$$

Now Eqn (6) can be evaluated for an average, passerine bird given only the bird's body mass and air speed (V), together with air density, air viscosity and the acceleration due to gravity (g). Equation (6) predicts the requirements of birds (and bats) flying at cruising speed in a wind tunnel within about ten to fifteen per cent (Tucker[8]).

POWER AVAILABLE FOR FLIGHT

Equation (6) describes the power required for flight but says nothing about the power available from the flight muscles. If the flight muscles can provide some maximum sustained power output $(P_{0,max})$, this value, together with the power required (P_0) from the flight muscles, defines the minimum (V_{min}) and maximum (V_{max}) air speeds at which the birds can fly (Fig. 1).

Now consider what happens to the relation between $P_{0,max}$ and P_0 as a bird gets larger. Suppose the bird's wing span increases. The mass of the wings increases as the wings become longer and thicker. The induced power changes since it depends both on wing span and total body mass. If more power is required to fly, the mass of the flight muscles would increase, which will cause an increase in body mass and in parasite power. Profile power will also change, since it depends upon both wing area and wing moment of inertia. Also, as body mass increases, the wing mass increases as well since stronger wings are required to support a larger body mass. Then the induced power changes with the consequences described previously in this paragraph. Obviously, any change in wing span requires adjustment in all of the terms of Eqn (1). The enlarged bird will fly only if the power available after all of the adjustments are made equals or exceeds the calculated power requirements.

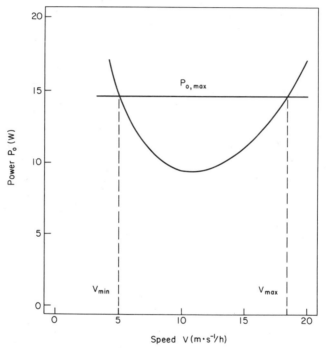

FIG. 1. Power required (curved line) for level, flapping flight, and maximum sustained power available ($P_{0,\,max}$) at different air speeds for the hypothetical bird described in Table I. The intersections of the power required and power available curves define the maximum and minimum speeds (V_{min}, V_{max}) between which the bird can carry out sustained flight.

When all the adjustments referred to above are made, equation (1) can be rewritten to yield an equation for $P_{0,\,max}$ of the following form:

$$P_{0,\,max} = [\phi_1(P_{0,\,max}, b, V) + \phi_2(P_{0,\,max}, V)][1 + \phi_3(b, V)] \qquad (10)$$

where ϕ_1, ϕ_2 and ϕ_3 indicate functions of the specified variables in air of constant density and viscosity. Equation (10) and the functions it contains are developed in the Appendix (see Eqn (31)). Solutions of Eqn (10) can be obtained for birds with a given wing span flying at different air speeds (Fig. 2) at sea level in air with density (ρ) of $1\cdot23 \, \text{kg m}^{-3}$ and viscosity (μ) of $17\cdot9 \times 10^{-6} \, \text{N s m}^{-2}$.

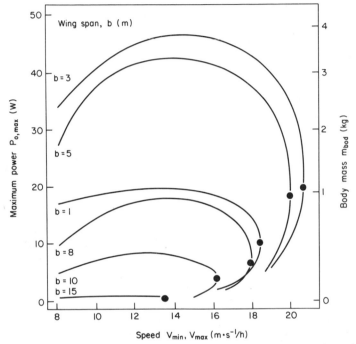

FIG. 2. Maximum power and body mass (exclusive of the wings) at various values of maximum speed for birds with different wing spans. The dots indicate the highest possible maximum air speed for each wing span.

The curves in Fig. 2 indicate that birds with a range of $P_{0,\text{max}}$ values, all with the same wing span, could fly, but at different maximum speeds. Or, since total body mass is an increasing function of $P_{0,\text{max}}$ (Eqns (14) and (20)) birds with a range of body sizes, all with the same wing span, could fly, but with different maximum speeds.

To simplify this analysis, I will consider birds of different wing spans that have values for $P_{0,\text{max}}$ and total body mass such that V_{max} is the highest possible for a particular wing span. The appropriate combinations of $P_{0,\text{max}}$ and V_{max} are indicated by dots in Figure 2. The evolution of birds with a given wing span but either heavier or lighter than the birds indicated by dots (i.e., with greater or smaller $P_{0,\,max}$ values) is of course possible. Heavier birds would fly more slowly and require more power than if they had the mass corresponding to the dot, but there might be offsetting advantages associated with greater mass. For example, the bird would be more robust and could accommodate heavier accessories such as a large beak or large leg muscles. On the other hand, lighter birds than those indicated by the dot fly more slowly but use less power.

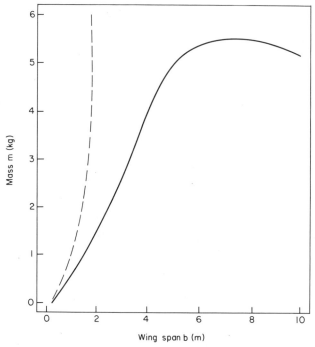

FIG. 3. Total masses of birds with different wing spans (solid curve), corresponding to the dots in Fig. 2. The dashed curve (calculated from equation (7)) indicates the mean masses of living birds with different wing spans.

Let us now look at certain morphological and flight characteristics of the birds described by dots in Fig. 2 (including a bird with a span of 0·25 meters that is not shown in Fig. 2). These characteristics comprise total body mass (Fig. 3), the ratio between wing mass and total body mass (Fig. 4), the maximum rate of climb, the maximum climb angle and the maximum acceleration (Fig. 5). The calculations on which these figures are based are described in the Appendix.

Few living birds that have wing spans of 2 m or more indulge in steady flapping flight, and they are more heavily built than their hypothetical counterparts (Fig. 3). If the present analysis is correct, they are flying at $P_{0,\,max}$ values that are near the maximum $P_{0,\,max}$ value for their wing spans (Fig. 2), rather than the maximum V_{max} values assumed here.

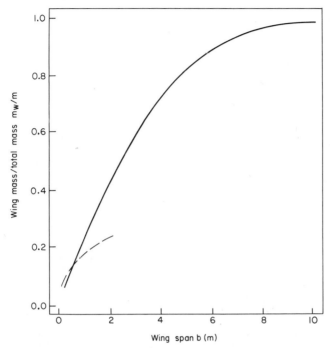

FIG. 4. The ratios of wing mass to total mass of birds with different wing spans, corresponding to the dots in Fig. 2. The dashed curve (calculated from Eqn (7) and the equation given in Table I) indicates the mean ratio for living birds.

Figure 4 indicates that as the hypothetical birds have larger wing spans, their wings make up a larger proportion of their total body mass. As an extreme example, a hypothetical bird with a 10 m wing span, although theoretically capable of flapping flight, would have a minuscule body making up only two per cent of its total mass. Such a bird would be impractically delicate—its total mass, if spread uniformly over its wing area (assuming an aspect ratio of 12 and tissue density of $10^3\,kg\,m^{-3}$) would form a film less than one mm thick.

Figure 5 shows that hypothetical birds with large wing spans would have low rates of climb, low climb angles, and low accelerations. Such birds could fly only over large level plains or seas or in areas suitable for soaring. For example, a bird with a wing span of 3 m would, in still air, have a maximum rate of climb less than 0.4 m s^{-1} and a climb angle less than 2·3 degrees, and it could change its speed in 10 seconds by less than one-fifth of its top speed. Since winds in the atmosphere commonly have vertical components of velocity greater than one m s^{-1}, a large bird could gain altitude more than three times as fast by flying in updrafts than it could by flapping its wings in still air. In contrast, a small bird flying in the same updraft would change

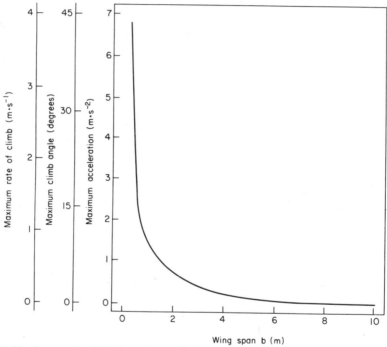

Fig. 5. Maximum rates of climb, maximum climb angles and maximum accelerations of birds in level flight with different wing spans, corresponding to the dots in Fig. 2. The single curve describes all these data approximately (within 5 % of the maximum ordinate value).

its maximum rate of climb by a factor of only 1·3. These considerations explain why larger flying birds spend much of their time on open plains, over water or over rough terrain where horizontal winds create updrafts.

The largest known wingspans for flying animals are claimed for the extinct reptilian pterosaurs. Bramwell[2] summarizes the aerodynamic characteristics of a *Pteranodon* with a wing span of 8·2 m; fossil remains claimed to represent a wing span of 15·5 m have recently been found (Lawson[5]). The latter figure is somewhat uncertain, since it is based on an extrapolation of the length of the humerus, which makes up only 6% of the estimated wing length. If creatures with such large spans really existed, steady flapping flight would be impractical for them. They could carry enough muscle mass for flapping flight only if they were extremely delicately built, with a total mass of about 5 kg of which only 16 g would comprise the body, exclusive of the wings, for a span of 15 m. This body mass would be half that of a house sparrow, and such a creature could flap steadily only if the boundary layer over the wings and body could be kept laminar. Any attempt to carry a bulkier body with more powerful muscles would result in the animal being too heavy to fly by flapping.

In summary, it is conceivable that flapping-flight animals very much larger than those alive today could be built from existing biological materials. However, these animals would have low rates of climb, acceleration and angles of climb that would restrict them to open areas or soaring conditions. The extreme delicacy of their construction would certainly render them susceptible to destruction by atmospheric turbulence as Bramwell[2] suggests.

APPENDIX

Birds' wings may be represented as a pair of tapered, cantilever beams, of constant width h and maximum depth l, in which each element of the beam length (wing span, b) supports the same load per unit length, and the stress (s) in the beams is constant. The total load on the wings is proportional to the mass of the body (m_{bod}) and

$$m_{bod} \propto \frac{shl^2}{b} \tag{11}$$

(Horton[4]). The mass of the wings (m_w) depends on their dimensions,

$$m_w \propto hlb. \tag{12}$$

Bone seems to have about the same tensile and compressive strengths for vertebrates of all sizes, so it is reasonable to assume that the stress in the beams representing the wings is the same for birds of all sizes. If

$$h \propto b \tag{13}$$

then combining Eqns (11) and (12) yields

$$m_w \propto m_{bod}^{1/2} b^2. \tag{14}$$

Since $b \propto m^{1/3}$ for living birds (Eqn 7), substituting into Eqn (14) yields

$m_w \propto m^{1 \cdot 17}$ approximately, which is close to the empirical relation of $m_w \propto m^{1 \cdot 11}$ (Greenewalt[3]).

Power Output of Flight Muscles

The maximum power output of a muscle is a product of four factors:

$$P_{0,\,max} = sAdf \tag{15}$$

where s is the maximum stress the muscle is able to exert, A is the cross-sectional area of the muscle, d is the distance of contraction, and f is the frequency of contraction. Since s is approximately constant for muscle (Alexander[1]),

$$P_{0,\,max} \propto Adf. \tag{16}$$

I assume that d is proportional to the wing span b, that

$$f \propto 1/b, \tag{17}$$

which fits the data for birds (Greenewalt[3]), and that the bodies of birds are isometric so that

$$A \propto m_{bod}^{2/3}. \tag{18}$$

Incorporating these assumptions into Eqn (15) yields

$$P_{0,\,max} \propto A \tag{19}$$

and

$$m_{bod} \propto P_{0,\,max}^{3/2}. \tag{20}$$

Induced Power

The induced power output ($P_{0,\,in}$) of the wings is given by Eqn (2) (Tucker[8]); for constant air density (ρ) and gravitational force (g) this implies

$$P_{0,\,in} = \frac{2g^2}{0 \cdot 7\pi\rho} \cdot \frac{(m_w + m_{bod})^2}{b^2 V}. \tag{21}$$

Since wing mass (m_w) is a function of body mass (m_{bod}) and wing span (Eqn (14))

$$P_{0,\,in} = \frac{2g^2}{0 \cdot 7\pi\rho} \cdot \frac{(k_1 m_{bod}^{1/2} b^2 + m_{bod})^2}{b^2 V} \tag{22}$$

where k_1 is a constant. Since m_{bod} is a function of $P_{0,\,max}$ (Eqn (20)), we have

$$P_{0,\,in} = \frac{2g^2}{0 \cdot 7\pi\rho} \cdot \frac{(k_2 P_{0,\,max}^{3/4} b^2 + k_3 P_{0,\,max}^{3/2})^2}{b^2 V} \tag{23}$$

where k_2 and k_3 are also constants.

Parasite Power

The parasite power output drags the bird's body through the air and is given by

$$P_{0,\,par} = \tfrac{1}{2}\rho C_D S V^3 \tag{24}$$

(Tucker[8]) where S is the cross-sectional area of the bird's body. The drag coefficient (C_D) can be assumed to be related to Reynolds number (Re) by the proportionality

$$C_D \propto (\text{Re})^{-1/2} \tag{25}$$

as long as the boundary layer remains laminar (Tucker[8]) where

$$\text{Re} = \frac{\rho l V}{\mu}, \tag{26}$$

l is some linear dimension of the bird's body, and μ is the air viscosity. For isometric bodies

$$l \propto S^{1/2} \tag{27}$$

$$S \propto A \tag{28}$$

so that

$$P_{0,\,par} \propto A^{3/4} V^{5/2} (\rho\mu)^{1/2}. \tag{29}$$

Since $P_{0,\,max} \propto A$ (Eqn (19))

$$P_{0,\,par} \propto P_{0,\,max}^{3/4} V^{5/2} (\rho\mu)^{1/2}. \tag{30}$$

Profile Power

Profile power is dissipated as the wings move through the air and is given by Eqn (4).

Maximum Power Output

The maximum power output can now be expressed by summing the induced power output (Eqn (23)), the parasite power output (Eqn (30)), and the profile power output (Eqn (4)) each evaluated at V_{max}. Thus

$$P_{0,\,max} = \left[\frac{2g^2 (k_2 P_{0,\,max}^{3/4} b^2 + k_3 P_{0,\,max}^{3/2})^2}{0 \cdot 7 \pi \rho b^2 V_{max}} + k_5 P_{0,\,max}^{3/4} V_{max}^{5/2} (\rho\mu)^{1/2} \right]$$
$$\times (1 + k_4 \, \text{Re}^{-1/2}). \tag{31}$$

There remains to evaluate the constants in Eqn (31). To achieve this, I will replace certain variables with appropriate values for a bird with a wing span of 1 m (Table I). These values will be indicated by appending the subscript O to the symbol for the variable.

TABLE I. Characteristics of a hypothetical bird with a wing span of 1 m flying at sea level. Parasite power is calculated from Eqn (3), and maximum power output from the sum of Eqns (2), (3) and (4).

Quantity	Symbol	Value
Air density	ρ	$1 \cdot 23 \text{ kg m}^{-3}$
Air viscosity	μ	$17 \cdot 9 \times 10^{-6} \text{ N s m}^{-2}$
Wing span	b_0	1 m
Total mass	m_0	$0 \cdot 751 \text{ kg}$
Body mass	$m_{\text{bod},O}$	$0 \cdot 612 \text{ kg}$
Wing mass*	$m_{w,O}$	$0 \cdot 139 \text{ kg}$
Parasite power	$P_{\text{par},O}$	$7 \cdot 94 \text{ W}$
Maximum power output	$P_{0,\text{max},O}$	$14 \cdot 6 \text{ W}$
Assumed maximum speed	$V_{\text{max},O}$	$18 \cdot 2 \text{ m s}^{-1}$

* $m_w = 0 \cdot 191 \text{ m}^{1 \cdot 11}$, derived from Greenewalt.[3]

Thus, from Eqns (14) and (20)

$$k_2 = \frac{m_{w,O}}{P_{0,\text{max},O}^{3/4} b_0^2} ; \tag{32}$$

from Eqn (20)

$$k_3 = \frac{m_{\text{bod},O}}{P_{0,\text{max},O}^{3/2}} ; \tag{33}$$

from Eqn (30)

$$k_5 = \frac{P_{0,\text{par},O}}{P_{0,\text{max},O}^{3/4} V_{\text{max},O}^{5/2} (\rho_0 \mu_0)^{1/2}} . \tag{34}$$

Finally

$$P_{0,\text{pr},O} = 471 \left(\frac{\mu_0}{\rho_0 m^{1/3} V_O} \right)^{1/2} (P_{0,\text{par},O} + P_{0,\text{in},O}) \tag{35}$$

(Tucker[9]). This means that, since

$$m_O^{1/3} = b_0 / 1 \cdot 1 \tag{36}$$

(Tucker[8]) when $\text{Re} = \rho b V / \mu$,

$$P_{0,\text{pr},O} = 494 (R_{e,O})^{-1/2} (P_{0,\text{par},O} + P_{0,\text{in},O}), \tag{37}$$

or, from Eqn (4), $k_4 = 494$.

The values in Table I, together with Eqns (32), (33) and (34) yield the following values for the constants of Eqn (31):

$$k_2 = 0.0186, \qquad k_3 = 0.0109, \qquad k_4 = 494, \qquad k_5 = 0.160.$$

REFERENCES

1. Alexander, R. M. Muscle performance in locomotion and other strenuous activities. *In* "Comparative Physiology". (Bolis, L., Schmidt-Nielsen, K., and Maddrell, S. H. P., Eds). North Holland, Amsterdam. pp. 1–21 (1973).
2. Bramwell, C. D. Aerodynamics of *Pteranodon. Biol. J. Linn. Soc.* 3, 313–328 (1971).
3. Greenewalt, C. H. Dimensional relationships for flying animals. *Smithsonian Misc. Collections* 144, 1–40 (1962).
4. Horton, H. L. (Ed.) "Machinery's Handbook". Industrial Press, Inc., New York (1970).
5. Lawson, D. A. Pterosaur from the latest Cretaceous of West Texas: discovery of the largest flying creature. *Science*, 187, 947–948 (1975).
6. Pennycuick, C. J. Power requirements of horizontal flight in the pigeon. *J. Exp. Biol.* 49, 527–555 (1968).
7. Pennycuick, C. J. The mechanics of bird migration. *Ibis*, 111, 525–556 (1969).
8. Tucker, V. A. Bird metabolism during flight: evaluation of a theory. *J. Exp. Biol.* 58, 689–709 (1973).
9. Tucker, V. A. Energetics of natural avian flight. *In* "Avian Energetics", (Paynter, R. A., Jr., Ed.) Publ. Nuttall Ornith. Club, No. 15, 298–328 (1974).
10. Tucker, V. A. Aerodynamics and energetics of vertebrate fliers. *In* "Swimming and Flying in Nature", Vol. 2 (Wu, T. Y., Brokaw, C. J. and Brennen, C., Eds.). Plenum Press, New York, pp. 845–867 (1975).

PART V

30. The Sustained Power Output from Striated Muscle.

TORKEL WEIS-FOGH† AND R. McN. ALEXANDER

Department of Zoology, University of Cambridge, England
and
Department of Pure and Applied Zoology, University of Leeds, England.

Note: This paper grew from a talk given in the general discussion at the Symposium by Torkel Weis-Fogh. It developed in discussion between us at and after the Symposium, and was almost ready for submission at the time of his death. However he did not see the final draft so I cannot be certain that every passage would have met with his full approval.

R.McN.A.

ABSTRACT

The myofibrils of vertebrate striated muscle and of some insect muscles exert isometric stresses in the region of $0.4 \ MNm^{-2}$, and the fastest known muscles have intrinsic speeds in the region of $25 \ s^{-1}$. Muscles with these properties making repeated contractions can deliver power up to a maximum of about $500 \ W \ (kg \ myofibril)^{-1}$, or $250 \ W \ (kg \ muscle)^{-1}$ for aerobic muscles containing an appropriate fraction of mitochondria. This maximum power can be delivered only at repetition frequencies above about $10 \ s^{-1}$: at lower repetition frequencies less power can be delivered, even when the intrinsic speed is $25 \ s^{-1}$. Muscles which operate at frequencies below $10 \ s^{-1}$ tend to have lower intrinsic speeds, so the power actually delivered by them is lower than the theoretically attainable maximum for the frequency in question.

INTRODUCTION

The functions of many muscles require more or less regular cycles of contraction and relaxation. This is true of many muscles used for running, swimming and flight and also of respiratory muscles and heart muscle. This

† Deceased.

paper is concerned with the power output, averaged over a complete cycle of contraction and relaxation, which can be sustained by such muscles. Our purpose is to define the main parameters which limit this power output and to estimate, on the basis of available physiological data, the maximum power output attainable. A simplified approach is adopted to establish the principles and obtain an approximate numerical result.

The power output of a muscle is limited by the rate at which its metabolic processes can produce ATP and by the rate at which its myofibrils can transduce chemical energy to mechanical work. The metabolic and mechanical processes will have to be considered separately. They occur in separate fractions of the volume of the muscle. Consider a muscle which has a fraction q of its volume filled by myofibrils and the remaining fraction $(1 - q)$ filled by mitochondria and other structures concerned with the production of ATP. Let the maximum mechanical power which the myofibrils can deliver in repetitive contractions be $\Pi_{f, \max}$ per unit volume of myofibril. Let the maximum possible rate of production of ATP, per unit volume of mitochondria etc., be just sufficient to support a mechanical power output $\Pi_{m, \max}$. Then the mechanical power output per unit volume of complete muscle, P, cannot exceed $q\Pi_{f, \max}$ and cannot exceed $(1 - q)\Pi_{m, \max}$. In an ideally proportioned muscle the maximum power per unit volume will be given by

$$P_{\max} = q\Pi_{f, \max} = (1 - q)\Pi_{m, \max}. \tag{1}$$

Elimination of q from this equation yields

$$P_{\max} = \Pi_{m, \max}\Pi_{f, \max}/(\Pi_{m, \max} + \Pi_{f, \max}). \tag{2}$$

$P_{\max}$, $\Pi_{m, \max}$ and $\Pi_{f, \max}$ are powers per unit volume. They can be converted to power per unit mass by division by the density ρ which is assumed to be the same for both fractions of the muscle. The density of mammal muscles and presumably of the other muscles with which we will be concerned is about 1060 kg m^{-3}.[25]

The next two sections of this paper are attempts to obtain realistic values for $\Pi_{f, \max}$ and $\Pi_{m, \max}$.

THE MAXIMUM POWER OUTPUT OF THE MYOFIBRILS

The properties of the myofibrils.

The stress developed by a muscle contracting isometrically (i.e. without being allowed to shorten) depends on the length at which it is held. Fig. 1(a) is based on data for a frog leg muscle[15] but can be taken as typical of the muscles which have been investigated.

Consider a muscle whose myofibrils develop their maximum isometric stress σ_0 at length l_0, and a smaller isometric stress σ_l at some other length l. These stresses are defined in terms of the cross-sectional area of myofibrils: the corresponding stresses for the whole muscle are $q\sigma_0$ and $q\sigma_l$. For any length l $(> l_0)$ there is a corresponding length l' $(< l_0)$ at which the same

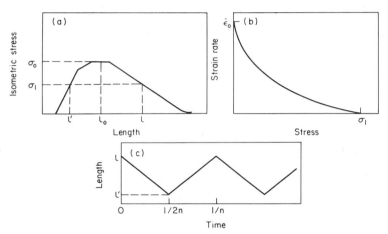

FIG. 1. Graphs illustrating assumptions made about the properties of the muscle and its mode of operation. (a) shows the relationship between the stress developed in isometric contraction, and muscle length. (b) shows the relationship between strain rate, in isotonic contraction, and stress at length l. (c) shows the assumed pattern of shortening and elongation. Further explanation is given in the text.

isometric stress σ_l is developed. It will be assumed that every contraction considered in this paper starts and ends at such a corresponding pair of lengths. This assumption seems reasonable but we have no evidence as to whether it is realistic. Contraction from length l to length l' involves total strain $\varepsilon_{tot} = (l - l')/l_0$. If a contraction starts at the maximum length at which any stress can be exerted and continues to the minimum length at which stress can be exerted the total strain will have its maximum possible value, ε_{max}.

Figure 1(a) would be broadly similar in shape to a half cycle of a cosine curve, if it were symmetrical instead of being skew. The relationship between σ_l and ε_{tot} derived from it is represented very closely by the equation

$$\sigma_l = \sigma_0 \cos\left(\pi\varepsilon_{tot}/2\varepsilon_{max}\right) \tag{3}$$

If the muscle contracted infinitely slowly from length l to length l' it could exert at every intermediate length the appropriate isometric stress and the work done per unit volume of myofibrils would be W_{max}, where

$$W_{max} = -\int_{l}^{l'} \sigma_l \cdot \mathrm{d}l/l_0. \tag{4}$$

By inserting the value of σ_l from (3) and integrating we find

$$W_{max} = (2\sigma_0 \varepsilon_{max}/\pi) \sin(\pi \varepsilon_{tot}/2\varepsilon_{max}). \tag{5}$$

This is the maximum work per unit volume obtainable for given ε_{tot}, because less stress is exerted during shortening at a finite rate than in isometric contraction. The rate of shortening can be expressed as the strain rate $\dot{\varepsilon}$, which is the rate of shortening in muscle lengths per unit time, $(\mathrm{d}l/\mathrm{d}t)/l_0$. It has been shown for a wide variety of muscles that the stress σ developed at length l during shortening at strain rate $\dot{\varepsilon}$ can be represented very accurately by Hill's equation, as generalized by Abbott and Wilkie[1]

$$\sigma = (b\sigma_l - \dot{\varepsilon}a)/(b + \dot{\varepsilon}) \tag{6}$$

where a and b are constaints for the muscle in question, with dimensions of stress and strain rate, respectively. This relationship is illustrated by Fig. 1(b). The isometric stress σ_l is only developed when the strain rate is zero and the maximum strain rate $\dot{\varepsilon}_0$ is only attained when the stress is zero. $\dot{\varepsilon}_0$ is often referred to as the intrinsic speed of the muscle. It can be shown by putting $\sigma = 0$ in Eqn (6) that $\dot{\varepsilon}_0 = b\sigma_l/a$.

It will be convenient in the equations which follow to use a dimensionless quantity $p = \sigma_l/a$, which will be called the characteristic stress ratio. Note that $\dot{\varepsilon}_0 = pb$. It will also be convenient to assume that p is constant for a given muscle, which is not strictly true. This assumption simplifies the mathematics and it will be shown later that the errors it introduces are not serious. By inserting p and $\dot{\varepsilon}_0$ in Eqn (6) we obtain

$$\sigma = \sigma_l(\dot{\varepsilon}_0 - \dot{\varepsilon})/(\dot{\varepsilon}_0 + p\dot{\varepsilon}). \tag{7}$$

Equations (3) and (7) describe well the behaviour of muscles in tetanic contractions, and it will be assumed that they also describe the behaviour of the muscles we are considering in the living animal.

We will assume further that shortening and re-extension of the muscle occur as shown in Fig. 1(c). The repetition frequency is n so the time occupied by a complete cycle of shortening and extension is $1/n$. It is assumed that half this time is occupied by shortening and half by extension. This seems reasonably realistic: the forward and back strokes of a leg are often very roughly equal in duration, as are the up and down strokes of a wing, the systole and diastole of a heart, and so on. It is also assumed that shortening occurs at a constant speed. The total strain ε_{tot} occurs in time $1/2n$ so the (constant) strain rate is given by

$$\dot{\varepsilon} = 2n\varepsilon_{tot}. \tag{8}$$

To make the shortening speed constant the muscle must be activated rapidly at the end of the extension phase, and the level of activation must remain high throughout shortening and then decline rapidly to zero.

The optimum strain rate.

In this section we will derive an expression for the optimum strain rate which will enable a muscle to deliver most power at a given repetition frequency. The expression will be used to calculate values of $\Pi_{f,\,max}$.

If Eqn (7) holds σ/σ_l will be constant throughout a contraction at constant strain rate, and the work done in each contraction per unit volume of myofibrils will be

$$W = (\sigma/\sigma_l)\, W_{max} \tag{9}$$

where W_{max} is the work done in a very slow contraction involving the same total strain. Equation (7) gives us an expression for (σ/σ_l) and Eqn (5) gives an expression for W_{max}. Inserting these values in Eqn (9) and noting that $\varepsilon_{tot} = \dot{\varepsilon}/2n$ (from Eqn (8))

$$W = \frac{2\sigma_0 \varepsilon_{max}(\dot{\varepsilon}_0 - \dot{\varepsilon})}{\pi(\dot{\varepsilon}_0 + p\dot{\varepsilon})} \sin\left(\frac{\pi\dot{\varepsilon}}{4n\varepsilon_{max}}\right). \tag{10}$$

This amount of work is done in each contraction and the repetition frequency is n so the power output per unit volume of myofibrils is

$$\Pi_f = nW = \frac{2n\sigma_0 \varepsilon_{max}(\dot{\varepsilon}_0 - \dot{\varepsilon})}{\pi(\dot{\varepsilon}_0 + p\dot{\varepsilon})} \sin\left(\frac{\pi\dot{\varepsilon}}{4n\varepsilon_{max}}\right). \tag{11}$$

The value of Π_f depends on the value of $\dot{\varepsilon}$. It is small when $\dot{\varepsilon}$ is small and also when $\dot{\varepsilon}$ approaches its maximum possible value, $\dot{\varepsilon}_0$. It has a maximum value at some intermediate value of $\dot{\varepsilon}$ which will be called $\dot{\varepsilon}_{opt}$. Differentiation of Eqn (11) with respect to $\dot{\varepsilon}$ gives

$$\frac{d\Pi_f}{d\dot{\varepsilon}} = \frac{2n\sigma_0 \varepsilon_{max}}{\pi(\dot{\varepsilon}_0 + p\dot{\varepsilon})} \left[\frac{\pi(\dot{\varepsilon}_0 - \dot{\varepsilon})}{4n\varepsilon_{max}} \cos\left(\frac{\pi\dot{\varepsilon}}{4n\varepsilon_{max}}\right) - \frac{\dot{\varepsilon}_0(1 + p)}{(\dot{\varepsilon}_0 + p\dot{\varepsilon})} \sin\left(\frac{\pi\dot{\varepsilon}}{4n\varepsilon_{max}}\right) \right]. \tag{12}$$

When $\dot{\varepsilon} = \dot{\varepsilon}_{opt}$ and Π_f is maximal, $d\Pi_f/d\dot{\varepsilon}$ must be zero. By setting $d\Pi_f/d\dot{\varepsilon} = 0$ in (12) we find

$$(4n\varepsilon_{max}/\pi)\tan(\pi\dot{\varepsilon}_{opt}/4n\varepsilon_{max}) = (\dot{\varepsilon}_0 + p\dot{\varepsilon}_{opt})(\dot{\varepsilon}_0 - \dot{\varepsilon}_{opt})/\dot{\varepsilon}_0(1 + p). \tag{13}$$

The following dimensionless quantities are defined so that conclusions obtained from Eqns (11) and (13) can be the more easily expressed in general form.

(i) *The relative repetition frequency* $(n\varepsilon_{max}/\dot{\varepsilon}_0)$. This is a dimensionless number which relates the repetition frequency to properties of the muscle.

(ii) *The relative shortening rate* $(\dot{\varepsilon}/\dot{\varepsilon}_0)$. This is the strain rate expressed as a fraction of the intrinsic speed.

(iii) *The amplitude fraction* $(\varepsilon_{tot}/\varepsilon_{max})$. This is the total strain which occurs in each contraction expressed as a fraction of the maximum possible strain.

(iv) *The characteristic power number* $(\Pi_f/\sigma_0\dot{\varepsilon}_0)$. This dimensionless number relates the power per unit volume of myofibrils to the maximum isometric stress and the intrinsic speed of the muscle. The instantaneous power output per unit volume of myofibrils with stress σ and strain rate $\dot{\varepsilon}$ is $\sigma\dot{\varepsilon}$. However the characteristic power number is always much less than 1 because no muscle can achieve stress σ_0 and strain rate $\dot{\varepsilon}_0$ simultaneously (see Fig. 1(b)) and because in the conditions of working which have been defined the muscle is contracting for only half the time.

Equation (13) can be solved graphically to find the relative shortening rate $(\dot{\varepsilon}_{opt}/\dot{\varepsilon}_0)$ which gives maximum power output for any combination of relative repetition frequency $(n\varepsilon_{max}/\dot{\varepsilon}_0)$ and characteristic stress ratio (p). Results are given in Table I. Division of the relative shortening rate by the relative repetition frequency gives $(\dot{\varepsilon}/n\varepsilon_{max})$ which equals $2\varepsilon_{tot}/\varepsilon_{max}$ (from Eqn (8)). Amplitude fractions obtained in this way are given in Table II.

Tables I and II show that to achieve the maximum power output possible at low relative repetition frequencies, a muscle should adjust its relative shortening rate to a low value such that the amplitude fraction is 1 (i.e. the muscle should use the whole of its useful range of lengths). At high relative

TABLE I. Optimum relative shortening rates $(\dot{\varepsilon}_{opt}/\dot{\varepsilon}_0)$, for given relative repetition frequencies $(n\varepsilon_{max}/\dot{\varepsilon}_0)$ and characteristic stress ratios (p).

$n\varepsilon_{max}/\dot{\varepsilon}_0$	$\dot{\varepsilon}_{opt}/\dot{\varepsilon}_0$			
	$p = 1$	$p = 3$	$p = 5$	$p = 10$
0·01	0·02	0·02	0·02	0·02
0·03	0·06	0·05	0·05	0·05
0·10	0·17	0·15	0·14	0·12
0·30	0·33	0·27	0·24	0·20
1·00	0·40	0·33	0·28	0·23

TABLE II. Optimum amplitude fractions ($\varepsilon_{tot}/\varepsilon_{max}$) for given relative repetition frequencies ($n\varepsilon_{max}/\dot\varepsilon_0$) and characteristic stress ratios (p).

$n\varepsilon_{max}/\dot\varepsilon_0$	Optimum ($\varepsilon_{tot}/\varepsilon_{max}$)			
	$p = 1$	$p = 3$	$p = 5$	$p = 10$
0·01	1·00	1·00	0·95	0·90
0·03	0·93	0·90	0·87	0·80
0·10	0·83	0·75	0·69	0·60
0·30	0·55	0·45	0·40	0·33
1·00	0·20	0·16	0·14	0·11

repetition frequencies, however, it should adjust its relative shortening rate to 0·2–0·4 (the precise value depending on the characteristic stress ratio; see Table I) and the amplitude fraction will be small. The tables also show the transition between these extremes. Note that Hill's[18] conclusion that maximum power output is obtained at relative shortening rates around 0·3 applies to repetitive contraction only if the relative repetition frequency is high.

Table III shows maximum characteristic power numbers obtained from Eqn (11) by inserting the values of optimum relative shortening rate given in Table I. Note that these numbers are high and relatively constant (for given p) at relative repetition frequencies above about 0·3, but are lower at lower frequencies. To obtain from a muscle a power output near the maximum of which it is capable it must be operated at a reasonably high repetition frequency.

It has been assumed that the characteristic stress ratio $p(=\sigma_l/a)$ is constant. For real muscles, a is nearly constant but σ_l changes during shortening.[1, 6]

TABLE III. Maximum characteristic power number ($\Pi_{f, max}/\sigma_0\dot\varepsilon_0$) obtainable for given relative repetition frequencies ($n\varepsilon_{max}/\dot\varepsilon_0$) and characteristic stress ratios (p).

$n\varepsilon_{max}/\dot\varepsilon_0$	$\Pi_{f, max}/\sigma_0\dot\varepsilon_0$			
	$p = 1$	$p = 3$	$p = 5$	$p = 10$
0·01	0·006	0·006	0·006	0·005
0·03	0·017	0·015	0·014	0·012
0·10	0·044	0·035	0·029	0·021
0·30	0·074	0·050	0·039	0·026
1·00	0·085	0·055	0·042	0·027

What errors are introduced by assuming p is constant and assigning to it the value p_0 ($=\sigma_0/a$)? At low relative repetition frequencies the error is negligible because the work done in each contraction is approximately equal to W_{max} (Eqn 5) which is independent of p. At high relative repetition frequencies the error is negligible because the amplitude fraction is small so that contraction is limited to the range of lengths at which $p \simeq p_0$. At intermediate relative repetition frequencies appreciable errors must occur but we find them acceptable because our calculations are in any case only approximate and because the assumption which introduces the errors simplifies the mathematical treatment.

Numerical values for the power output of the myofibrils

Data assembled by Close[12] shows that ($1/p_0$) is 0·25–0·30 for various fast vertebrate striated muscles, about 0·15 for slow twitch muscles and 0·10 for a very slow tortoise muscle. It is about 0·9 for rabbit heart muscle.[14] Most striated muscles probably have characteristic stress ratios between 1 and 10 ($1/p_0$ between 0·10 and 1). This range of values accordingly appears in Tables I to III.

Most measurements of the maximum isometric stress developed by frog limb muscles have given values between 0·2 and 0·35 MNm^{-2} (see Ref. 11). These are stresses based on total cross-sectional area (i.e. they are values of $q\sigma_0$). The muscles concerned consist mainly of white fibres with only a small proportion of mitochondria[34] so q is probably more than 0·8 and σ_0 must be around 0·25–0·4 MNm^{-2}. Rat fast twitch and slow twitch muscles have both been found to develop maximum isometric stress ($q\sigma_0$) about 0·3 MNm^{-2} [12,43]. If their composition is similar to that of mouse muscles,[33] σ_0 must be between 0·35 and 0·4 MNm^{-2}. Locust wing muscle develops a maximum isometric stress ($q\sigma_0$) of about 0·16 MNm^{-2} at 11°C[39] and its mitochondria occupy about 0·3 of its cross-sectional area so $q \simeq 0·7$.[9] Hence $\sigma_0 \simeq 0·23$ MNm^{-2} at 11°C, and there are strong indications that much higher stresses would be exerted at higher temperatures.[39] These and other data[3] lead us to suggest $\sigma_0 \simeq 0·4\ MNm^{-2}$ as a typical value for striated muscles. Some obliquely striated muscles with long sarcomeres can develop higher isometric stresses but cannot deliver particularly high power because their intrinsic speeds are relatively low.[8,32]

Whereas σ_0 is probably fairly uniform for a wide range of vertebrate and insect striated muscles, the intrinsic speed is very variable. The highest intrinsic speeds which have been observed seem to be about 25 s^{-1} for a mouse finger muscle and a rat eye muscle at 35°C.[12,13] These muscles might have a slightly higher temperature in the body and consequently be a little faster. Locust wing muscle has an intrinsic speed of 6–9 s^{-1} at 30°C but often operates at or above 35°C when it would presumably be faster.[10]

Faster muscles may be discovered but for the present it seems likely that 25 s^{-1} is close to the maximum intrinsic speed attainable in Nature.

For a muscle of this speed, capable of developing a maximum isometric stress of 0.4 MNm^{-2}, $\sigma_0 \dot{\varepsilon}_0$ would be 10 MW m^{-3}. If its characteristic stress ratio lay between 3 and 4 (the usual range for fast muscles,[12]) $\Pi_{f,\,max}$ at high repetition frequencies would be about $0.05 \sigma_0 \dot{\varepsilon}_0$ or 0.5 MW m^{-3} (Table III). The maximum power output per unit mass of myofibrils, $\Pi_{f,\,max}/\rho$ would then be about 500 W kg^{-1}.

THE MAXIMUM POWER OUTPUT OF THE METABOLIC APPARATUS

We will now estimate $\Pi_{m,\,max}$, the maximum mechanical power output which can be supported by unit volume of the metabolic fraction of muscle. In our present state of knowledge we can only do this by reference to the measured power output of muscles of known composition.

We will consider the wing muscles of the hummingbird *Amazilia* and the locust *Schistocerca*. These are aerobic muscles, and the mitochondria account for most of the metabolic fraction. In both cases the mitochondria are densely filled with cristae (much more densely than the mitochondria of many other tissues) and it seems likely that their rate of ATP production, per unit volume, may be near the maximum attainable.

When hummingbirds hover their wing muscles deliver at least 150 W kg^{-1}.[20, 42] This value is calculated from the mean metabolic rate during prolonged hovering, assuming that mechanical energy is generated with efficiency 0.2. Much higher metabolic rates occur briefly during hovering and the wing muscles can probably deliver 300 W kg^{-1} or even more. The myofibrils and mitochondria each occupy about half the volume of the muscle ($q \simeq 0.5$[21]) so the power output (Π_m/ρ) which can be supported by unit mass of mitochondria is at least 300 and probably 600 W kg^{-1}.

Locust wing muscle delivers 80 W kg^{-1} in level flight.[38, 40, 41] Flying locusts can generate lift up to twice body weight, when the wing muscles deliver 170 W kg^{-1}.[40] The mitochondria occupy 0.3 of the volume of the muscle[9] so the corresponding value of the mitochondrial power output Π_m/ρ is $170/0.3 = 570 \text{ W kg}^{-1}$.

These data lead us to suggest that the maximum attainable value of Π_m/ρ for aerobic muscles is probably in the region of 500 W kg^{-1}.

We lack adequate data for calculating $\Pi_{m,\,max}$ for anaerobic muscles. It can probably be much higher than $\Pi_{f,\,max}$, for some white muscles such as the wing muscles of domestic fowl contain very small proportions of sarcoplasm.[29] If this is the case P_{max} for anaerobic muscles may be only a little less than $\Pi_{f,\,max}$ (Eqn 2).

TABLE IV. Estimates of the power output of muscles in strenuous activities. Examples involving high (power output/muscle mass) have been selected from the data available. Asterisks (*) indicate power outputs which can be maintained only for a few seconds and presumably depend partly on anaerobic metabolism. Daggers (†) indicate power outputs calculated from oxygen consumption on the assumption of an efficiency of 0·2. Question marks (?) indicate estimates based on incomplete data.

	Body mass kg	Activity	Muscles	Repetition frequency, s^{-1}	Power output, W	Muscle mass, kg	Power/muscle mass $W\,kg^{-1}$	References
Insects								
Schistocerca	2×10^{-3}	flight (extra load)	flight muscles	19	0·06	$3·6 \times 10^{-4}$	170	38, 42
Drosophila	2×10^{-6}	hovering flight	flight muscles	240	$5·6 \times 10^{-5}$	$3·6 \times 10^{-7}$	160	38, 42
others		flight	flight muscles				70–360†	41
Birds Vultur	8	fast flight	flight muscles	2·5			54*	24
Columba	0·4	fast climbing flight	flight muscles	9·4	20*	0·09	220*	27, 28
Melopsittacus	0·035	fast flight	flight muscles	14	6†	9×10^{-3}	150†	37
Amazilia	5×10^{-3}	hovering	flight muscles	35	0·23(–0·45?)	$1·5 \times 10^{-3}$	150(–300?)	42
Bats								
Pteropus	0·78	fast flight	flight muscles		10†	0·09	110†	35, 46
Phyllostomus	0·093	fast flight	flight muscles		2·0†	0·01	200†	35, 46
Dolphin (Tursiops)	89	fast swimming	swimming muscles	4?	1660*	15	110*	2, 16
Man	70	cycling	leg muscles	1·5?	1000*	12?	80*	45
	70	violent exercise	heart	3	11	0·25	45	7, 22

THE MAXIMUM POWER OUTPUT OF WHOLE MUSCLE

Estimates of $\Pi_{f,\,max}$ and $\Pi_{m,\,max}$ derived from the preceding sections can be used with Eqn (2) to estimate P_{max}. Figure 2 has been prepared in this way. Values of $\Pi_{f,\,max}$ have been taken from Table III for $p = 3$ and $\sigma_0\dot\varepsilon_0 = 10\ \mathrm{MW\,m^{-3}}$ (see above). We note that $\varepsilon_{max} = 1\cdot1$ for a frog leg muscle[15]), $0\cdot5$ for a rat eye muscle[3] and $0\cdot8$ for lo⌐ ᵗ flight muscle.[39] In preparing Fig. 2, ε_{max} has been taken to be $1\cdot0$. The ᵤ   ᵢity of the muscle, ρ, has been assumed to be $1060\ \mathrm{kg\,m^{-3}}$[25] $(\Pi_{m,\,max}/\rho)$ has been assumed to be $500\ \mathrm{W\,kg^{-1}}$ for aerobic muscle. The graphs indicate that very fast aerobic muscles should be able to deliver about $250\ \mathrm{W\,kg^{-1}}$ and that very fast anaerobic muscles with low proportions of sarcoplasm may be able to deliver nearly $500\ \mathrm{W\,kg^{-1}}$, provided in each case that they operate at repetition frequencies above about $10\ \mathrm{s^{-1}}$. At lower frequencies less power can be delivered.

Table IV shows some of the highest known power outputs for muscles in intact animals. Those which involve repetition frequencies above $10\ \mathrm{s^{-1}}$ are all delivered by muscles working aerobically and are reasonably near the

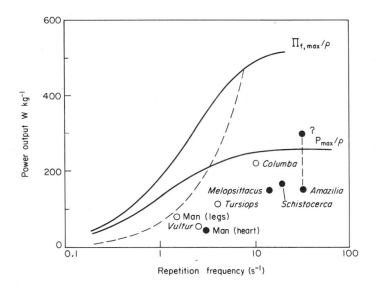

Fig. 2. Graphs showing the maximum obtainable mechanical power output per unit mass of myofibrils $(\Pi_{f,\,max}/\rho)$ and the maximum per unit mass of aerobic muscle (P_{max}/ρ) estimated as explained in the text, plotted against repetition frequency. The broken line shows values which Π_f/ρ would have if $\varepsilon_0 = 3\cdot3n$. Some observed power outputs taken from Table IV are also shown. Filled symbols indicate aerobic power outputs.

theoretical limit for aerobic muscles at high frequencies ($250 \, \text{W kg}^{-1}$, see Fig. 2). Some of those which involve lower repetition frequencies are from aerobic and some from anaerobic working but all fall well below the theoretical limits for the frequencies concerned. This will be discussed in the next section of this paper.

Insect fibrillar muscle was not considered in the theoretical sections of this paper because Hill's equation is inappropriate to its oscillatory manner of working (see Ref. 44). However, *Drosophila* and some of the "other insects" of Table IV have fibrillar flight muscles and it appears that their power outputs are similar to that of other fast striated muscles. One would expect $\Pi_{m, \max}$ to be about the same for them as for the other fast muscles. However the maximum difference in stress between contraction and relaxation of *Lethocerus* flight muscles is only about $0.2 \, \text{MNm}^{-2(30)}$ which is lower than the values of σ_0 obtainable from vertebrate striated muscle.

THE RELATIONSHIP BETWEEN REPETITION FREQUENCY AND INTRINSIC SPEED

In a few cases the intrinsic speed $\dot{\varepsilon}_0$ of a muscle and the repetition frequency n at which it operates are both known (Table V). It appears that muscles

TABLE V. The intrinsic speeds of muscles ($\dot{\varepsilon}_0$) and repetition frequencies (n) at which they work. In the cases of the mouse, rat and cat limb muscles the frequency given is the stride frequency at the trot/gallop transition. For *Schistocerca*, *Pecten* and *Octopus* it is the observed frequency of the wing beat or of swimming movements. For the rabbit heart muscle it is the resting pulse frequency. EDL, extensor digitorum longus. An asterisk indicates an estimate based on data for a related species of similar size.

	Body mass, kg	Muscle	n, s^{-1}	Ref.	$\dot{\varepsilon}_0$, s^{-1}	Ref.	$n/\dot{\varepsilon}_0$
mouse	0·025	EDL	7·5	17	22	12	0·34
		soleus	7·5	17	11	12	0·68
rat	0·25	EDL	5·5	17	17	12	0·32
		soleus	5·5	17	7	12	0·79
cat	2·5	quadriceps	4·0*	17	11	12	0·36
		soleus	4·0*	17	5	12	0·80
rabbit	2	heart	3·3	31	3	14	1·1
Schistocerca	0·002	wing	18	40	9	10	2
Pecten		striated adductor	2·5*	31	3	31	0·8
Octopus		funnel retractor	1*	36	2·4	23	0·4

which operate at low frequencies have relatively low intrinsic speeds so that in every case the range of repetition frequencies at which high power output is required is above $0.3\,\dot{\varepsilon}_0$ (i.e. $\dot{\varepsilon}_0 \leqslant 3.3n$). If ε_{max} is about 1 as seems to be typical for fast striated muscles (see above) the relative repetition frequency is 0·3 or more, so the characteristic power number must be near its maximum value (Table III). However, power output is proportional to the product of characteristic power number and intrinsic speed and the power obtainable from these muscles is lower than it would be if they had higher intrinsic speeds. The limb muscles of cats, for instance, could deliver more power if they had the same intrinsic speeds as the corresponding muscles in mice (Table V).

Because of their relatively low intrinsic speeds muscles which operate at low frequencies deliver less power than is indicated by the continuous lines in Fig. 2. These lines assume $\dot{\varepsilon}_0 = 25\,s^{-1}$. The broken line shows values of Π_f/ρ for $\dot{\varepsilon}_0 = 3.3n$. No corresponding line has been drawn for aerobic power output because we do not know how Π_m is related to the frequency at which the muscle operates.

It follows from Eqn (7) and Table I that muscles delivering maximum power at relative repetition frequencies above 0·3 will develop stresses below $0.4\,q\sigma_0$ (for $p = 3$). A typical striated muscle capable of developing $0.3\,MNm^{-2}$ in isometric contraction would be expected to exert about $0.1\,MNm^{-2}$ when delivering maximum power in these circumstances. It is therefore unexpected to find stresses around $0.3\,MNm^{-2}$ acting in the limb muscles of a dog taking off for running jumps.[4] The muscles concerned include the extensors of the hip which shorten throughout the period of contact of the foot with the ground so that there could be no question of the high stresses occurring while the muscle was extending (i.e. when $\dot{\varepsilon}$ was negative). The dog was travelling at $6\,m\,s^{-1}$, about 1·5 times its trot/gallop transition speed. This observation suggests either that dog limb muscles are capable of isometric stresses well above $0.3\,MNm^{-2}$, or that their intrinsic speeds are higher than Table V suggests, or both. Our understanding of the power output of muscles in this and other cases could be improved by more studies in which the performance of the same muscle was investigated both in natural activities of the intact animal and in physiological experiments with the excised muscle.

REFERENCES

1. Abbott, B. C. and Wilkie, D. R. The relation between velocity of shortening and the tension-length curve of skeletal muscle. *J. Physiol., London.* **120**, 214–223 (1953).
2. Alexander, R. McN. "Animal Mechanics". Sidgwick and Jackson, London (1968).

3. Alexander, R. McN. Muscle performance in locomotion and other strenuous activities. *In* "Comparative Physiology". 1–22. (Bolis, L., Maddrell, S. H. P. and Schmidt-Nielsen, K., eds). Amsterdam, North Holland.
4. Alexander, R. McN. Mechanics of jumping by a dog. *J. Zool., Lond.* **173**, 549–573 (1974).
5. Ashhurst, D. E. The fibrillar flight muscles of giant water-bugs: an electron-microscope study. *J. Cell. Sci.* **2**, 435–444 (1967).
6. Bahler, A. S., Fales, J. T. and Zierler, K. L. The dynamic properties of mammalian skeletal muscle. *J. gen. Physiol.* **51**, 369–384 (1968).
7. Bell, G. H., Davidson, J. N. and Emslie-Smith, D. "Textbook of Physiology and Biochemistry" 8th Ed. Churchill Livingstone, Edinburgh (1972).
8. Bennet-Clark, H. C. The energetics of the jump of the locust *Schistocerca gregaria*. *J. exp. Biol.* **63**, 53–83 (1975).
9. Bücher, T. Formation of the specific structural and enzymic pattern of the insect flight muscle. *Biochem. Soc. Symp.* **25**, 15–28 (1965).
10. Buchthal, F., Weis-Fogh, T. and Rosenfalck, P. Twitch contractions of isolated flight muscle of locusts. *Acta physiol. scand.* **39**, 246–276 (1957).
11. Calow, L. J. and Alexander, R. McN. A mechanical analysis of a hind leg of a frog. *J. Zool., Lond.* **171**, 293–321 (1973).
12. Close, R. I. Dynamic properties of mammalian skeletal muscles. *Physiol. Rev.* **52**, 129–197 (1972).
13. Close, R. I. and Luff, A. R. Dynamic properties of inferior rectus muscle of the rat. *J. Physiol., Lond.* **236**, 259–270 (1974).
14. Edman, K. A. P. and Nilsson, E. Relationships between force and velocity of shortening in rabbit papillary muscle. *Acta physiol, scand.* **85**, 488–500 (1972).
15. Gordon, A. M., Huxley, A. F. and Julian, F. J. The variation in isometric tension with sarcomere length in vertebrate muscle fibres. *J. Physiol, Lond.* **184**, 170–192 (1966).
16. Gray, J. Studies in animal locomotion. VI. The propulsive powers of the dolphin. *J. exp. Biol.* **13**, 192–199 (1936).
17. Heglund, N. C., Taylor, C. R. and McMahon, T. A. Scaling stride frequency and gait to animal size: mice to horses. *Science,* **186**, 1112–1113 (1974).
18. Hill, A. V. The heat of shortening and dynamic constants of muscle. *Proc. R. Soc. B.* **126**, 136–195 (1938).
19. Jensen, M. Biology and physics of locust flight. III. The aerodynamics of locust flight. *Phil. Trans. B.* **239**, 511–552 (1956).
20. Lasiewski, R. C. Oxygen consumption of torpid, resting, active and flying hummingbirds. *Physiol. Zool.* **36**, 122–140 (1963).
21. Lasiewski, R. C., Galey, G. R. and Vasques, C. Morphology and physiology of the pectoral muscles of hummingbirds. *Nature, Lond.* **206**, 404–405 (1965).
22. Lind, A. R. and McNicol, G. W. Muscular factors which determine the cardio-vascular responses to sustained and rhythmic exercise. *Canad. med. Assoc. J.* **96**, 706–713 (1967).
23. Lowy, J. and Millman, B. M. Mechanical properties of smooth muscles of cephalopod molluscs. *J. Physiol., Lond.* **160**, 353–363 (1962).
24. McGahan, J. Flapping flight of the Andean condor in nature. *J. exp. Biol.* **58**, 239–253 (1973).
25. Méndez, J. and Keys, A. Density and composition of mammalian muscle. *Metabolism,* **9**, 184–188 (1960).

26. Moore, J. D. and Trueman, E. R. Swimming of the scallop, *Chlamys opercularis* (L). *J. exp. mar. Biol. Ecol.* **6**, 179–185 (1971).
27. Pennycuick, C. J. Power requirements for horizontal flight in the pigeon. *J. exp. Biol.* **49**, 527–555 (1968).
28. Pennycuick, C. J. and Parker, G. A. Structural limitations on the power output of the pigeon's flight muscles. *J. exp. Biol.* **45**, 489–498 (1966).
29. Pepe, F. A. Structure of the myosin filaments of striated muscle. *Progr. Biophys.* **22**, 77–96 (1971).
30. Pringle, J. W. S. and Tregear, R. T. Mechanical properties of insect fibrillar muscle at large amplitudes of oscillation. *Proc. R. Soc. Lond. B.* **174**, 33–50 (1969).
31. Prosser, C. L. (ed.) "Comparative Animal Physiology" 3rd Ed. Saunders, Philadelphia (1973).
32. Rüegg, J. C. Contractile mechanisms of smooth muscle. *Symp. Soc. exp. Biol.* **22**, 45–66 (1968).
33. Shafiq, S. A., Gorycki, M. A. and Milhorat, A. T. An electron microscope study of fibre types in normal and dystrophic muscles of the mouse. *J. Anat.* **104**, 281–293 (1969).
34. Smith, R. S. and Ovalle, W. K. Varieties of fast and slow extrafusal muscle fibres in amphibian hind limb muscles. *J. Anat.* **116**, 1–24 (1973).
35. Thomas, S. P. Metabolism during flight in two species of bats. *J. exp. Biol.* **63**, 273–294 (1975).
36. Trueman, E. R. and Packard, A. Motor performances of some cephalopods. *J. exp. Biol.* **49**, 495–507 (1968).
37. Tucker, V. A. Respiratory exchange and evaporative water loss in the flying budgerigar. *J. exp. Biol.* **48**, 67–87 (1968).
38. Weis-Fogh, T. Fat combustion and metabolic rate of flying locusts (*Schistocerca gregaria* Forskal) *Phil. Trans. B.* **237**, 1–36 (1952).
39. Weis-Fogh, T. Tetanic force and shortening in locust flight muscle. *J. exp. Biol.* **33**, 668–684 (1956).
40. Weis-Fogh, T. Biology and physics of locust flight, VIII. Lift and metabolic rate of flying locusts. *J. exp. Biol.* **41**, 257–271 (1964).
41. Weis-Fogh, T. Diffusion in insect wing muscle, the most active tissue known. *J. exp. Biol.* **41**, 229–256 (1964).
42. Weis-Fogh, T. Energetics of hovering flight in hummingbirds and in *Drosophila*. *J. exp. Biol.* **56**, 79–104 (1972).
43. Wells, J. B. Comparison of mechanical properties between slow and fast mammalian muscles. *J. Physiol., Lond.* **178**, 252–269 (1965).
44. White, D. C. S. and Thorson, J. The kinetics of muscle contraction. *Progr. Bioph. mol. Biol.* **27**, 173–255 (1973).
45. Wilkie, D. R. Man as an aero engine. *J. Roy. aero. Soc.* **64**, 477–481 (1960).
46. Wimsatt, W. A. "Biology of Bats". Academic Press, New York and London (1970).

INDEX

Atropine, 91

Babies, human, oxygen consumption, 31
Bacteria
    efficiency, 239
    flagella, 234, 239, 244
    locomotion, 233–4
    Reynolds number, 205
    size of, 2
    speed, 237
    wave length, 237
Baluchitherium, 4
Barracuda, 362
    gill area, 61
    swimming speed, 222
Bats, 406–7
    flight muscle, 411, 520–1
    power output, 13, 409
    wing
        area, 423
        loading, 369
Beam
    cantilever, 505
    loading, 164
Bee
    blood sugar, 16
    size of, 2
Beetle, click, *see* Click beetle
Beetle, water, *see* Water beetle
Bending moment
    limb, 175–6, 178, 180, 182
    beam, 164
    cilium, 252
    wing, 373, 403, 412–3, 415–6, 418
    strength, limb, 155, 162
    stresses, 173
    wave, cilia, 246
Bernoulli's equation, 288, 572
Beroe, 248–51, 253–4
Biceps brachii muscle, 46–7
Bidessus, 270–1, 279
Biped, locomotion, 93, 95, 98–9, 103–4
    see also Birds, running; Kangaroo;
            Man
Birds
    aquatic, 368
    brain, size of, 6, 7

flight, 41–8, 365–404, 406–7, 421–443, 479–509
    muscles, 520–1
    gallinaceous, wing area, 423–5
    intermittent flight, 437–443
    leg bones, 171, 174–5
    metabolic rate, 11, 12
    non-passerine, energy saving, 440
        resting metabolic power, 133
    passerine, energy saving, 440
        metabolic rate, 11
    power, for flight, 409, 432
        for running, 133
    of prey, 368–9, 495
    running, 93, 99
        incremental cost of transport, 131
        oxygen consumption, 129–30
    skeleton, mass, 180
    temperature, 10
    wing, area, 422–6
        bones, 171, 175
    see also specific names
Blackbird, 442, 481–2, 488–9, 491
Blackcap, 442
Blepharisma, 248–51, 253–4
Blood cells, red, size, 30
Blood volume, 15
Bluefish, swimming, 90
Bobwhite
    incremental cost of transport, 131
    oxygen consumption, 129
Body
    mass, 218–22
    shape
        fish, 303–6
        water beetles, 272, 274
Boeing 747, 369
Bohr effect, 15
Bonasa, 425
Bone
    fracture, 155
    human, 160–1
    mineral content, 162
    stiffness, 162
Bounces, in hopping and running, 114
Boundary layer, 215–6, 361
    laminar, 221, 224–5, 317–20, 328–9, 486, 505, 507
    reattachment, *see* Flow, reattachment
    separation, *see* Flow, separation